60/61 Advances in Polymer Science

Fortschritte der Hochpolymeren-Forschung

Liquid Crystal Polymers II/III

Editor: M. Gordon
Guest Editor: N. A. Platé

With Contributions by
M. G. Dobb, H. Finkelmann, J. Grebowicz,
J. E. McIntyre, N. A. Platé, G. Rehage,
V. P. Shibaev, B. Wunderlich

With 121 Figures and 41 Tables

Springer-Verlag
Berlin Heidelberg New York Tokyo
1984

ISBN-3-540-12994-4 Springer-Verlag Berlin Heidelberg New York Tokyo
ISBN-0-387-12994-4 Springer-Verlag New York Heidelberg Berlin Tokyo

Library of Congress Catalog Card Number 61-642

Typesetting: Th. Müntzer, GDR; Offsetprinting: Br. Hartmann, Berlin;
Bookbinding: Lüderitz & Bauer, Berlin

2154/3020-543210

Editors

Editorial

With the publication of Vol. 51, the editors and the publisher would like to take this opportunity to thank authors and readers for their collaboration and their efforts to meet the scientific requirements of this series. We appreciate our authors concern for the progress of Polymer Science and we also welcome the advice and critical comments of our readers.

With the publication of Vol. 51 we should also like to refer to editorial policy: *this series publishes invited, critical review articles of new developments in all areas of Polymer Science in English (authors may naturally also include works of their own).* The responsible editor, that means the editor who has invited the article, discusses the scope of the review with the author on the basis of a tentative outline which the author is asked to provide. Author and editor are responsible for the scientific quality of the contribution; the editor's name appears at the end of it.
Manuscripts must be submitted, in content, language and form satisfactory, to Springer-Verlag. Figures and formulas should be reproducible. To meet readers' wishes, the publisher adds to each volume a "volume index" which approximately characterizes the content.

Editors and publisher make all efforts to publish the manuscripts as rapidly as possible, i.e., at the maximum, six months after the submission of an accepted paper. This means that contributions from diverse areas of Polymer Science must occasionally be united in one volume. In such cases a "volume index" cannot meet all expectations, but will nevertheless provide more information than a mere volume number.

From Vol. 51 on, each volume contains a subject index.

Editors Publisher

Table of Contents

Thermotropic Mesophases and Mesophase Transitions of Linear, Flexible Macromolecules

Bernhard Wunderlich and Janusz Grebowicz*
Department of Chemistry, Rensselaer Polytechnic Institute Troy, New York 12181, USA

The field of mesophases is subdivided into six different phases: liquid crystals, plastic crystals, condis crystals and the corresponding LC, PC, and CD glasses. Liquid and plastic crystals are the traditional phases with positional and orientational disorder, respectively. Condis crystals are conformationally disordered. On hand of tables of thermodynamic transition parameters of small and large molecules it is shown that the orientational order in liquid crystals is only a few per cent of the total possible, while the positional order in plastic crystals is virtually complete. Condis crystals have a wide variety of different degrees of conformational disorder. The glass transitions of all mesophases are similar in type. Macromolecules in the liquid crystalline state produce high orientation on deformation. Plastic crystals consist always of small molecules. Condis crystals may under proper conditions anneal to extended chain crystals.

* On leave from the Center of Molecular and Macromolecular Studies, Polish Academy of Sciences, Łodz, Poland

Advances in Polymer Science 60/61

1 Introduction

A question that has persisted throughout the development of science has been: "What is the structure of the condensed phase?" In the earliest stage of knowledge about atoms backed by experiment, Dalton states in 1808 [1)]: "The essential distinction between liquids and solids, perhaps consists in this, that heat changes the figure of arrangement of the ultimate particles of the former continually and gradually, whilst they retain their liquid form; whereas in the latter, it is probable, that change of temperature does no more than change the size, and not the arrangement of the ultimate particles." This microscopic description of a solid as a material with fixed atomic arrangement is still valid today. The "ultimate particles" one would describe today as motifs, i.e. as the atoms, ions, molecules or parts of molecules which represent the crystallographic repeating units. The "change in size" one relates today to the asymmetric increase in vibrational amplitude. When asking about the types of arrangement possible in a solid, we can make the obvious subdivision into order and disorder. This leads to the distinction of crystalline and amorphous solids (glasses). The crystals have long range order, the glasses do not. In this review we want to concentrate on the in-between states of order, those of "mesomorphic order".[1] Mesomorphic is a term proposed by Friedel in 1922 [2)] for the materials of "middle" (Grk. mesos) "form" (Grk. morphe). Because of their in-between microscopic structure, mesophase crystals (Grk. phasis = appearance) may be in-between liquid and solid in macroscopic appearance. It thus becomes difficult to identify mesophases as solid or liquid.

To give a base for discussion, one can make use of the following operational definition of liquids and solids:[2] A solid is either a crystal (stable below its melting or crystal-to-crystal transition temperature) or a glass, which changes at its glass transition temperature to a liquid. The glass is an amorphous solid, metastable relative to the crystalline state. Both, the melting temperature T_m and the glass transition temperature T_g, are easily identified by thermal analysis, for example. Figure 1 shows thermal analysis traces through the melting and glass transitions. The transition temperatures T_m and T_g provide thus a precise, operational definition for solids and liquids.

Turning briefly to the possible variety of chemical substances, we find that since the discovery of linear macromolecules some 60 years ago, it is possible to subdivide all molecules into three major classes [3)]:

1. Small molecules
2. Flexible macromolecules[3]
3. Rigid macromolecules.

Small molecules are held in the solid state only by weak forces and melt at relatively low temperature without losing their molecular integrity. Often these molecules

1 The field of "lyotropic" liquid crystals will not be covered extensively. For references see Sect. 5.1

2 It is not possible to ask for a mechanical criterion, since some liquids have, for example, higher shear viscosity than some crystals

3 A macromolecule is defined according to Staudinger [4)] as a molecule of more than 1000 atoms

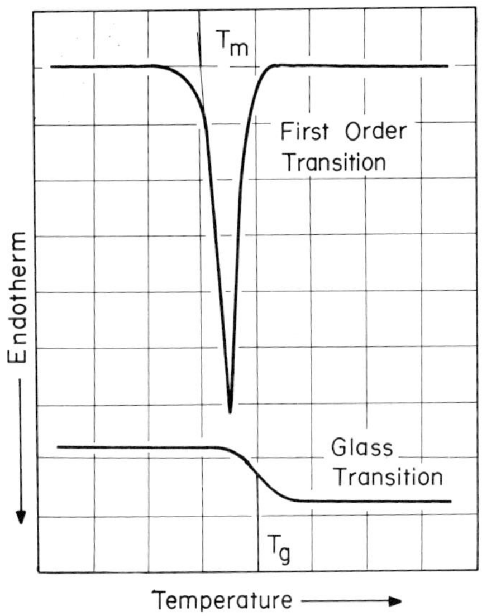

Fig. 1. Schematic thermal analysis results on fusion and devitrification (top and bottom curves, respectively). The area under the peak in the top curve represents the heat of fusion H_f and can be used to calculate the entropy of fusion $\Delta S_f = \Delta H_f/T_m$ in case of equilibrium melting. The increase in heat capacity ΔC_p at T_g is related to the number of motifs that gain mobility

can even be found in the gaseous state. Flexible macromolecules always have linear portions as the major property-determining structure. The flexibility of these linear chain portions allows the macromolecules to melt without loss of molecular integrity. The rigid macromolecules can exist only in the solid state. Melting occurs with loss of molecular integrity. Strong bonds must be broken to gain the mobility necessary for the liquid state.

Mesophase materials are possible for all three classes of molecules. Biological mesophases were already discovered in the middle of the 19th Century. Reinitzer later described the special two-stage melting of cholesteryl benzoate. These materials were then named liquid crystals by Lehmann in 1904 [5]. Small molecule mesophase materials will be referred to from time to time in this review as reference materials.

Mesophase materials of linear, flexible macromolecules have gained attention only more recently, when it was found that parallel molecular orientation is easily achieved in some of these mesophases. This orientation can lead to high modulus and tensile strength [6]. Presently there exists a certain amount of confusion in the literature about the description, properties, and nomenclature of these macromolecular mesophases and their place in the arrangement of all matter. Even for the better understood small-molecule mesophases there are some problems in the description of glasses and in the separation of orientational and conformational disorder. Also, the distinction between mesophases based on molecular structure and on supermolecular structure is not always made. We will try in this review to clarify some of these points.

Mesophases of rigid macromolecules have not found much attention as such. There should be an increase in mechanical properties due to partial order. One may expect that more attention will be paid to these materials in the future, as the mesophases of the other two classes of molecules are better understood.

2 Description of the Mesophases

The various types of mesophases have usually been treated separately, and rarely did researchers review more than one mesophase at a time. One of the few descriptions of liquid *and* plastic crystals was given by Smith [7]. We will expand on this attempt and try to combine the description of all three major types of mesophase order, namely:

1. Liquid crystals
2. Plastic crystals and
3. Condis crystals.

The term "condis crystal", which is a contraction of the term "conformationally disordered crystal", was coined to designate the most important mesophase for flexible, linear macromolecules. We are not aware of prior naming of this class of mesophases[4].

Before detailing the mesophases and their transitions, one must delineate the three conventional states that form the limits between which the mesophases are found. The first limiting state is the fully ordered crystal. It shows long-range, three-dimensional order. The amorphous solid (glass) has only short-range order. Going from the crystal to the isotropic liquid, one loses order and gains molecular or conformational mobility at the melting temperature. For glasses, there is no change in order, but similar gain in molecular or conformational mobility occurs at the glass transition temperature.

The melting process can be characterized by the entropy change on fusion [8]. Thermal analysis of fusion is illustrated in Fig. 1. We can break the overall entropy of fusion ΔS_f into the three approximate parts: positional, orientational, and conformational entropy of fusion

$$\Delta S_f = \Delta S_{pos} + \Delta S_{or} + \Delta S_{conf} \,. \qquad (1)$$

Crystals of spherical motifs such as noble gases and metals show only positional disordering. It will be shown in Sect. 5.2 that for almost spherical motifs (i.e. plastic crystals) the same rule applies. Independent of the molecular size, ΔS_f is usually between 7 and 14 J/(K mol) [9]. This is the rule of Richards [10], analogous to Trouton's rule for entropies of evaporation. Similarly, one finds for crystals which separate on melting into small, non-spherical, rigid motifs, that the entropy of fusion consists of positional and orientational contributions. In this case one finds a ΔS_f between 30 and 60 J/(K mol) [9] as first observed by Walden [11]. Again, ΔS_f is size-independent. As soon as a molecule can show conformational disordering, it has additional contributions to ΔS_f. This conformational contribution is dependent on the number of bonds around which conformational disordering is possible. The paraffins represent a well documented series of molecules with increasing fractions of conformational

4 The need ultimately to include conformational disorder in the system of mesophase materials has already been pointed out by Smith (Ref. [7], p. 193) and becomes obvious when reading discussions of the behavior of typical condis crystals (see, for example, Ref. [108])

entropy of fusion [9]. The practically spherical methane shows a final ΔS_f of 10.3 J/(K mol) which is largely positional. Nonspherical ethylene which undergoes no conformational isomerism, has a ΔS_f of 32.2 J/(K mol), consisting of positional and orientational contributions. The straight-chain n-decane has a ΔS_f of 118.2 J/(K mol) because of its large conformational entropy in the melt. Camphor, which is nearly spherical and rigid, in turn, is similar in numbers of atoms to n-decane ($C_{10}H_{16}O$ versus $C_{10}H_{22}$), but has a ΔS_f of only 14.3 J/(K mol). It cannot contain conformational or orientational contributions. Linear, flexible macromolecules, finally, have, because of their large size, only negligible positional and orientational contributions per mole of repeating units. Their entropy of fusion is dominated by conformational contributions. Analysis of equilibrium data on fusion of many macromolecules has revealed that ΔS_{conf} is about 9.5 J/(K mol) [12] for each mole of bonds around which conformational freedom is attained on fusion.

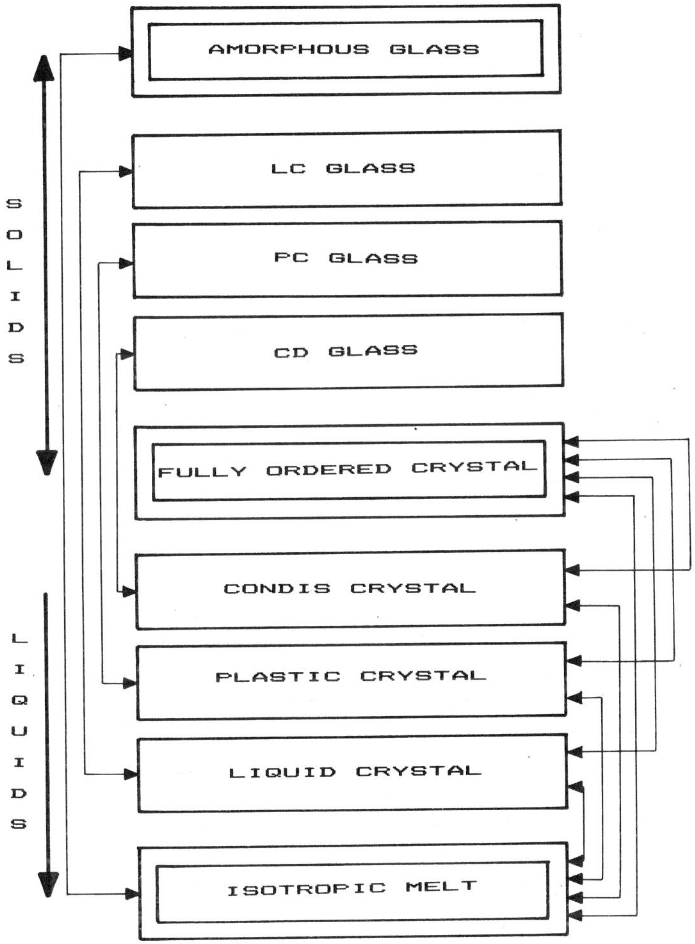

Fig. 2. Schematic diagram of the relationship between the three limiting phases (double outline) and the six mesophases. The top five phases are solid. The bottom four phases show increasing mobility i.e. liquid-like behavior

The devitrification process shows no change in entropy. Thermally, an increase in heat capacity is observed as is illustrated in Fig. 1. The exact temperature range of this increase is dependent on cooling and heating rate. The glass transition is dynamic in nature. Empirically, one could observe that molecules of similar size and number of motifs that start moving on devitrification have similar increases in heat capacity, ΔC_p. For small motifs one finds a ΔC_p of about 11.3 J/(K mol) [13]. For larger motifs, such as phenylene or larger ring structures, the contribution may be two to three times this amount [14]. As one can judge from entropy increase on fusion the gain of disorder, one can judge from heat capacity increase on devitrification the gain of mobility.

Based on this simplified description of the melting and glass transitions, it is possible to propose six major types of mesophases. Figure 2 shows these mesophases in relation to the glass, crystal, and melt. First, it is possible to keep orientational order, but lose positional order. These "positionally disordered" crystals or "orientationally ordered" liquids [7] are widely known as liquid crystals [15]. The name liquid crystal was given because of the obvious, liquid-like flow of these materials. By now it is too late to try to change the nomenclature, especially when the possible new names would be cumbersome [7].

Next is a group of materials with a mesophase which shows "orientational disorder", but positional order. These materials are widely known as plastic crystals [16] because of the ease of deformation of such crystals. Again, this name is well accepted and much less cumbersome than "orientationally disordered" crystals [7].

The third group of mesophase materials represents the "conformationally disordered" crystals, called condis crystals. The physical properties of condis crystals, which largely maintain positional and orientational order, change in much too subtle a way from the fully ordered crystals so that a common property could be attached to their name.

The differences between the three mesophase crystals are largely based on the geometry of the molecules. We can expand on the summary of Smith [7] who compared liquid and plastic crystals and states: The molecules of liquid crystals always have a rigid, mesogenic group which is rod- or disc-like and causes a high activation energy to rotational reorientation.[5] The molecules of plastic crystals, in contrast, are compact and more globular, so that their reorientation is not opposed by a high activation barrier. The condis crystals, in turn, consist of flexible molecules which can undergo relatively easily hindered rotation to change conformation without losing positional or orientational order.

While all three mesophases have some degree of long range order in common with the crystal, they also have some degree of non-vibrational motion in common with the liquid. As a result, all mesophases must show, just as the liquid, a glass transition if crystallization to full order can be avoided on cooling. This fact has often been overlooked as a key identifier for the mesomorphic state. At temperatures below the respective glass transitions one can thus have three further mesophases. Besides the "normal" amorphous glasses there are:

5 Note that this is only true if the liquid crystalline state is caused by the molecular structure and not its superstructure (see Sect. 5.1)

1. Positionally disordered glasses (LC-glasses);
2. Orientationally disordered glasses (PC-glasses);
3. Conformationally disordered glasses (CD-glasses).

Because all glasses are clearly solids, the names "liquid crystal glass" or "plastic crystal glass" are awkward. The terms LC-, PC- and CD-glass thus stand for: glass obtained by quenching a liquid crystal, plastic crystal, or condis crystal, respectively.

The transitions between the bottom five phases of Fig. 2 may occur close to equilibrium and can be described as thermodynamic first order transitions (Ehrenfest definition [17]). The transitions to and from the glassy states are limited to the corresponding pairs of mobile and solid phases. In a given time frame, they approach a second order transition (no heat or entropy of transition, but a jump in heat capacity, see Fig. 1).

3 Thermodynamics of Mesophase Transitions

The major transitions among the three limiting phases and the six mesophases were already indicated by the double arrows in Fig. 2. To minimize confusion in the designation of the transition temperatures involving various types of crystals, we have adopted a system which uses a subscript indicating the direction of change. T_i is the final change to the isotropic melt, T_d, the disordering from a more ordered crystal, T_o, the ordering to a more ordered crystal. T_m is reserved for the complete melting of the fully ordered crystal. The various glass transitions are labelled T_g with an appropriate second subscript, when needed: T_{ga}, T_{gl}, T_{gp}, and T_{gc}.

Flexible, linear macromolecules frequently show on cooling only partial crystallization. Such materials can often be described by the crystallinity model (typical crystallinities of 30–90%) [18,19]. Two of the questions in the description of macromolecular mesophases must thus be: what is the crystallinity of the sample? and: which phases are present?

The formation of macromolecular liquid crystals usually seems to be virtually complete. Also, when the mesogenic group is sufficiently mobile, there is only little supercooling of the transition to the liquid crystalline state. Plastic and condis crystals, on the other hand, are expected to behave more like the fully ordered, macromolecular crystals, i.e., partial crystallinity is expected in these cases.

Liquid crystal forming materials have thus five possible pure phases (fully ordered crystal, LC-glass, amorphous glass, liquid crystal, and isotropic melt). In addition, there may be four two-phase materials (fully ordered crystal and LC-glass, fully ordered crystal and amorphous glass, fully ordered crystal and isotropic melt, and fully ordered crystal and liquid crystal). The correlations between these phases are shown in Fig. 3.

Each of the glass transitions has the appearance as given in Fig. 1. From the magnitude of the increase in heat capacity at the transition it should be possible to obtain information on the total amount of mobility gained, as outlined in Sect. 2. For well characterized systems, the glass transitions can also be used for the approximate determination of the weight fraction of crystallinity w^c

$$\Delta C_p = (1 - w^c)\,\Delta C_p^0 \qquad (2)$$

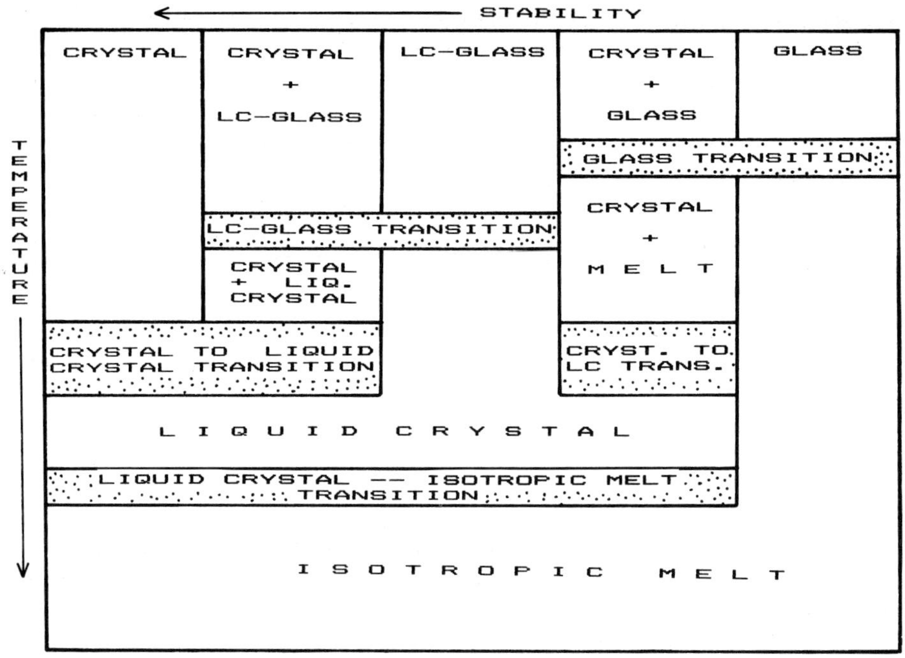

Fig. 3. Schematic diagram of the possible states of a liquid crystal forming material. Only the left side of the diagram corresponds to equilibrium. The states towards the right are increasingly metastable or unstable, but are often reached for flexible, linear macromolecules

where ΔC_p is the measured increase in heat capacity and ΔC_p^0 is the heat capacity increase for the glass transition of the pure amorphous glass or mesophase glass. The small amount of information available seems to indicate that there is not much difference between the ΔC_p^0 for the amorphous glass and the LC- and PC-glasses [20, 21].

The transition from a fully ordered crystal to a liquid crystal occurs at temperature T_d (disordering transition). This is a first order transition, which on thermal analysis resembles complete melting (T_m), i.e. the transition from the fully ordered crystal to the isotropic melt as described in Fig. 1 and Sect. 2. The heats of transition are usually only slightly smaller than for true melting, which is also indicated by the observation that liquid crystals have a specific volume close to that of the isotropic liquid [7]. Most of the change in interaction takes place at this disordering transition. The discussion of the entropy of transition will allow one to get more insight into the amount of disordering (see Sect. 5.1). In partially crystalline samples the measured heat of fusion ΔH_f can be used to estimate the weight fraction of crystallinity w^c

$$\Delta H_f = w^c \, \Delta H_f^0 \,. \tag{3}$$

The heat of fusion of the fully crystalline sample ΔH_f^0 must be known for this determination. It can be obtained by calibration of samples with widely different crystallinities in conjunction with Eq. 2, or by parallel use of a different crystallinity measuring technique, such as density or X-ray diffraction analysis [18]. Besides lowering the heat of transition, partial crystallization leads usually to a lowering of the melting temperature and a broadening of the melting range. The major factors causing these changes in melting temperature are small crystallite size and the degree of interconnection between the phases. Lowering in melting temperatures due to small crystal size may range from a few kelvins to the glass transition temperature. A melting point elevation is also possible by strain in the molten or liquid crystalline phase. Details about this topic are given in Ref. [3] and must be taken into account in the discussion of transitions of flexible, linear macromolecules. In many cases, crystalline regions are only 2 to 10 nm in at least one dimension, quite in contrast to the liquid crystalline domains, which are macroscopic in magnitude.

The transition of the liquid crystal to the isotropic melt, T_i, is governed largely by entropy. It was shown, for example, by Flory [22] that a phase change of this type could occur just due to the shape of the molecule. Since liquid crystalline behavior can be caused by a relatively large rod-like or disc-like mesogenic group [15], we expect for the entropy of transition at most the size-independent orientational contribution of about 15–45 J/(K mol). Usually the transition entropy is much smaller, indicating that the orientational order in liquid crystals is quite imperfect (see Sect. 5.1). Because of the existence of an entropy of transition, one finds a first order transition at T_i. For flexible, linear macromolecules it is initially surprising that the transition is frequently close to equilibrium, i.e. supercooling is small. The heat of transition is also practically the same for all thermal histories (no crystallinity effect).

As a result of the strong drive to equilibrium, it is usually difficult to quench the isotropic melt to an amorphous glass when liquid crystal formation is possible. Extraordinary quenching techniques may be needed. Once produced, the amorphous state loses its metastability on heating above the glass transition T_{ga}. The melt is quite unstable, so that it may not be possible to keep the melt from changing to the mesophase [21].

The arrow at the top of Fig. 3 indicates schematically the increase in stability of the various phase combinations. Exact placement of the three middle groups of phases would have to take into account the level of crystallinity of the second and fourth material. The temperatures of the transitions are given in probable relative sequence. In some cases it was found, however, that T_{ga} and T_{gl} are close in temperature [20] in others, they are the order of magnitude of 50 K apart [21].

The possible transitions of plastic and condis crystal-forming materials are shown in Fig. 4. For plastic crystals, this diagram is fully based on information on low molecular weight materials. No flexible, linear macromolecules which resemble plastic crystalline behavior have been reported (see Sect. 5.2.3). Similarly, little attention has been paid in the past to conformationally disordered mesophases in small molecules. In fact, some of the plastic crystals of larger organic molecules may actually be condis crystals (see Sects. 5.2.2 and 5.3.3). Since the positional order is preserved in both plastic and condis crystals, the possible phase relations are similar. The major difference from the liquid crystals is the possibility of partial mesophase formation.

There are again five pure phases possible (fully ordered crystal, mesophase glass, amorphous glass, mesophase, and isotropic melt). In addition, there is the possibility of seven two-phase materials (fully ordered crystal and mesophase glass, fully ordered crystal and amorphous glass, mesophase glass and amorphous glass, mesophase

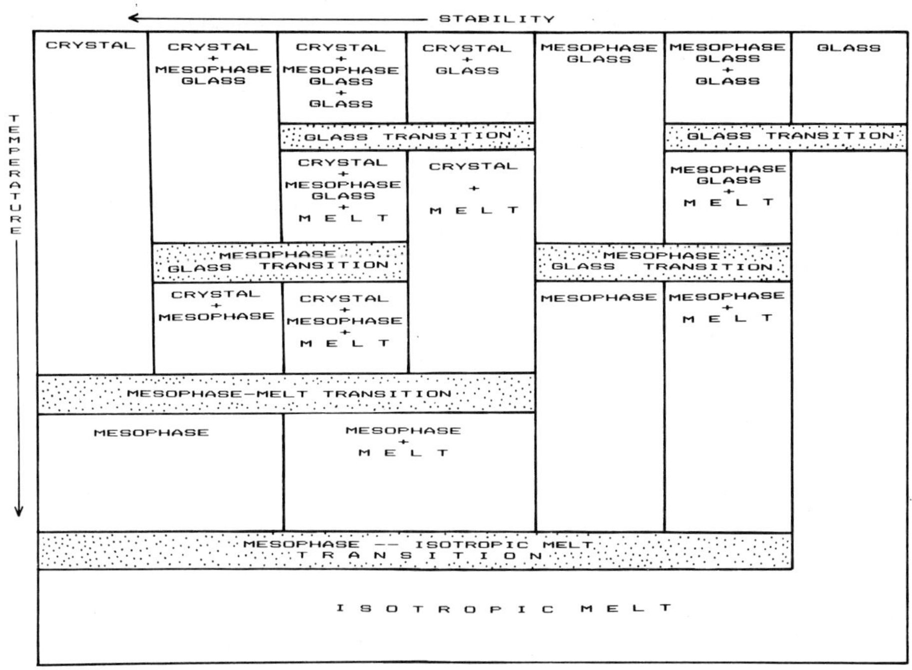

Fig. 4. Schematic diagram of the possible states of plastic or condis crystal forming materials. Only the left side of the diagram corresponds to equilibrium. The states towards the right are increasingly metastable or unstable, but are often reached for flexible, linear macromolecules

glass and isotropic melt, fully ordered crystal and mesophase, fully ordered crystal and melt, mesophase and melt). Furthermore, it is possible to have 3 three-phase materials (fully ordered crystal with mesophase glass and amorphous glass, fully ordered crystal with mesophase glass and isotropic melt, fully ordered crystal with mesophase and isotropic melt).

The various transitions can again be analyzed as discussed for the liquid crystals. The entropy of transition of the plastic crystals to the isotropic melt is often close to the expected, size-independent positional entropy gain of 7–14 J/(K mol). This is an indication of rather high positional order. The equilibrium phases are represented by the left side of Fig. 4. Towards the right side, the diagram shows decreasing thermodynamic stability. The exact order is set by the crystallinity. For plastic crystals the glass transition was shown to be similar in ΔC_p to the amorphous glass [23]. No information is available as yet on condis crystals. The glass transition temperatures are expected to show greater variation than in the liquid crystal case. It will be suggested in Sect. 5.3 that some CD-glasses may have glass transitions above the mesophase to isotropic-phase transition or decomposition temperature. If observable, the glass transitions may be used to get some information on the weight fraction of crystallinity using Eq. (2).

The crystal-mesophase transitions at T_d are first order transitions. In the case of plastic crystals, they show usually the major heat of transition. In the condis crystals the magnitude of the transitions depends on the number of bonds which gain conformational freedom. A wider variation in heats and entropies of transition is thus expected.

4 Kinetics of Mesophase Transitions

The kinetics of the mesophase transitions can be discussed by comparison with the crystallization-melting and the vitrification-devitrification transitions of the three limiting phases of Fig. 2. Except for the crystallization kinetics, the literature on the kinetics of phase transitions is limited. For the mesophases, information is scarce.

Looking first at the glass transition kinetics of amorphous glasses, one finds that the glass transition temperature decreases approximately logarithmically with cooling rate. For poly(methyl methacrylate) [24] and polystyrene [25] the following equations have been found (heating rate q in K/min, T_g in K)

$$T_g(\text{PMMA}) = 383 + 4.23 \log q \,, \tag{4}$$

$$T_g(\text{PS}) = 372.5 + 4.03 \log q \,. \tag{5}$$

Both of these polymers do not have any reported mesophase. Since the glass transition is a freezing process on a 3 to 10 mobile unit (bead) scale, there are neither phase boundaries nor nucleation barriers for the process.

Measurements on cooling are often difficult to carry out, so that heating experiments are usually substituted. These may lead, however, to hysteresis phenomena if heating and cooling are not carried out at similar rates, or if annealing occurred close to the glass transition temperature before analysis. Figure 5 shows typical apparent heat

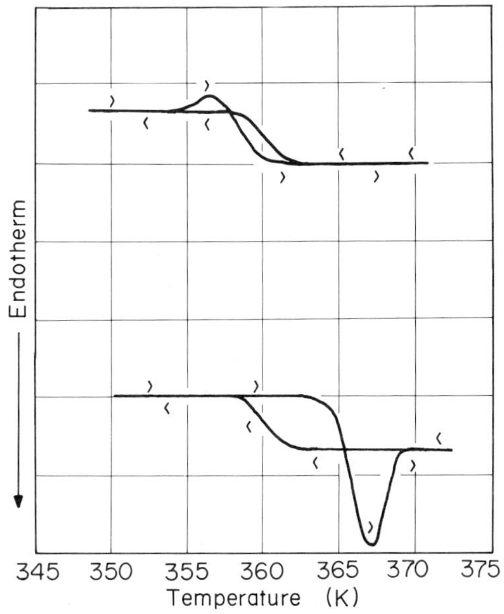

Fig. 5. Schematic of the heat capacity on cooling and heating at largely different rates in the glass transition region. Top Curves: Cooling at 5.0 K/min, heating at 0.25 K/min. Bottom Curves: Cooling at 5.0 K/min, heating at 150 K/min. (Hypothetical polymer, time dependence as in polystyrene)

capacities on heating and cooling through the glass transition region at largely different cooling and heating rates. No systematic study of the hysteresis effects on mesophase glasses has been reported, but it was observed in our laboratory that for liquid crystal glasses of two small molecules (azoxybenzene type [26]), a macromolecule with side-chain mesogenic groups [21], and a macromolecule with main chain mesogenic groups with flexible spacers [27] the hysteresis was strongly reduced. This can only mean that the glass transition is less time dependent than is observed for amorphous glasses. It must be mentioned, however, that the poly(ethylene terephthalate-*co*-oxybenzoate) liquid crystals showed a reduction in hysteresis only for partial crystallinity of the fully ordered type [20]. Under the condition of partial crystallinity many non-mesophase, flexible, linear macromolecules will show a reduction in hysteresis [28]. Much more detailed experimental information is necessary before an interpretation of these observations will be possible.

The crystallization transition is much better understood. It is a two step process which involves nucleation, followed by crystal growth [19]. The nucleation is often heterogeneous, meaning that nuclei exist already in the melt in the form of foreign particles. The crystal nuclei become active over a narrow time span as soon as the crystallization temperature is reached. Crystal growth has a constant rate in the direction normal to the crystal surface as long as the temperature is constant and growth is not impeded by mass or heat transport. Until the first crystals impinge on each other, the overall crystal growth is described by the free growth equation

$$v^c = NG(vt)^3 \tag{6}$$

where v^c is the volume fraction crystallinity, N the number of crystal nuclei per unit volume, G a geometric factor ($4\pi/3$ for spheres), v is the linear crystal growth rate, and t, time. As soon as crystals impinge on each other, the growth slows. Mathematically this impingement is corrected for by the Poisson [29] equation, usually called the Avrami equation [30]

$$1 - v^c = e^{-Kt^3} \tag{7}$$

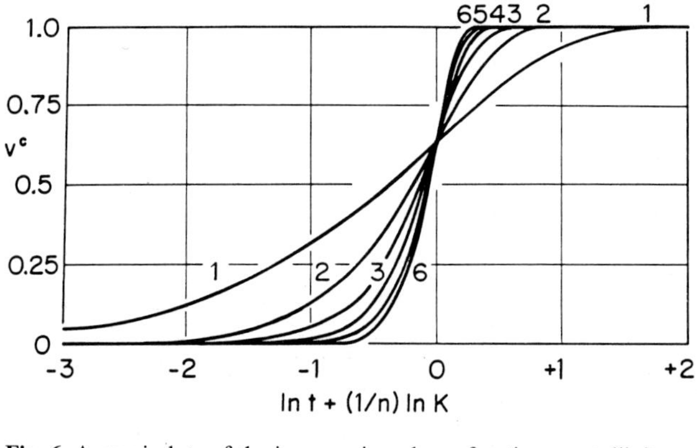

Fig. 6. Avrami plots of the increase in volume fraction crystallinity as a function of time for Avrami exponents n from 1 to 6 (see Eq. (7)). For partial crystallization v^c would be normalized to the ultimate crystallinity reached. K represents the geometry and time constants and nucleation terms [19]

where K is a combination of all constant terms in eq. 6 ($K = NGv^3$). Differences in nucleation rate and morphology change the exponent of t in Eq. (7). A detailed discussion of the various exponents and also the limitations of the application of Eq. (7) for polymeric materials are given in Ref. [19]. Figure 6 shows typical Avrami plots for the crystallization kinetics with various exponents. Turning to mesophases, we find that available experimental data can be fitted to Eq. (7). The conclusion drawn from these experiments is that transitions from the isotropic melt to the mesophase and from the mesophase to the crystal are also controlled by nucleation and growth.

Price and Wendorff [31] and Jabarin and Stein [32] analyzed the solidification of cholesteryl myristate. Under equilibrium conditions it changes at 357.2 K from the isotropic to the cholesteric mesophase and at 352.9 K to the smectic mesophase (see Sect. 5.1.1). At 346.8 K the smectic liquid crystal crystallized to the fully ordered crystal. Dilatometry resulted in Avrami exponents of 2, 2, and 4 for the respective transitions. The cholesteric liquid crystal has a second transition right after the relatively quick formation of a turbid homeotropic state from the isotropic melt. It aggregates without volume change to a spherulitic texture. This process was studied by microscopy [32] between 343 and 355.2 K and revealed another nucleation controlled process with an Avrami exponent of 3.

Price and Wendorff found Avrami exponents of 4 for the mesophase to fully ordered crystal transition of cholesteryl nonanoate and acetate [33, 34]. Details on the mesophase to fully-ordered-crystal transition were obtained by measuring nucleation rates and linear crystal growth at various temperatures using microscopy [35]. Over 6.5 to 10 K temperature ranges these transitions showed unchanged linear crystal growth rates, quite in contrast to crystallization from the isotropic phase. Using standard crystalli-

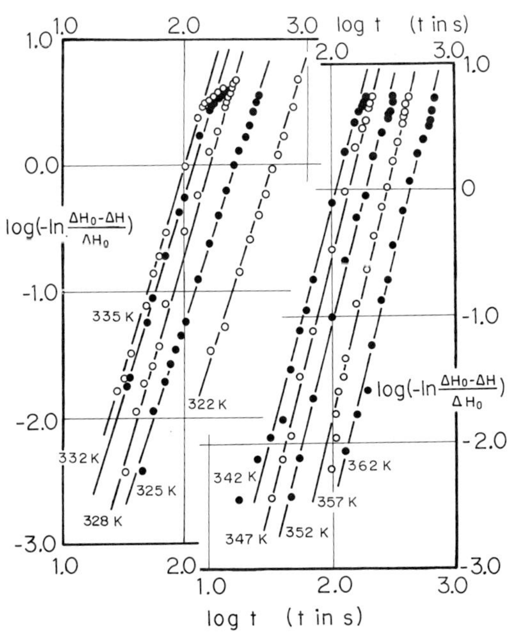

Fig. 7. Double logarithmic Avrami plots for the crystallization of liquid crystalline poly(oxy-2,2′-dimethylazoxybenzene-4,4′-diyloxydodecanedioyl) to the fully ordered state between 322 and 362 K. Average of the Avrami exponent 3.5 ± 0.3. Diagram courtesy J. Wiley and Sons [27]

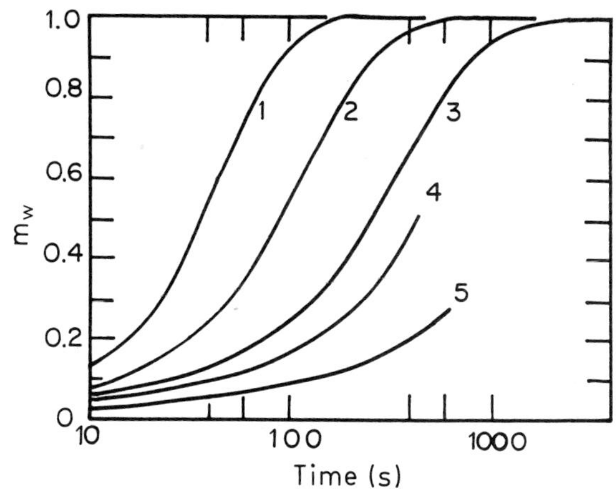

Fig. 8. Superheating of polyethylene extended chain crystals. Curves: 1) at 421.7 K, 2) at 419.2, 3) at 417.7 K, 4) at 416.7 K, 5) at 414.7 K. The equilibrium melting temperature is 414.6 K. Drawn after Ref. [40]. ^{m}w is the weight fraction molten, obtained by isothermal calorimetry

zation theory, the nucleation was deduced to be time dependent and the nuclei are thought to be of small size (0.65 to 1.46 nm radius).

Adamski and Klimczyk analyzed cholesteryl pelargonate [36] and caproate [37] liquid crystal to fully-ordered-crystal transitions over a temperature range of about 25 K. Again, the appearance of the fully ordered crystals was that of a spherulitic superstructure. The nucleation was time dependent, and the linear growth rate of the spherulites decreased with decreasing temperature by a factor 1/2 to 1/3, in contrast to the nonanoate and acetate. The Avrami exponent was close to 4 as judged from the measurement of the crystallized volume in the field of view under the microscope.

The kinetics of transition from the liquid crystal to the fully ordered crystal of flexible, linear macromolecules was studied by Warner and Jaffe [38] on copolyesters of hydroxybenzoic acid, naphthalene dicarboxylic acid, isophthalic acid, and hydroquinone. The analytical techniques were optical microscopy, calorimetry and wide angle X-ray diffraction. Despite the fact that massive structural rearrangements did not occur on crystallization, nucleation and growth followed the Avrami expression with an exponent of 2. The authors suggested a rod-like crystal growth.

The mesophase to fully ordered crystal transition in the flexible spacer, main chain, mesogenic polymer is closely similar to the low molecular weight materials. Calorimetry and light scattering revealed a spherulitic morphology after athermal (time independent) nucleation [27]. A typical example is shown in Fig. 7. The Avrami exponent varies between 3 and 4 over a 40 K wide temperature range which includes the maximum of linear crystal growth rate. The linear growth rates are comparable to polymers of slow crystallization [such as poly(ethylene terephthalate) or isotactic polystyrene].

Little information is available for the crystallization of condis crystals from the isotropic melt, but even less difference from the crystallization of fully ordered crystals is expected. It will be shown in Sect. 5.3 that one possible special effect of crystallization of macromolecular condis crystals is the ability to chain extend after initial crystallization. This process has been analyzed to some degree and occasional low Avrami exponents have been reported [39]. No information is available for the kinetics of

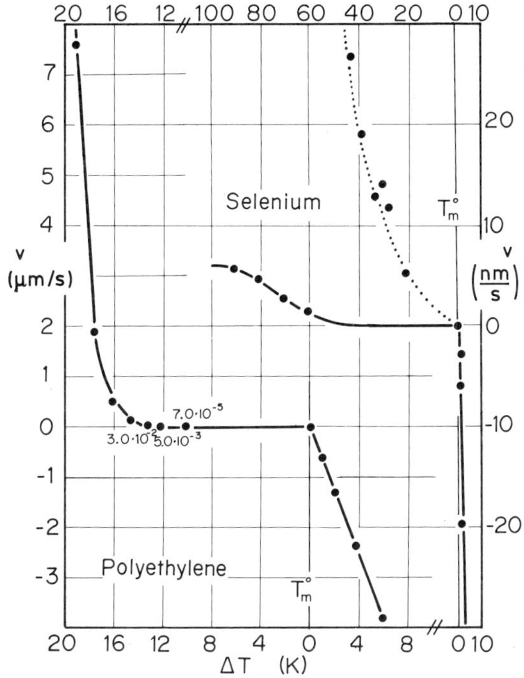

Fig. 9. Melting kinetics and crystallization kinetics of polymeric selenium (right) and polyethylene (left). The equilibrium melting temperatures are 494.2 and 414. 6K. The dotted curve indicates that on crystallization of the macromolecule from small molecules Se_2 there is no "molecular nucleation" necessary as in the melt crystallization (see also ref. 43 for a more detailed discussion of Se crystallization and melting). Drawn after Ref. [41]

the condis to fully ordered crystal transition. Similarly, solid-solid transitions in crystals are not very frequently studied.

The reverse of the crystallization kinetics, the melting kinetics, has also received only little attention. Even for fully ordered crystals transforming to the isotropic liquid, only little information is available. The superheating of a crystal on melting is different from the supercooling on crystallization because nuclei for melting are always present in the form of defects, edges or corners of the crystal. Melting thus starts instantaneously and the kinetics can only be followed experimentally if the linear melting rate is slower than the addition of the heat of transition. Figure 8 illustrates the melting kinetics of large crystals of polyethylene [40]. The similarity to the melting kinetics is seen by comparing Fig. 8 to Fig. 6. Figure 9 shows complete phase-transition kinetics when traced through the melting temperature for selenium and polyethylene. Typical for macromolecules is the break in the kinetics which is linked to the need of molecular nucleation [19,41]. We are not aware of any experiments on the kinetics of any of the crystal-mesophase or mesophase-isotropic melt transition except for the melting of condis crystals of polytetrafluoroethylene [42]. In this latter case the results are, as expected, similar to the fully ordered crystals [19].

5 Mesophases and Their Transitions

5.1 Liquid Crystals and LC-Glasses

5.1.1 Molecular Structure

In liquid crystals or LC-glasses one looks for orientational order and an absence of three-dimensional, long-range, positional order. In liquid crystals, large scale molecular motion is possible. In LC-glasses the molecules are fixed in position. The orientational order can be molecular or supermolecular. If the order rests with a supermolecular structure, as in soap micelles and certain microphase separated block copolymers, the molecular motion and geometry have only an indirect influence on the overall structure of the material.

In this review we are mainly concerned with thermotropic materials, i.e. with liquid crystals and LC-glasses which do not contain a solvent. The transitions of the macromolecular, thermotropic liquid crystals are governed then by temperature, pressure and deformation. In lyotropic liquid crystals and LC-glasses a solvent or dispersing agent is present in addition. The transitions then also become concentration dependent.

Supermolecular level liquid crystals are not of major concern in this review. In fact, they form a class of materials which should best be separated from the "normal liquid crystals". Although there are structural similarities to the molecular liquid crystals, molecular motion and transition behavior of these liquid crystals based on supermolecular structure is completely different.

Molecular-level orientational order can best be achieved with mesogens which are rods, laths, or discs. These shapes have a large length (or diameter) to width (or thickness) ratio R. Table 1 contains a listing of a series of typical liquid crystals. A larger collection can be found in Ref. [44]. The thinnest possible rods are strings of conjugated trans double bonds which have a diameter of about 0.4 nm (Table 1, entry 1, trans,trans-nona-2,4-dienoic acid). The most common small-molecule liquid crystals have p-connected phenylene groups or other ring structures, as shown in entries 2–7 of Table 1. The ratio R for 4,4-dibenzylideneaminobiphenyl is, for example, five. Entry 10 in Table 1 is an example of a more disc-like mesophase [45]. Larger disc-like molecules are found in coal-tar and petroleum pitches after aromatic polymerization on heating to between 600 and 750 K [46].

Macromolecules can form similar liquid crystals if the mesogenic groups are linked by flexible spacers. The mesogenic groups can be attached to macromolecules as side-chains or as part of the main-chain. Detailed descriptions of such materials are given below. The mobility of the mesogenic groups in the liquid-crystal is based on conformational motion in the spacers. For full characterization the degree of participation of the nonmesogenic spacers in the transitions must be known.

Larger mesogens are produced by rigid, linear macromolecules such as all-aromatic para-connected polyamides and polyesters. Molecules of this type can at best form LC-glasses. Because of the large size of the mesogenic group, which may be the whole macromolecule, the decomposition temperature is lower than the glass transition temperature. A solvent may lower the glass transition temperature so that a lyotropic liquid crystal results. The best known example of this type of liquid crystal is the spinning solution of poly(p-phenylene terephthalamide). Examples of more flexible macromolecules which can become rigid by formation of a secondary structure are

Table 1. Examples of Liquid Crystals

1. Nona-2,4-dienoic acid (nematic)	$(T_d = 296\ K,\ T_i = 322\ K)$ $CH_3-(CH_2-)_3\overset{H}{C}=\overset{H}{C}-\overset{H}{C}=\overset{H}{C}-\overset{\overset{O}{\Vert}}{C}-OH$
2. 4,4′-Dibenzylideneaminobiphenyl (nematic)	$(T_d = 507\ K,\ T_i = 533\ K)$ 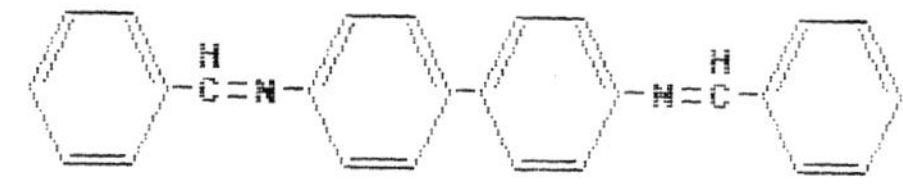
3. 4-p-Methoxybenzylidene-aminobiphenyl (nematic)	$(T_d = 435\ K,\ T_i = 447\ K)$ 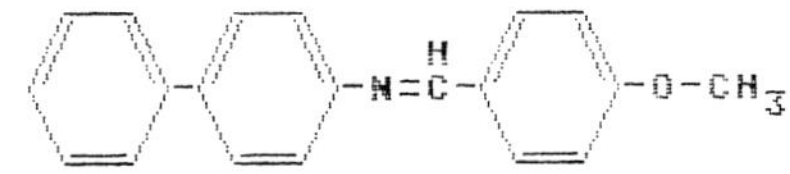
4. Anisal-p-aminoazobenzene (nematic)	$(T_d = 424\ K,\ T_i = 458\ K)$ 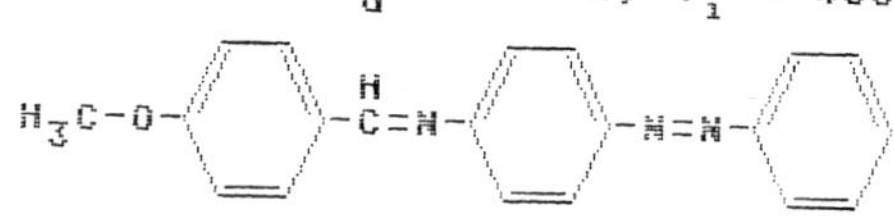
5. p-Azoxyanisole (nematic)	$(T_d = 389\ K,\ T_i = 406\ K)$ 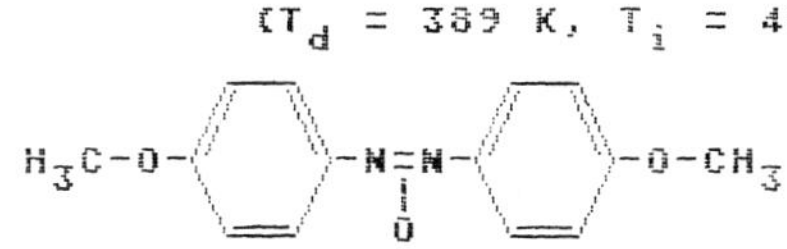
6. Ethyl-p-azoxycinnamate (smectic)	$(T_d = 414\ K,\ T_i = 537\ K)$ 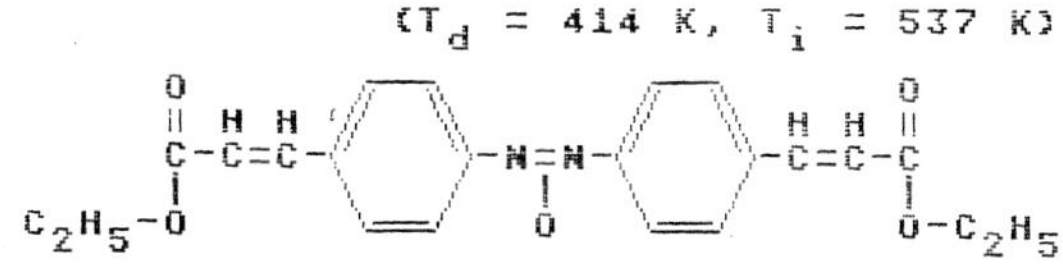
7. 4′-Methyloxybiphenyl-4-carboxylic acid (nematic)	$(T_d = 531\ K,\ T_i = 573\ K)$ 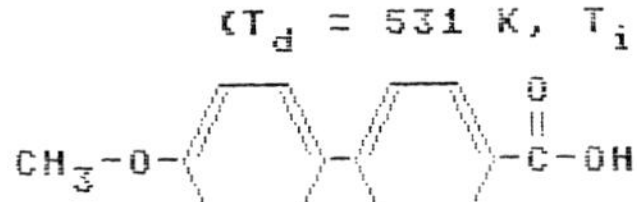
8. p-n-Propoxybenzoic acid (nematic)	$(T_d = 418\ K,\ T_i = 427\ K)$ 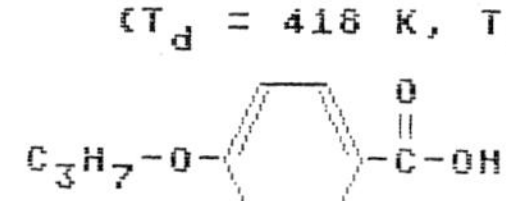

Table 1 (continued)

9. Cholesterol esters (cholesteric)

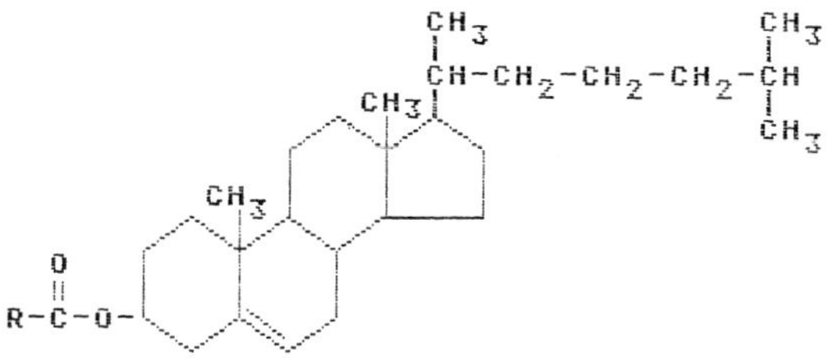

10. Benzenehexa-n-heptanoate (discotic)

(T_d = 354 K, T_i = 359 K)

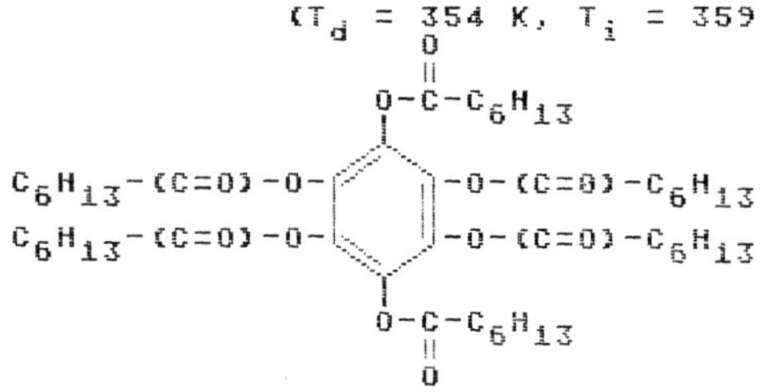

polypeptides and poly(amino acid)s. When dispersed in a suitable solvent, these molecules form relatively rigid alpha-helices of about 2 nm width and perhaps 20–50 nm length [47]. Even larger mesogens are found in desoxyribonucleic acids and certain viruses, which may form lyotropic liquid crystals with water as dispersing agent [48]. The tobacco mosaic virus has, for example, a length of 300 nm and a diameter of 20 nm and thus reaches almost macroscopic dimensions. De Gennes [49] suggested that silicate glass fibers of the correct R may form liquid crystals with macroscopic mesogens for the proper dispersing agent. Similarly, one can reach macroscopic disc dimensions by going to increasingly larger graphitic structures. The larger mesogens need in all cases a liquid dispersing agent to keep the system above the glass transition.

Mesophases of supermolecular structure do not need a rigid mesogen in the constituent molecules. For many of these materials the cause of the liquid crystalline structure is an amphiphilic structure of the molecules. Different parts of the molecules are incompatible relative to each other and are kept in proximity only because of being linked by covalent chemical bonds. Some typical examples are certain block copolymers [50], soap micelles [51] and lipids [52]. The overall morphology of these substances is distinctly that of a mesophase, the constitutent molecules may have, however, only little or no orientational order. The mesophase order is that of a molecular superstructure.

Block copolymers of sufficiently large and incompatible sequences of repeating units undergo microphase separation. Because of the geometric restriction caused by the needed alignment of the junctions between the different blocks, the phases separate into regular arrangements of lamellae, rods, or even spheres. The type of geometry and the choice of matrix phase is mainly dependent on concentration and block length. Within the microphases there is no special order beyond the normal amorphous or semicrystalline structure found in the corresponding homopolymers. Except for size effects, the melting and glass transition of the block copolymers corresponds to the constitutent homopolymers [3, 53]. The mesophase is based on the microphase regions and not on the constituent molecules. Since the molecules cross the phase boundary, it is sufficient to have one of the blocks become glassy to prevent the complete sample from flowing.

Soaps and lipids are small, amphiphilic molecules. They consist of more or less straight-chain, flexible hydrocarbon portions of 8 to 20 carbon atoms attached to a polar group. The polar groups are water soluble, but are kept by the non-polar groups at the interface. As a result, micelles are formed which can, as in the block copolymers, be lamellar, rod-like, or spherical. The reason for micelle formation is the incompatibility of the two parts of the molecule, not the presence of a rigid molecular mesogenic group. Although the translational and orientational motions are limited by the attachment to the polar interface, there is, at sufficiently elevated temperature, conformational mobility. The constituent molecules may undergo their normal glass transition, crystallization, or ordering to a condis crystals. They influence, in this way, the overall mesophase properties indirectly. Since most of these systems have only been analyzed over the limited temperature range possible in water, little is known about the changes caused by transitions of the constituent molecules. A detailed description and classification of the polymorphs of lipids was given by Charvolin and Tardieu [54].

Returning to the molecules with rod-, lath-, or disc-like mesogens, one can further classify various states of order. Broadly, one distinguishes between nematic, smectic, and discotic liquid crystals or LC-glasses (nematic from the Grk. nema = thread, smectic from the Grk. smegma = soap, and discotic from the Grk. diskos = disc). Figure 10 illustrates schematically the arrangements of the mesogens in the three types of mesophases. For large, rod-like mesogens the additional term canonic (Grk. kanon = rod) was suggested by F. C. Frank (see 15c), it has, however, not found wide acceptance.

The nematic phase, N, shows no long range order of the centers of gravity. A correlation exists only in the direction of the mesogenic groups. Positive and negative directions within the molecules (polarity), if they exist, are not distinguished. The low molecular mass nematics flow like liquids and it takes only minor forces to achieve alignment of the axes of the mesogens in a given direction. Nematics are uniaxially positive and have a schlieren texture which commonly shows thread-like defects [2] (disclinations). A thermodynamically equivalent phase to the classical, nematic phase N is the cholesteric phase CH. It results if a mesogen is chiral (i.e. it has a distinguishable mirror image) or if a chiral molecule is dissolved in a nematic liquid crystal. The basically nematic structure of the cholesteric phase contains a helical distortion of the orientation direction of much longer periodicity than the molecular dimension. Typical examples of N and CH are given in Table 1. The cholesteryl esters gave the name to the CH-phases.

The smectic phases are commonly divided into seven basic polymorphs [7, 55]. They all show layered structures of the mesogens (see Fig. 10). The layers can slide

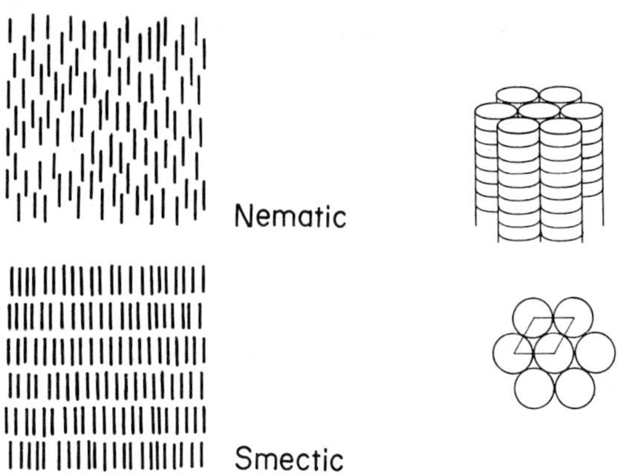

Fig. 10. Schematic representation of nematic, smectic and discotic mesophases

more or less freely relative to each other. The layer spacing is easily measured by X-ray diffraction. The smectics are thus more ordered than the nematics. If both a smectic and a nematic phase are stable for a given material, the smectic phase will occur at lower temperature in accord with its higher degree of order. In the smectics A and C the layers behave as two-dimensional liquids with the mesogens normal (A) or tilted (C) relative to the layer plane. In smectics B additional order occurs inside the layer planes, observable by X-ray diffraction. Smectic B is even more ordered than A and C. In tetramorphic liquid crystals the transition sequence is: fully ordered crystal to smectic B, to smectic C, to smectic A, to nematic, to isotropic liquid. A typical example is terephthal-bis-(p-butylaniline) with the respective transition temperatures 386, 417, 445, 473, and 509 K. Smectics D, E, F and G are less frequent and more complex. They appear only as low temperature polymorphs in addition to at least one of the other smectic phases [55]. As the degree of order increases in the smectic phases, it becomes more difficult to distinguish them from fully ordered crystals [7].

The discotic phases can show also a complex polymorphism. Nematic and cholesteric-like, low viscosity phases have been reported recently. In these, the director vector is perpendicular to the plane of alignment of the flat molecules [56] in contrast to the normal nematics and cholesterics where it is parallel to the molecular axis. Most frequently, however, discotics form columnar arrangements as shown in Fig. 10. The order within the columns may change from liquid to quasi-crystalline. The columns are then packed in hexagonal or tetragonal coordination, but are free to slide in the direction parallel to their axes [57]. The viscosity of these more ordered discotics is considerably higher than the nematic discotics.

5.1.2 Transitions of Small Molecules

For liquid crystals of small molecules only the equilibrium transitions have received attention (left side of Fig. 3). In addition to the basic transitions indicated in Fig. 2, the polymorphic transitions have to be added which normally are also first order

Table 2. Examples of Heats and Entropies of Transition of Nematic, Smectic, and Discotic Mesophases

Nematic[a]:	T_i (K)	ΔH_{total} (kJ/mol)	ΔS_{total} [J/(K mol)]	ΔS_i [J/(K mol)]
p-Azoxyanisole	391.4	31.0	80.5	1.82
p-n-Propoxybenzoic acid	426.7	27.2	65.7	5.9
N-(p-methoxyphenyl)-α-(p-ethoxyphenyl)nitrone	411.2	38.4	91.7	1.8
4-Benzylideneamino-4′-methoxybiphenyl	448.7	35.5	79.4	0.87
Diethylazoxy-α-methylcinnamate	414	35.0	91.8	2.93
4,4′-Di-n-butoxybenzene	409.9	22.3	59.1	3.2
4-Picolideneamino-4′-methoxybiphenyl	454.5	36.5	78.3	0.77
p-Methoxybenzylidene-(n-propyl-p-aminocinnamate)	407.2	27.6	78.2	1.05
p-Cyanobenzylidene-(n-propyl-p-aminocinnamate)	436.2	24.9	66.5	1.55
p-Cyanobenzylidene-p-aminoanisole	397.4	26.7	65.6	1.59
Cholesterylformate	370.2	22.1	60.2	1.09
Smectic[a]:				
4,4′-Di-n-nonoxybenzene	396.6	45.8	128.2	18.3
p-n-Nonoxybenzoic acid	418.2	37.7	101.7	10.1
N-(p-methoxyphenyl)-α-(p-duodecoxyphenyl)nitrone	400.2	60.4	157.8	5.5
Diduodecylazoxy-α-methylcinnamate	360	75.1	212.9	23.4
Discotic[b]:				
Benzenehexa-n-heptanoate	359.3	66.37	209.91	59.9

[a] Data from Ref. [59]; [b] Data from Ref. [60]

transitions. A particularly nice example is given in the scanning calorimeter trace of Fig. 11 for the transitions of bis(4′-n-octyloxybenzal)-1,4-phenylenediamine. One observes that the transitions into the mesophases have considerably less tendency to supercool than the transitions into the fully ordered crystal [58]. A collection of thermal data on about 200 mesophases, which form the basis of much of the following brief discussion, has been published by Barrall and Johnson [59]. Table 2 displays some of these data.

The transition of a nematic phase to the isotropic melt can be seen to produce only a small increase in entropy. It is much less than expected for a full entropy increase normally assigned to orientational disordering on melting [see Eq. (1), 20 to 50 J/(K mol)]. Figure 11 and Table 2 indicate that ΔS_i is only a few percent of the total entropy of fusion. Adding all the mesophase transitions listed in Fig. 11, an entropy increase of 41.4 J/(K mol) is found which would be in accord with the gaining of full orientational freedom of the mesogenic group, distributed over all mesophase transitions. The remaining 45.6 J/(K mol) of transition entropy for the change from fully ordered crystal to mesophase is more than enough to account for the loss of positional order [7–14 J/(K mol)], but not enough to account for the additional conformational disordering of the alkyl chains [9–10 J/(K mol methylene), i.e., about 150 J/(K mol)].

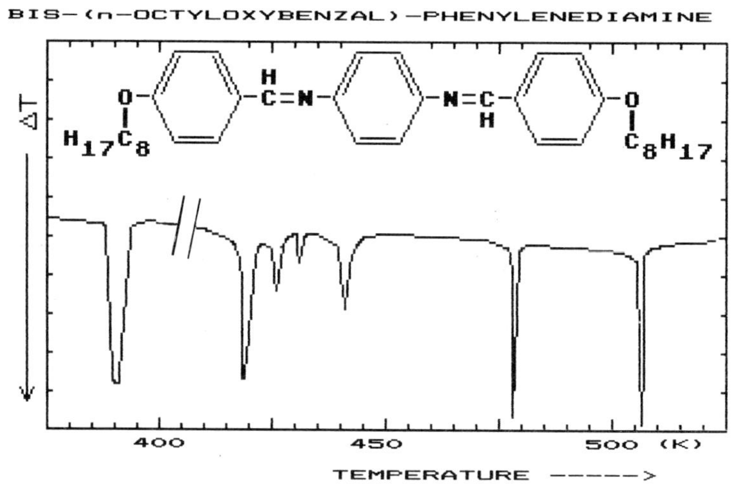

Fig. 11. DSC-trace of bis-(4′-n-octyloxybenzal)-1,4-phenylenediamine at 5 K/min. The lowest temperature transition is the fully ordered crystal-to-smectic transition, followed by four transitions between the five smectic polymorphs. Next is the smectic-nematic transition, followed by the nematic-to-isotropic liquid transition. Drawn after data by Petrie [58]

Temperature K	Enthalpy kJ/mol	Entropy J/(K mol)
388	17.6	45.6
415	6.7	16.3
422	2.1	5.0
428	0.4	1.0
437	3.2	7.1
476	3.6	7.5
504	2.3	4.6

One must conclude that either there are additional lower solid-solid transitions, or that the alkyl chains are poorly ordered in the crystal.

The influence of the alkyl chains on the mesophase transitions has been studied for several homologous series [59]. In all nematics there is hardly any influence of the first 5 to 7 CH_2-groups on ΔS_i. For longer sequences, ΔS_i seems to increase by 0.1 to 0.7 J/(K mol), hardly enough to account for more than a fraction of the conformational entropy change possible. The alkyl chains must thus have practically full conformational freedom in the nematic phase. As the alkyl chains increase in length, smectic phases become possible. In line with the greater order in the smectic phase, the increase in ΔS_i per methylene is now larger [up to 2 J/(K mol)], but still far from the full gain of conformational freedom. For very long chains, the mesophase transitions may become substantial, but are still too small to account for much of the conformational entropy of fusion. For 4,4′-di-n-octadecylazoxybenzene, for example, there are the smectic phases B and C observed, with a combined entropy of transition to the isotropic phase of 86.6 J/(K mol) which must be compared with a total entropy of fusion of about 300 J/(K mol) and a possible conformational entropy of fusion of the two octadecyl groups alone of about 350 J/(K mol).

The discotic example of Table 2 is of particular interest since the homologous benzenehexa-n-hexanoate shows polymorphism, but has no discotic mesophase. The successive solid-solid transitions of the benzenehexa-n-hexanoate had been interpreted quantitatively in terms of conformational melting of the paraffinic chains, progressing from the periphery of the molecule. The overall entropies of all transitions of benzenehexa-n-heptanoate are too low to account for the change from fully ordered molecules to the random structure of the melt. The discrepancy has been estimated in this case to be about 140 J/(K mol) [60]. Since the heat capacity has been measured for the hexanoate and the heptanoate from 13 K to beyond the melting temperatures, it could be shown that the entropies of the melts have the normal progression for increasing numbers of methylene groups. The mesophase-forming compound has a higher heat capacity in the fully ordered crystal, i.e. it gains entropy before the mesophase transitions. This observation points to a need for full thermal analysis of mesophase materials. It indicates clearly that disorder can be gained not only in transitions, but also well below the transition temperatures. Between polymorphic mesophase transitions abnormally high heat capacities have frequently been observed. The observations on the discotic materials indicate that such high heat capacities may occur even at lower temperatures.

LC-glasses are known for quite some time, but few detailed studies have been undertaken. Vorlaender [61] wrote in 1933: "... we shall recognize one day that the crystalline-liquids can assume any imaginable degree of viscosity which might be conceived also as hardness." He proceeded to list six possible LC-glasses. Particularly p-N-diethylamino-benzal-1-aminonaphthalene-4-ethylbenzoate was described to give on rapid cooling with ice a nearly clear, hard and brittle glass which could be scratched and split with a needle. The glass crystallizes only on annealing at room temperature for two or three days. The latter suggests a T_g of about 300 to 350 K (T_d = 461–462 K,

Table 3. Glass Transitions of LC-Glasses[a]

	T_g (K)	ΔC_p [J/(K mol)]	Beads	Ref.
N-(2-hydroxy-4-methoxybenzylidene)-4'-butylaniline (nematic)	204	107	7	[64]
p-butyl-p'-methoxyazoxybenzene (nematic)	207.6	129	7	[26]
eutectic mixture of p-butyl- and p-ethyl-p'-methoxyazoxybenzene (nematic)	208.1	102	6	[26]
cholesteryl hydrogenphthalate (cholesteric)	295	180	10	[63]
cholesteryl-2,4-dichlorobenzoate (cholesteric)	283	171	9	[65]

[a] For a summary of lyotropic polypeptide glasses see: E. T. Samuels, Liquid crystalline order in polypeptides in A. Blumstein, ed. "Liquid Crystalline Order in Polymers". Academic Press, New York, NY 1978. For the description of glasses of dicholesterylesters of dicarboxylic acids see: D. Gross, Z, Naturforsch. *B27*, 472 (1972)

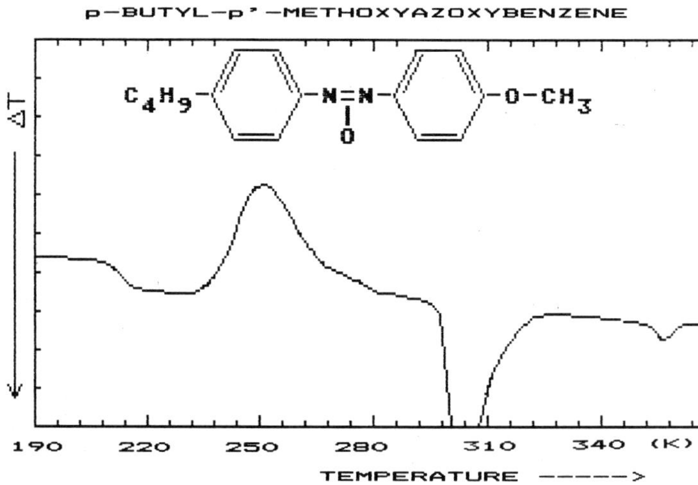

Fig. 12. DSC-trace of the LC-glass of p-butyl-p′-methoxyazoxybenzene. The glass was produced by cooling at 50 K/min. Heating rate 50 K/min. Drawn after Ref. [26]

Temperature K	Enthalpy kJ/mol	Entropy J/(K mol)	ΔC_p J/(K mol)
207.6	—	—	129[a]
250	—10.3[b]	—	—
296.5	13.4	42.5	—
352.5	0.6	1.6	—

[a] Extrapolated to completely glassy.
[b] Increase in crystallinity from 19 to 100%

$T_i = 470$ K). Sorai and Seki [62,63] proposed more recently the term glassy liquid crystal. They based their analysis on a more detailed thermodynamic study.

Table 3 contains a listing of several LC-glasses and their transition parameters. Figure 12 shows a scanning calorimeter trace of quenched p-butyl-p′-methoxy-azoxybenzene LC-glass with a glass transition at 207.6 K. The metastable liquid crystal above the glass transition temperature crystallizes at about 250 K to the fully ordered crystal. At 296.5 K the transition to the stable liquid crystal occurs, and at 352.5 K, one sees the transition to the isotropic melt. In line with the observation of a fast transition of an isotropic melt to the liquid crystal, it has not yet been possible to produce an amorphous glass for any small molecule which is able to form liquid crystals.

The increases in heat capacity listed in Table 3 reveal a contribution of all mobile structural motifs (called beads), mesogenic groups as well as alkyl pending groups, to the glass transition. This makes the increase in heat capacity of LC-glasses at T_g similar to the increase in heat capacity observed for comparable amorphous glasses. Quite different to this "normal" glass transition are the first order mesophase transitions which were described above. They show a deficit in the entropy of disordering, which was made up by an increased heat capacity between and before the transitions.

From the data in Fig. 12 one can find, for example, that the total entropy of fusion [46.8 J/(K mol)] is enough to account only for the positional and orientational entropy of fusion. The alkyl groups would have required an additional 40 to 50 J/(K mol) for conformational disordering. The large increase in heat capacity at the glass transition, however, points to all groups gaining mobility at T_g, i.e. all alkyl units are frozen below T_g.

The time dependence of the glass transition was studied by Sorai and Seki [64] who estimated an activation enthalpy of 75 kJ/mol for N-(2-hydroxy-4-methoxy-benzylidene)-4′-butylaniline, a value not much different from the 2,3; 2,5; and 4,3 isomers of the same compound which do not form mesophases. For the azoxybenzene LC-glasses of Table 3 little time dependence was suggested by a missing hysteresis [26].

Glasses have also been reported for nematic p-methoxybenzylidene-p-n-butyl-aniline [66] and some commercial liquid crystals [67] and for smectic ethyl-(p-anisol-amino)-cinnamate [68], p-n-hexyloxybenzylidene-p-aminobenzoic acid (T_g = 376 K) and p-n-nonyloxybenzylidene-p-aminobenzoic acid (T_g = 203 K) [69].

5.1.3 Macromolecules with Side-Chain Mesogenic Groups

Schematically the arrangement of a macromolecule with side-chain mesogenic groups in the liquid crystalline state is shown in Fig. 13. Flexible spacers give the mesogenic group its mobility. It is of interest to note that in such liquid crystals all positional mobility of the mesogen is based on conformational motion of the flexible spacer and backbone. Restricting this mobility either prohibits ordering, or freezes the order into the glassy state.

There exist large numbers of vinyl and related monomers which contain mesogenic groups. Blumstein and Hsu list over 100 such monomers in a recent review [70]. On polymerization, the resulting macromolecules contain these mesogenic groups as side chains. Perfect retention of nematic, smectic, and cholesteric order of the monomer could, however, only be achieved on polymerization in the presence of cross-linking agents [71]. These products are then of necessity LC-glasses. Little is known about their transition behavior.

Earlier studies of macromolecules with flexible, non-mesogenic side-chains had revealed that the transition properties change with increasing side-chain length from typical polymeric behavior to that of the small molecule which corresponds to the side chain [72,73]. With short side-chains and proper backbone geometry (tacticity),

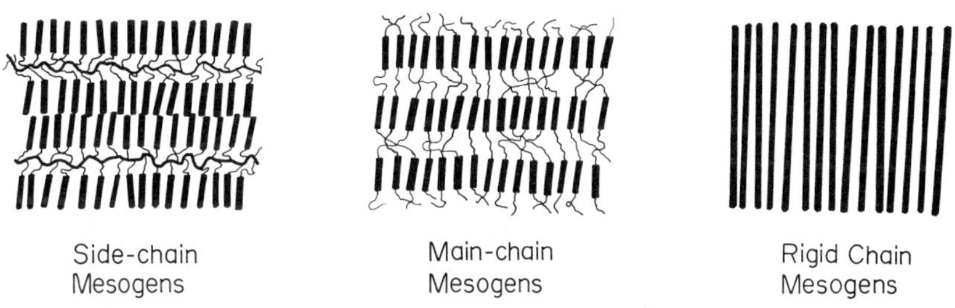

Fig. 13. Schematic representation of macromolecular liquid crystals. A smectic structure is assumed for all cases

Table 4. Transition Temperatures of Biphenyl Side-Chain Macromolecules[a)]

Monomer	T_g (K)	T_i (K)
$CH_2{=}CH{-}COO{-}C_{12}H_9$	383	553
$CH_2{=}CCH_3{-}COO{-}C_{12}H_9$	432	no mesophase
$CH_2{=}CH{-}C_{12}H_9$	413	no mesophase
$CH_2{=}CH{-}OCO{-}C_{12}H_9$	387	no mesophase
$CH_2{=}CH{-}COO{-}CH_2{-}C_{12}H_{19}$	323	no mesophase

a) Data from Ref. [77)]

a unique crystal, encompassing backbone and side-chain is possible. For longer, flexible side-chains, usually of more than 8 to 10 carbon atoms, it becomes increasingly possible to overcome the restrictions imposed by the backbone. The side-chains can now crystallize largely independent of the backbone chain. In these cases the transition temperatures, crystallization kinetics, crystal structures, crystallinities, and heats of transitions are closely related to the small molecule analogue. It was thus no surprise that mesogens attached to a polymeric backbone through flexible spacers are similar in behavior to low molecular weight liquid crystals [74–76)]. More surprising are the properties of a series of monomers which have the potential mesogen bonded directly to the backbone [70, 77)]. In these cases the appearance of a mesophase is not predictable and seems to depend on the specific combination of steric and polar characteristics of main and side-chain. For example, biphenyl groups in close proximity of the main chain cause a mesophase only in the case of poly(4-biphenyl acrylate) as is shown in Table 4. The closely similar poly(4-biphenyl methacrylate), poly(vinyl-4-biphenyl) and the poly(vinyl ester) do not lead to a mesophase. The last entry of Table 4, poly(methylene-4-biphenyl acrylate), has even a lower glass transition due to the extra methylene group, but still it does not lead to a mesophase. Also, none of the monomers of the polymers of Table 4 show mesophase behavior.

The transition behavior of a number of liquid crystals with side-chain mesogens is summarized in Table 5. The most obvious feature of macromolecular liquid crystals is the frequent absence of fully ordered crystals at low temperatures. If fully ordered crystals are observed, crystallization is incomplete, i.e. the observed phase states are to be described by an area on the right side of Fig. 3. Glass transitions, which were hard to find in low molecular weight liquid crystals (see Table 3), are now prominent.

The magnitude of the increase in heat capacity at the glass transition of the LC-glass is similar to that expected for fully amorphous glasses and also similar to that of equivalent low molecular mass LC-glasses. Figure 14 shows a comparison of a scanning calorimeter trace of amorphous, glassy poly(acryloyloxybenzoic acid) (entry 2 of Table 5) with the mesophase glass. The amorphous liquid is metastable and changes irreversibly at about 440 K to the mesophase. The mesophase glass, shown in the bottom curve, can be seen to have about the same heat capacity increase, but a glass transition temperature shifted by about 60 K. The other available data on heat capacity increases at the glass transition suggest also a "normal" gain of motion at the transition. All LC-glasses show the behavior of typical one-phase materials. Backbone and side-chain do not move independently. The closer the mesogen is to the main-chain, the higher is the glass transition temperature.

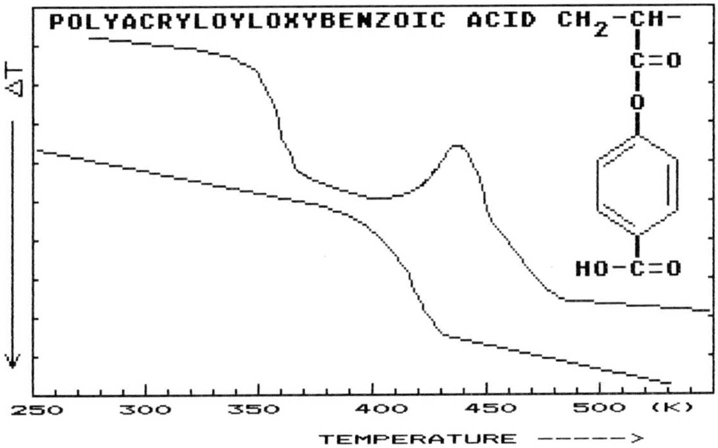

Fig. 14. DSC-traces of glasses of a macromolecule with a mesogen in the side-chain. Amorphous glass (top curve) and the corresponding LC-glass (bottom curve) drawn after data from Ref. [21]. Heating rates 50 K/min. The exotherm indicates the irreversible transition to the mesophase

Information on the crystal to liquid crystal transitions is scarce and is to be treated with caution since partial crystallization is prominent and polymorphism of the smectic phase is frequent. Only the data on poly(acryloyloxybenzoic acid) (entry 2 of Table 5) have been extrapolated to 100% crystallinity. As with the low molecular weight liquid crystals, the total heat of transition is lower than expected for fully ordered crystals. Various combinations of two phase structures as suggested by Fig. 3 could be produced for the poly(acryloyloxybenzoic acid) [21].

The transition to the isotropic phase shows less gain in entropy than expected for the fully oriented mesogen alone. The mesophase must thus contain considerable disorder, even for the mesogens. As in the low molecular weight materials, there is some indication that larger alkyl groups increase the transition entropy, but much less than expected for fully crystallized methylenes. All entropies of transition above 11 J/(K mol) are found for mesogens with pentyl or hexyl tails (entries 11, 14, 15, 17, 21, and 25 of Table 5). The nematic mesophases (entries 6, 8, 9, and 16 of Table 5) show a distinctly lower transition entropy than the smectic phases and seem less affected by long methylene sequences (compare entries 8 and 9 with 16). Occasional high transition entropies for nematic (entry 2), or low transition entropies for smectic materials (entries 20 and 24) may warrant reexamination of the structure assignment and/or the calorimetry.

The structure of liquid crystals of macromolecules with side-chain mesogens has been reviewed extensively (Refs. [6, 70, 73, 75, 77, 86]). The structure consists generally of double layers of parallel or antiparallel mesogens in an overall smectic structure. In contrast to the small crystalline areas of macromolecules, the ordered regions of the mesophases are many micrometers in size. The special stiffness of the macromolecules with mesogens directly attached to the main-chain leads often to a larger supercooling for the otherwise quite reversible isotropic-liquid to liquid-crystal transition [77, 80]. Hardy et al. proposed for a series of nematic p-alkoxyphenyl esters of polyacryloyloxybenzoic acid an irregular helical structure as the mesogen [87]. This

Table 5. Transition Parameters of Mesophases of Macromolecules with Side-Chain Mesogens

Macromolecule[a] Backbone/Side-chain	T_g (K)	ΔC_p J/(K mol)	T_d K	ΔH_d kJ/mol	T_i K	ΔS_i J/(K mol)	Ref.
Polyacrylates: $[CH_2—CH(COOR)—]$							
1. $—C_6H_4—C_6H_5$ (smectic)	383	(35)	—	—	553	8.1	[77]
	375	—	—	—	540	8.8	[78]
2. $—C_6H_4—COOH$ (nematic?)	408	43	585	22	—	(8)	[21]
3. $—C_6H_4—CH{=}N—C_6H_5$ (smectic)	365	—	—	—	530	5.0	[78]
4. $—C_6H_4—CH{=}N—C_6H_4—O—C_2H_5$ (smectic)	360	—	—	—	540	3.9	[78]
Polyacrylamide: $[CH_2—CN(CONHR)—]$							
5. $—C_6H_4—C_6H_5$ (smectic)	—	—	—	—	538	3.6	[80]
Polymethacrylates: $[CH_2—CCH_3(COOR)—]$							
6. $—C_6H_4—CH{=}N—C_6H_4—O—C_2H_5$ (nematic)	470	—	—	—	555	1.9	[78]
7. $—C_6H_4—O—CO—C_6H_4—O—C_6H_{13}$ (smectic)	453	—	—	—	520	8.0	[84]
8. $—(CH_2—)_2O—C_6H_4—C_6H_4—O—CH_3$ (nematic)	393	—	—	—	425	2.1	[81]
9. $—(CH_2—)_2O—C_6H_4—CO—O—C_6H_4—O—CH_3$ (nematic)	374	—	—	—	394	2.1	[76]
10. $—(CH_2—)_2O—C_6H_4—CO—O—C_6H_4—O—C_3H_7$ (smectic)	393	—	—	—	402	8.8	[76]
11. $—(CH_2—)_2O—C_6H_4—CO—O—C_6H_4—O—C_6H_{13}$ (smectic)	373	—	—	—	413	11.7	[76]
12. $—(CH_2—)_6O—C_6H_4—C_6H_4—OCH_3$ (smectic)	—	—	392	4.6	409	6.3	[81]
13. $—(CH_2—)_6O—C_6H_4—C_6H_4—O—C_2H_5$ (smectic)	—	—	407	4.3	435	10.2	[81]
14. $—(CH_2—)_6O—C_6H_4—O—C_5H_{11}$ (smectic)	353	—	—	—	402	17.7	[81]
15. $—(CH_2—)_6O—C_6H_4—C_6H_4—O—C_6H_{13}$ (smectic)	—	—	—	—	432	23.0	[81]
16. $—(CH_2—)_6O—C_6H_4—CO—O—C_6H_4—O—CH_3$ (nematic)	368	—	—	—	378	2.3	[76]
	365	(211)	—	—	375	2.9	[82]
17. $—(CH_2—)_6O—C_6H_4—CO—O—C_6H_4—O—C_6H_{13}$ (smectic)	333	—	—	—	388	19.2	[76]
18. $—C_6H_4—(CH_2—)_6CO—O$-cholesteryl (smectic)	—	—	—	—	455	9.7	[83]
19. $—C_6H_4—(CH_2—)_{12}CO—O$-cholesteryl (smectic)	—	—	—	—	441	7.1	[83]

Polymethacrylamides: $[CH_2{-}CCH_3(CONHR){-}]$							
20. $-(CH_2-)_5CO-O-C_6H_4-O-CO-C_6H_4-O-C_6H_{13}$ (smectic)	—	—	412	6.4	546	1.4	[84]
21. $-(CH_2)_{11}CO-O-C_6H_4-O-CO-C_6H_4-O-C_6H_{13}$ (smectic)	—	—	412	5.8	433	11.2	[84]
22. $-(CH_2)_{11}CO-O$-cholesteryl (smectic)	393	—	(330, 346)	—	453	4.6	[79]
Polymethylethylsiloxanes: $[SiCH_3(CH_2{-}CH_2R]{-}O{-}]$							
23. $-CH_2-O-C_6H_4-CO-O-C_6H_4-OCH_3$ (nematic)	288	107	—	—	334	2.3	[85]
24. $-CH_2-O-C_6H_4-CO-O-C_6H_4-CN$ (smectic)	293	78	—	—	334	1.9	[85]
25. $-CH_2-O-C_6H_4-CO-O-C_6H_4-O-C_6H_{13}$ (smectic)	288	46	—	—	385	12.5	[85]
26. $-CH_2-CO-O$-cholesteryl (smectic)	318	67	—	—	388	3.6	[85]

[a] All C_6H_4 are p-connected phenylene groups, cholesteryl = $C_{27}H_{45}$

puts the director vector practically at right angles to the mesogen. Such structure would represent the link between liquid crystals with side-chain mesogens and macromolecules which can assume a rigid conformation due to secondary structures, such as are found in poly(amino acids), for example. The polyacryloyloxybenzoates seem to be able to retain thermotropic character. In contrast to the methacrylates with longer flexible spacers between mesogen and backbone-chain, the glass transition increases in the acryloyloxybenzoic acid derivatives from 333 K for the phenyl ester to 463 K for the p-butoxyphenyl ester. The influence of the main chain can also be seen on the macromolecules with cholesteric mesogens in the side-chain. All of these are smectic, rather than cholesteric [79]. Only on disordering the placement of the side-chains by copolymerization is a cholesteric mesophase possible [83].

As shown above, by shortening the flexible spacer, it is possible to produce rather rigid chains which should probably better be described as main-chain mesogens. Mesophase superstructures can also be achieved with flexible side-chains (Shibaev et al. [88]). Lyotropic systems of this type are given by Finkelmann et al. [89].

5.1.4 Macromolecules with Main-chain Mesogenic Groups

The schematic arrangement of a macromolecule with main-chain mesogenic groups is shown in Fig. 13. Again, the flexible spacers permit, as in the side-chain case, mobility of the mesogens through conformational motion in the spacer. In the side-chain case, the coupling to the main chain led frequently to smectic structures. In the main-chain case, there have only been nematic liquid crystals found to date. As the spacer decreases in length or flexibility, the glass transition temperature increases. The mesophase can then be brought into the liquid crystalline state only by a solvent (lyotropic liquid crystals). Of particular interest are naturally the materials at the borderline of thermotropic behavior.

Table 6 contains data on the transition parameters for a large series of main-chain mesogen macromolecules. Figure 15 shows scanning calorimetry data for poly(oxy-2,2′-dimethylazoxybenzene-4,4′-diyloxydodecanedioyl), entry 8 in Table 6. These data should be compared to data on low molecular mass p-butyl-p′-methoxyazoxybenzene which are shown in Fig. 12. Similar comparisons are available for two benzalazines (entries 14 and 16 of Table 6). Although the entropies of transition from the liquid crystal to the isotropic melt are small for all polymers listed; as expected, they are larger than those of the corresponding nematic small molecules [1.6, 3.35, and 3.26 J/(K mol), respectively for the small molecules corresponding to entries 8, 14, and 16]. The main-chain nematics seem to have a somewhat larger entropy of transition than the side-chain nematics for larger flexible spacers.

From entries 30 to 59 in Table 6 the typical odd-even alternations in transition parameters of homologous series are obvious [96]. They are based on the interruption (odd) or maintenance (even) of an overall linear propagation of a zig-zag-chain with changing methylene or oxygen groups. Entries 13 to 26 of Table 6 indicate that similar transition parameters are found for homo- and copolymers as long as the average flexible spacer length is maintained. Odd-even structures remain of effect (see entries 19 to 24). Phase diagrams of several copolymers have been reviewed [97].

The crystal to liquid crystal transition has been measured frequently, but not studied in detail. From the generally poor crystallization of the polyester types which make up the liquid crystals [3] one expects typical samples to have low crystal-

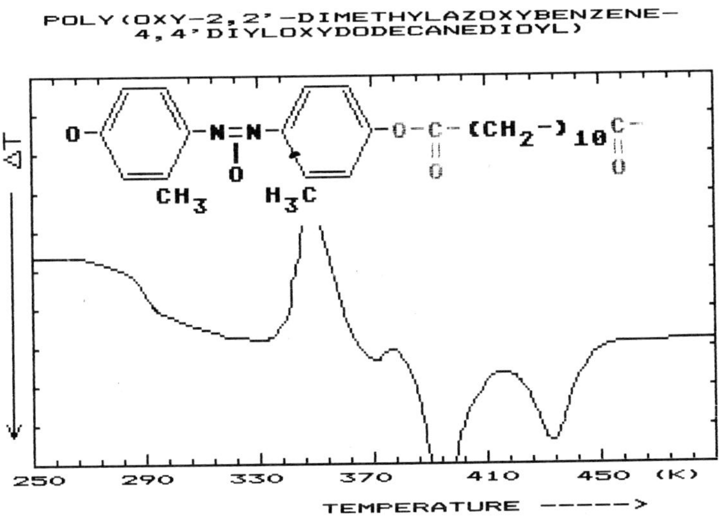

Fig. 15. DSC-trace of a mesophase macromolecule with flexible spacers in the main chain. Heating rate 50 K/min. Cooling rate 50 K/min. The mesophase glass crystallizes on heating at about 345 K and melts then at about 385 K. The reversible mesophase-isotropic transition is at about 320 K. Drawn after data from Ref. [27]

linity (10 to 50%). In very few cases has attention been paid to extrapolating the heats of transition to 100% crystallinity, but even then (entry 8 in Table 6), the heat of transition is much lower than expected for typical macromolecules [27]. For many well crystallizing macromolecules one finds at least 1 to 2 kJ heats of fusion per mol of C- or O-atom of the repeating unit. For example, poly(ethylene terephthalate) has, when extrapolated to 100% crystallinity, a heat of fusion of 26.9 kJ/mol; entry 8 in Table 6 for which the data are extrapolated to 100% crystallinity has only half this heat of fusion with more than double the number of large atoms. The low heats of transition are thus not only caused by low crystallinity, but also by rather poor crystal structures [27]. Much of the flexible spacers seem to be poorly ordered in the crystal which must accommodate the large and rigid mesogenic group. Many of the studied polyesters are also of rather low molecular mass [98]. The transition of the mesophase to the isotropic state was shown to be much more molecular mass dependent than the crystal to mesophase transition (Refs. [95] and [97], example: entry 11 of Table 6).

Glass transitions of the main-chain-mesogen liquid crystals have also not been studied in larger numbers. The example in Table 6 (entry 8) indicates contributions of all mobile groups, as was also found for the small molecules (Table 3) and the side-chain mesogen macromolecules (Table 5). For the mesogen alone, one would have expected only about 20 J/(K mol) increase in heat capacity at the glass transition [13,14]. A uniform, simultaneous freezing of large scale motion of spacer and mesogen is thus occurring at the glass transition.

Shortening the flexible spacer must ultimately lead to rigid chain macromolecules. Many examples of such macromolecules have been reviewed by Preston [99]. An interesting example is the series of poly(ethylene terephthalate-*co*-oxybenzoate)s

Table 6. Transition Parameters of Mesophases of Macromolecules with Main-Chain Mesogens

Macromolecule[a)]	T_g K	ΔC_p J/(K mol)	T_d K	ΔH_d kJ/mol	T_i K	ΔS_i J/(K mol)	Ref.
Azobenzene-type $[(CH_2-)_nO-CO-C_6H_3(CH_3)-N=N-C_6H_4-CO-O-]$[b)]							
1. n = 5	—	—	433	—	482	2.68	[90)]
2. n = 6	—	—	448	—	522	7.36	[90)]
3. n = 7	—	—	419	—	464	3.22	[90)]
4. n = 8	—	—	393	—	478	11.51	[90)]
5. n = 10	—	—	431	—	466	15.52	[90)]
$[(CH_2-CH_2-O-)_xCO-C_6H_4-N=N-C_6H_4-CO-O-]$							
6. n = 6 (x = 2)	—	—	410	11.9	461	5.9	[94)]
Azoxybenzene-type $[(CH_2-)_nO-CO-C_6H_3(CH_3)-N=NO-(CH_3)C_6H_3-CO-O-]$[c)]							
7. n = 7	—	—	385	—	429	4.31	[90)]
8. n = 10	288	220	391	13.5	418	9.76	[27)]
$[(CH_2-)_n-O-CO-(CH_3)C_6H_3-N=NO-C_6H_3(CH_3)-CO-O-]$[d)]							
9. n = 7	—	—	408	—	424	3.05	[90)]
$[(CH_2-CH_2-O-)_xCO-C_6H_4-N=NO-C_6H_4-CO-O-]$							
10. n = 6 (x = 2)	—	—	358	1.3	493	2.12	[94)]
11. n = 9 (x = 3)	—	—	367	4.2	433	1.06	[94)]
12. n = 12 (x = 4)	—	—	360	3.1	373	19.5	[94)]
Benzalazine-type $[(CH_2\text{-})_n-CO-O-C_6H_4-C(CH_3)=N-N=C(CH_3)-C_6H_4-O-CO-]$							
13. n = 6	—	—	511	10.6	578	19.4	[91)]
14. n = 8	—	—	476	7.95	529	16.7	[91)]
15. n = (6 + 10)/2[e)]	—	—	400	1.6	546	15.4	[91)]
16. n = 10	—	—	481	8.49	514	15.5	[91)]
17. n = 11	—	—	—	—	531	9.3	[91)]
18. n = (8 + 12)/2[e)]	—	—	—	—	528	8.6	[91)]
19. n = 12	—	—	—	—	545	18.0	[91)]
20. n = (10 + 14)/2[e)]	—	—	—	—	549	17.7	[91)]
21. n = (11 + 13)/2[e)]	—	—	—	—	525	9.4	[91)]
22. n = 13	—	—	—	—	509	9.5	[91)]
23. n = (11 + 15)/2[e)]	—	—	—	—	510	9.1	[91)]
24. n = (12 + 14)/2[e)]	—	—	—	—	529	17.8	[91)]

25. n = 14	—	—	—	—	516	17.9	[91]
26. n = (12 + 16)/2[e]	—	—	—	—	514	18.0	[91]
Aromatic polyesters $[(CH_2-)_nO-C_6H_4-CO-O-C_6H_4-O-CO-C_6H_4-O-]$							
27. n = 10	—	—	509	3.8	567	7.11	[92]
$[(CH_2-)_nO-C_6H_4-CO-O-C_6H_3(C_6H_5)-O-CO-C_6H_4-O-]$[f]							
28. n = 10	—	—	424	8.8	441	15.1	[92]
$[(CH_2)_nO-C_6H_4-CO-O-C_{10}H_6-C_{10}H_6-O-CO-C_6H_4-O-]$[g]							
29. n = 10	—	—	497	2.5	597	15.1	[92]
$[(CH_2-)_n\{O-C_6H_4-O-CO-C_6H_4-O-\}(CH_2-)_x\{O-C_6H_4-CO-O-C_6H_4-O-\}]$[h]							
30. n = 6, x = 2	—	—	572	2.8	606	22.4	[93]
31. n = 6, x = 3	—	—	546	10.8	556	6.7	[93]
23. n = 6, x = 4	—	—	544	8.2	580	15.2	[93]
33. n = 6, x = 5	—	—	510	6.0	543	11.7	[93]
34. n = 6, x = 6	—	—	530	6.2	555	24.1	[93]
35. n = 6, x = 7	—	—	504	6.8	532	12.8	[93]
36. n = 6, x = 8	—	—	507	7.4	535	27.2	[93]
37. n = 6, x = 9	—	—	478	6.6	518	18.6	[93]
38. n = 6, x = 10	—	—	488	6.2	512	16.3	[93]
39. n = 6, x = 12	—	—	474	9.1	487	11.3	[93]
40. n = 8, x = 2	—	—	534	2.5	569	18.0	[93]
41. n = 8, x = 3	—	—	502	6.8	521	11.5	[93]
42. n = 8, x = 4	—	—	524	1.3	563	16.4	[93]
43. n = 8, x = 5	—	—	494	8.1	522	13.6	[93]
44. n = 8, x = 6	—	—	506	8.8	532	21.0	[93]
45. n = 8, x = 7	—	—	477	6.0	511	15.5	[93]
46. n = 8, x = 8	—	—	481	6.8	509	25.3	[93]
47. n = 8, x = 9	—	—	455	6.5	492	23.6	[93]
48. n = 8, x = 10	—	—	462	6.1	494	37.0	[93]
49. n = 8, x = 12	—	—	455	9.9	478	20.3	[93]
50. n = 10, x = 2	—	—	506	4.2	565	26.2	[93]
51. n = 10, x = 3	—	—	485	10.9	505	16.3	[93]
52. n = 10, x = 4	—	—	496	3.8	539	14.9	[93]
53. n = 10, x = 5	—	—	469	5.1	501	17.2	[93]
54. n = 10, x = 6	—	—	494	8.6	525	28.1	[93]

Table 6 (continued)

Macromolecule[a)]	T_g K	ΔC_p J/(K mol)	T_d K	ΔH_d kJ/mol	T_i K	ΔS_i J/(K mol)	Ref.
55. n = 10, x = 7	—	—	454	7.1	485	17.6	93)
56. n = 10, x = 8	—	—	479	5.0	500	23.9	93)
57. n = 10, x = 9	—	—	453	6.1	476	20.0	93)
58. n = 10, x = 10	—	—	458	7.1	485	29.3	93)
59. n = 10, x = 12	—	—	448	5.5	466	19.4	93)
Diphenyldiurethane-type $[(CH_2-)_nO-CO-NH-C_{14}H_{12}-NH-CO-O-]$[i)]							
60. n = 5	—	—	521	10.8	527	26.2	95)
61. n = 6	—	—	493	4.2	507	14.0	95)
62. n = 8	—	—	504	17.1	510	11.4	95)
63. n = 10	—	—	475	1.2	494	16.9	95)
64. n = 12	—	—	447	0.4	459	15.5	95)
Copolyesters $[\{(CH_2-)_2O-CO-C_6H_4-CO-O-\}C_6H_4-CO-O]$							
65. 71/29[j)]	353	60.4	497	6.0	—	—	20)
66. 68/32[j)]	353	63.0	487	6.3	—	—	20)
67. 52/48[j)]	353	47.0	478	5.8	—	—	20)
68. 37/63[j)]	353	39.0	466	2.5	—	—	20)

a) All phenylenes are 1,4-connected in the backbone chain, all mesophases seem to be of the nematic type;
b) The methyl group is in the 2-position;
c) The methyl groups are in the 2,2′-positions;
d) The methyl groups are in the 3,3′-positions;
e) Copolyesters of equimolar amounts of two monomers of the indicated n's;
f) The phenyl group is in the 2-position;
g) $-C_{10}H_6-C_{10}H_6-$ = 1,1′-binaphthyl-4,4′-ylene;
h) Can be considered a strictly alternating copolymer. To compare with the other examples ΔH and ΔS could be divided by two and n taken as $(n + x)/2$;
i) $C_{14}H_{12}$ = 3,3′-dimethyl-4,4′-biphenyldiyl;
j) Mol-% ratio ethylene terephthalate to oxybenzoate

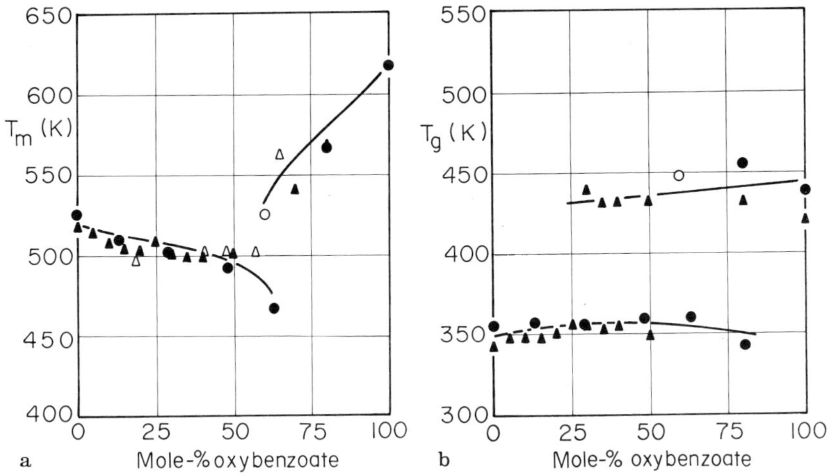

Fig. 16a and b. Phase diagrams of poly(ethylene terephthalate-*co*-oxybenzoate). **a.** Crystal melting temperatures; **b.** Glass transition temperatures. All samples with two glass transitions are most likely two-phase samples. The black circles up to 63 mol-% represent one-phase samples. Filled triangles: W. J. Jackson, Jr. and H. F. Kuhfuss, J. Polymer Sci., Polymer Chem. Ed. **14**, 2043 (1976). Open triangles: R. W. Lenz and K. A. Feichtinger, Polymer Preprints, Am. Chem. Soc. Div. Polymer Chem. *20*, 114 (1979). Filled circles: Ref. [20]. Open circle: J. Menczel and B. Wunderlich, J. Polymer Sci., Polymer Phys. Ed., *18*, 1433 (1980)

(entries 65–68 in Table 6). Their transitions to the isotropic state are already above the decomposition temperatures. Above 70 mol-% ethylene terephthalate the mesophase is lost. Slight changes in the random composition change the homogeneous polymer into a two-phase structure in which both phases have mesophase character, but widely differing glass transitions (differing by about 100 K). One expects, if small changes in composition can cause such large differences in the glass transition, a complicated interaction between the two components [20]. Sequences of the more flexible components may easily act as a solvent for the more rigid portions of the molecule. The phase diagram for crystallization shows no indication of the observed phase separation. Figure 16 compares the glass and melting transition diagrams of this system.

Only fragmentary details about the structure of the main-chain liquid crystals are known (for a review see Ref. [86]). Often condis crystals are confused with liquid crystals, and in many cases lyotropic liquid crystals are not separated from thermotropic materials. The problem is complicated since flexible chains, such as for example poly(gamma-benzyl glutamate) [47], can become rigid by a coil-to-helix transformation. Similarly, external stress or quenching can lead to incomplete orientation which may be described as a mesophase.

5.2 Plastic Crystals and PC-Glasses

5.2.1 Molecular Structure

In plastic crystals or PC-glasses one looks for positional order and an absence of orientational order. While for liquid crystalline order it is almost self-understood that the translational, liquid-like motion is cooperative, a special point must be made about the cooperative, orientational motion in plastic crystals. The molecular structure of the motif of the plastic crystal is close to spherical. But note, that several more disk-like molecules have been classified as plastic crystals. Their orientational disorder would then be one- or two-dimensional. From the van der Waals radii and the crystal structure one can deduce that the diameter of the equivalent true sphere or circular disc of the motif is generally by 15 to 20% larger than the available separation of motifs in the crystal, i.e. the motifs do not have enough room to rotate freely. The orientational motion requires a correlated movement between neighboring motifs. Also, one finds that the major volume change has usually occurred on changing from the fully ordered crystal to the plastic crystal. Only a minor additional change occurs on final transition to the isotropic melt. The need for cooperative motion persists thus also in the melt [100] and one expects a freezing of a plastic crystal to the glassy state (PC-glass) if on cooling the transition to the fully ordered crystal can be bypassed kinetically (see Fig. 2).

The structure of most plastic crystals is cubic [101], as one would expect for crystals of motifs of spherical shape (usually body or face centered). The softness of the crystals led to their discovery as a special state of matter by Timmermanns in the 1930's [5]. The softness is a direct consequence of the large number of slip-planes in close-packed structures and the relatively low attractive forces between the motifs. The frequent observation of ductility in metals has the same structural origin. An exceptionally plastic crystal is perfluorocyclohexane which is reported to flow under its own weight [16c]. The majority of plastic crystals are easily cut with a knife or extruded through a small hole. While liquid crystals showed as identifying optical property characteristic birefringence, plastic crystals are isotropic because of the highly symmetric cubic crystal structure. Only on transition to the frequently less symmetric fully ordered crystal does birefringence appear on cooling. Dunning [101] lists a brief description of the structure of over 60 plastic crystals. As a result of the orientational disorder and high mobility, X-ray diffraction of plastic crystals shows usually considerable background scatter and only a small number of reflections.

Although polymorphism in plastic crystals is less frequent than in liquid crystals, it does exist. Tetrakis(methylmercapto)methane, $C(SCH_3)_4$, for example, has four crystal modifications of which the three high temperature forms have a high degree of plasticity [100]. Also, it has been observed that plastic crystals are frequently mutually soluble [16b], a consequence of the less restrictive crystal structures. Phase separation of these solutions occurs often on transition to the fully ordered crystal, giving rise to quite complicated phase diagrams [102].

In view of the larger volume of plastic crystals, self-diffusion is much easier than in fully ordered crystals. The mechanism is similar to that in fully ordered crystals, not to that in liquids [7]. Much detailed information on the molecular motion has been gained by NMR studies [103].

5.2.2 Transitions

As with liquid crystals (Sect. 5.1.2), only the equilibrium transition to the fully ordered crystal and the transition to the isotropic melt have been studied extensively [16c, 102]. Both transitions are of the first order transition type, similar to that shown in Fig. 1. A series of transition data are collected in Table 7. Plastic crystals with motifs as small as

Table 7. Transition Parameters of Plastic Crystals

Substance	T_g K	ΔC_p J/(K mol)	T_d K	ΔH_d kJ/mol	T_i K	ΔS_i J/(K mol)	Ref.
N_2	—	—	35.6	0.23	63.1	11.4	16c)
CO	—	—	61.6	0.63	68.1	12.3	16c)
HCl	—	—	98.4	1.19	158.9	12.5	16c)
H_2S	—	—	(103.5, 126.0)	(1.54, 0.51)	187.2	12.7	16c)
CH_4	—	—	20.5	0.07	90.7	10.4	16c)
SiH_4	—	—	63.5	0.69	88.5	7.5	16c)
CF_4	—	—	76.2	1.48	89.5	7.8	16c)
CCl_4	—	—	225.5	4.59	250.3	10.0	16c)
$C(CH_3)_4$	—	—	140.0	2.58	256.6	12.7	16c)
2,3-dimethylbutane	76	51	136	6.43	145.1	5.5	104)
cyclohexanol	150	24	265	8.83	299	6.0	105)
cis-1,4-dimethyl-cyclohexane	94	56	172.5	8.26	223.3	7.4	105)
camphor	—	—	250.0	7.95	453.0	11.7	16c)

nitrogen and carbon monoxide have similar transition behavior as the large camphor molecule motifs. The entropies of transition to the isotropic state are often close to the expected value for positional disordering (see Sect. 2). This means that the plastic crystals have, when judged by the entropy of transition, high positional order. The liquid crystals achieved, in contrast, only a small fraction of the possible orientational order.

An inspection of the transition enthalpies to the plastic state reveals relatively small values, so that one may expect, as in the case of liquid crystals, that some of the enthalpy is gained continuously through increased heat capacity (see Sect. 5.1.2).

A transition to the plastic crystal is, however, not only geometry determined. This is illustrated by a comparison of adamantane (symmetric tricyclodecane) and the geometrically similar hexamethylenetetramine. The former is a typical plastic crystal, the latter not, because of its local polarity which hinders reorientation [16c]. More detailed theories and discussions of the phase changes have been given in Refs. [8, 16c], and [102].

The glass transition of PC-glasses was studied mainly in the laboratory of Seki. Figure 17 shows the thermal analysis curve of glassy 2,3-dimethylbutane [104]. Additional data on cyclohexanol [105] and cyclohexane [106] are shown in Table 7. Qualitative data on a series of other PC-glasses are given in Ref. [107]. The order of magnitude of the increase in heat capacity at the glass transition is the same as for amorphous glasses or LC-glasses. The residual entropy of 2,3-dimethylbutane at absolute zero is 7.4 J/(K mol), which would account for a random placement among 2–3 positions ($S = R \ln W$). This number is much smaller than the total of all possible orientational and conformational positions which may be estimated to be 12 [104]. It may be possible that the freezing of the cis and trans conformations alone accounts for the residual entropy [104]. In this case the PC-glass would be better described as a CD-glass. 2,3-dimethylbutane seems to be a case which might be either possessing a plastic or a condis crystal phase.

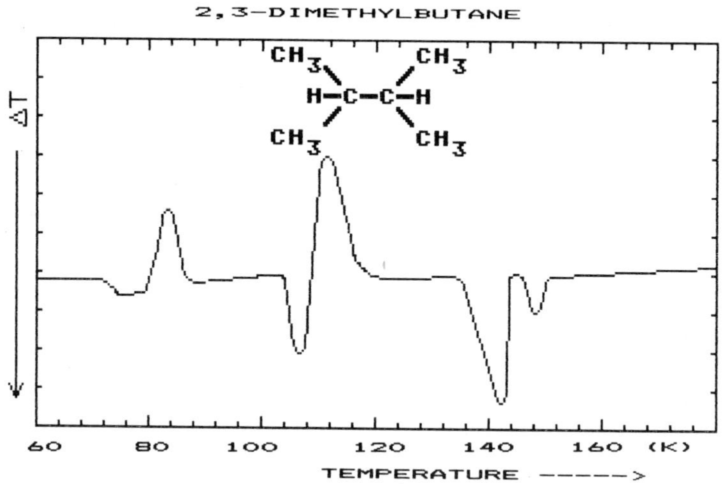

Fig. 17. DSC-trace of 2,3-dimethylbutane. The PC-glass transition occurs at 76 K. This is followed by the formation of a metastable crystal form (III), which transforms at about 110 K to the fully ordered crystal (II), which then changes to the plastic crystal (I) at 136 K and to the melt at 145 K (see also Table 7). Drawn after data of Ref. [104)]

Cyclohexanol, which has a lower change of heat capacity at the glass transition has also a low residual entropy of 4.7 J/(K mol) at zero kelvin for the PC-glass [105)]. Again, intramolecular effects (axial-equatorial conversions) alone could account for the residual entropy which would make cyclohexanol glass a CD-glass. The zero kelvin entropy of the PC-glass of cis-1,4-dimethylcyclohexane is 9.7 J/(K mol). A large amount of this zero entropy has been suggested to arise from the mixing of the d and l forms [106, 107)].

Although in the three investigated cases of PC-glasses it was not possible to quench the isotropic state to the glass, one expects for plastic (or condis) crystals a slower crystallization than in liquid crystals since a larger ordering is involved in their initial crystallization. As a result, it should be possible to observe amorphous glasses which would permit important comparisons. Also, it may be possible by solid state NMR of the glasses and plastic crystals to get more information on the actual motion.

5.2.3 Macromolecules

The question of "plastic" crystals of macromolecules has been brought up from time to time. Naturally, the condition of a globular molecule can in this case only be fulfilled by macromolecules which crystallize in a nonextended chain form. Globular proteins and viruses fall into this category [18)]. While it was shown that the more elongated shapes of these form liquid crystals (see Sect. 5.1.1), it has been demonstrated that sufficiently spherical shapes for orientational disordering are available. The presence of large amounts of water in the interstices of the crystals complicates the observation of truly plastic crystalline behavior. Most space groups observed do not have the high symmetry found in plastic crystals.

The examples of the extended-chain crystals, which from their thermal properties and mechanical properties have some similarities to plastic crystals, will probably all fall into the category of condis crystals. The need for differentiation of these molecules from true plastic crystals of small molecules was pointed out by Schneider et al. [108] who proposed the term "viscous crystal" for what is called condis crystal in this review.

5.3 Condis Crystals and CD-Glasses

5.3.1 General Description

In a condis crystal cooperative motion between various conformational isomers is permitted. In the CD-glass this motion is frozen, but the conformationally disordered structure remains. For a condis crystal it is not necessarily expected that all possible conformations can be reached, but all conformations of the same type are involved in the condis crystal motion. If conformational isomers of low energy exist which leave the macromolecules largely in a parallel, extended, low energy conformation, conditions for the formation of condis crystals are given. The conformational changes involve more or less hindered rotations about backbone bonds or side chain bonds and are thus the some degree related to the orientational motion in plastic crystals.

In order to find condis crystals, one looks for molecules with high temperature polymorphs that have high conformational entropies. For small molecules, there may be difficulties to distinguish between condis and plastic crystals, as is mentioned in Sect. 5.2.2. For macromolecules, there may be condis crystals of relatively rigid backbone or side-chain polymers that have, at higher temperature (or with the proper solvent), a liquid crystal polymorph. An example of a macromolecule that shows two mesophases is polyoxybenzoate (Sect. 5.3.2). Furthermore, one must distinguish crystals which develop practically independent motion of isolated side-groups from condis crystals. Such independent motion is excited over a wide temperature range and does not give rise to a mesophase transition. A typical example is the rotation of methyl groups in macromolecules. Such rotation starts gradually without a thermally noticeable transition. Overall, details of the condis state will have to be studied on a case by case basis (see Sects. 5.3.2 and 5.3.3).

No general rules about the entropy of transitions, as were found for liquid and plastic crystal transitions, can be set up for condis crystals. Two typical examples may illustrate this point. Polytetrafluoroethylene has a relatively small room-temperature transition-entropy on its change to the condis state and a larger transition entropy for final melting. Polyethylene has, in contrast, a higher condis crystal transition entropy than melting entropy (see Sect. 5.3.2).

Theories about the condis state can be based on the kink-model of the solid state [18]. Rotational isomers which do not disturb the parallel molecular arrangement too severely are introduced into the crystal. Pechhold and Blasenbrey [109] suggested, for example, for the condis crystals of trans-1,4-polybutadiene a cooperative, statistical treatment of pairs of straight and singly kinked chains. This model leads to large kink-blocks without torsion above a first order transition. On final melting these break into smaller kink blocks with torsional defects, jogs and folds [109]. Most likely the number of computed conformational defects is too small for the condis phase and a change of parameters may be needed. Baur [110] permits such a mixture of chains with larger numbers of defects (rotational isomers) which comes closer to

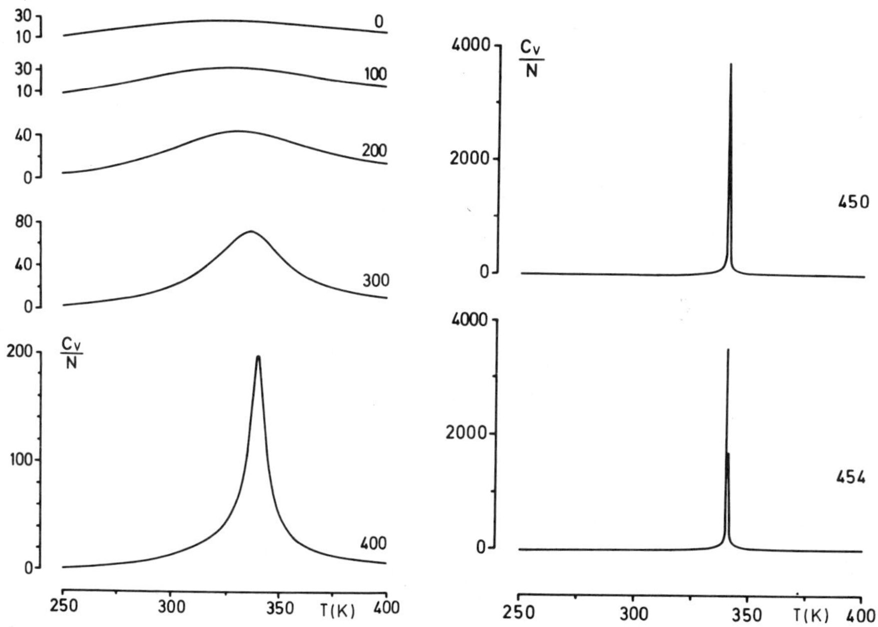

Fig. 18. Heat capacity of a cooperative system as a function of the excess energy on aggregation. The critical temperature of a first order transition is reached with the last curve (parameter = 454). The parameter 0 corresponds to an isolated hindered rotator. Curves after data of Ref. [110b)]

the condis crystal situation. The key of the theory is the evaluation of the number of different arrangements of the conformationally disordered chains at constant energy. This value is needed for the calculation of the partition function. The solution is closely related to the two dimensional Ising model with external field. In the condis crystal, the equivalent effect of the external field on the dipoles of the Ising model is the excess energy for the new rotational isomers. Since this model cannot be solved exactly, Baur used the Bragg-Williams and the quasi chemical methods for analysis. Figure 18 shows the effect of the introduction of new rotational isomers on the heat capacity as a function of the cooperative tendency of the chains with rotational defects. The latter is expressed by the excess energy of the asymmetrical fully-ordered-chain rotational-isomer-chain pairs over that of the symmetrical pairs. The temperature of the discontinuous transition depends, in addition, on the rotational isomer energy and the size parameters of the crystal.

5.3.2 Macromolecules

In this section information on possible condis states of the following macromolecules are reviewed: polyethylene, polytetrafluoroethylene, poly(vinylidene fluoride), polychlorotrifluoroethylene, polypropylene, trans-1,4-polybutadiene, cis-1,4-poly(2-methylbutadiene), polyoxybenzoate, poly(ethylene terephthalate), nylon, poly(diethyl siloxane), and polyphosphazene. There is no reason to assume that this selection is complete. Statton [111)] has shown, for example, already in 1959 on a list of 29 macromolecules that longitudinal and lateral disorder may exist. Similarly, textbooks [18, 112)]

Fig. 19. Fracture surface of extended chain polyethylene crystallized at about 480 MPa pressure at 493 K (21 h) followed by slow cooling (96% cryst.). Weight average molecular weight 24,100. Electron micrograph of a replica. Scale bar 1 micrometer. R. B. Prime, unpublished

contain several additional examples of disorder in macromolecular crystals. Many of these show conformational disorder and represent thus condis states. Often the disordered state is frozen (CD-glass) and the equilibrium condis crystal state is not reached before melting. Little further work has been done on these materials to show whether pressure or strain can stabilize a condis crystal below the melting temperature. The condis crystals described below have all been discussed in the literature as being some type of mesophase material, or as showing some behavior typical for condis crystals. Most of these macromolecules cannot be described as liquid crystals because of their chain flexibility. They do not possess mesogenic groups or shapes needed for the liquid or plastic crystalline state. From structural considerations alone, it is thus likely that the reported mesophases are condis states.

The special properties of a condis state of *polyethylene* were observed in our laboratory in 1964 when crystallizations were attempted at hydrostatic pressures above 300 MPa [113]. Extended chain crystals of close to 100% crystallinity resulted under these, otherwise more restrictive, crystallization conditions of elevated pressure [114, 115]. Figure 19 shows a typical fracture surface of such extended chain crystals. It was suggested from morphological evidence [116] that during the crystallization high mobility exists in the crystal phase which leads to the chain extension. Later [117] a special, highly symmetric (hexagonal) high pressure phase was discovered which was linked with the extended chain crystallization [118].

Figure 20 shows the phase diagram of polyethylene [119]. The existence range of the condis crystals increases with pressure and temperature. The enthalpy of the reasonably reversible, first order transition from the orthorhombic to the hexagonal condis phase of polyethylene is 3.71 kJ/mol at about 500 MPa pressure [121] which is about 80% of the total heat of fusion. The entropy of disordering is 7.2 J/(K mol), which is more than the typical transition entropy of paraffins to their high temperature

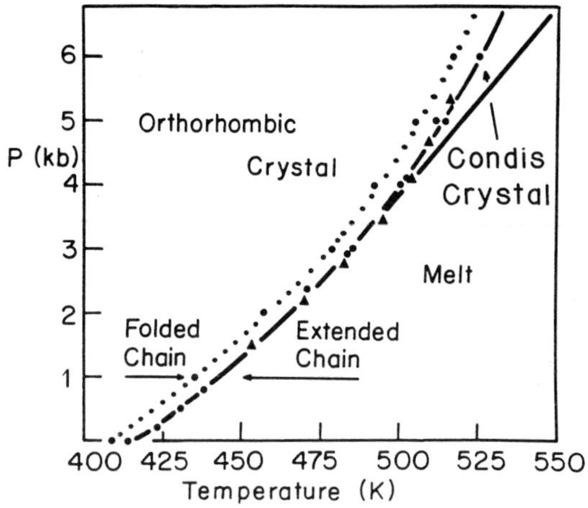

Fig. 20. Phase diagram of polyethylene. The difference between folded and extended chain crystal melting temperatures is largely a size effect (1 kbar = 100 MPa). Curve based on data of Ref. [119]

phase, 2.6 J/(K mol) for octadecane at about 170 MPa [120]. The volume change at the transition is 0.075 cm^3/g [122], which is about 8% of the crystalline volume. The suggestion that the high pressure phase of polyethylene is a mesophase (assumed at that time to be a nematic liquid crystal) was first made by Yasuniwa and Takemura [123] after optical microscopy at elevated pressure. In the meantime, it is proven that the disorder in polyethylene consists of statistically disordered conformations [120, 124], i.e. it represents, according to the definition in Sect. 2, a condis crystal. By Raman spectroscopy [125] evidence of many gauche conformations in the chain was brought (skeletal bands at 1067 and 1135 cm^{-1} are liquid-like). Similarly, ultrasonic studies [126] showed a decrease of the shear modulus to the liquid level at the transition to the condis phase (deduced from transverse velocity, little change was found in the longitudinal velocity which is related to the bulk modulus). All is in accord with the X-ray data which show an expansion normal to the chain axis and a contraction in the chain axis direction. Only sharp 100, 110, and 200 (equatorial) Bragg reflections have been observed, which indicate a good 2-dimensional lattice of the chains with disorder along the chains [124].

Polytetrafluoroethylene shows a complicated phase diagram of at least four polymorphs [3]. Of special interest to this discussion is phase I which exists at atmospheric pressure between about 303 K and the melting temperature (600 K). Starkweather [127] proposed that this phase I may be a smectic mesophase. This was concluded on the basis of the rheological behavior. Below about 300 K no flow could be detected within the instrumental limitations. As soon as phase I was formed, the apparent shear viscosity decreased rapidly with increasing temperature, to increase again on melting. The shear-rate dependence was found to be that observed for small molecule smectics. Based on the definitions given in Sect. 2, phase I should be a condis crystal. As polyethylene, polytetrafluoroethylene has no mesogenic groups and is flexible enough to show normal melting at 600 K. The melt crystallizes in a similar fashion as polyethylene at elevated pressure; i.e. the condis crystal, once grown in a chain folded macroconformation, can thicken to extended chain crystals of high crystallinity [128, 129]. This ease of chain extension and the low shear viscosity point to high mobility in the crystalline phase. The isolated polytetrafluoroethylene molecule has four rotational isomers [130]. The trans(+) and the trans(—) isomers have a $\pm 15°$ rotation about the backbone bonds from the strictly planar zig-zag conformation. The planar zig-zag conformation itself is sterically hindered due to the size of the fluorine atoms. The potential barrier between the two trans-isomers is only about 1.7 kJ/mol. The two other isomers are gauche-isomers at $\pm 123°$ rotation angles with potential energy minima 4–5 kJ/mol higher than the trans minima. At low temperature, the rigid crystal is triclinic (form II). It has a helical 1*13/6 conformation consisting within any one crystal domain of either all trans(+) or trans(—) conformations [18]. At about 295 K, this structure changes to a 1*15/7 helix in a solid-solid transition (form IV, trigonal). At an about 10 K higher temperature, this new helix changes to the condis crystal (crystal form I) with almost no further volume change (about 0.08 % [131]). The condis crystals show sharp X-ray diffraction spots only on the equator. The combined volume change of both room temperature transitions is 1.3 % [131], the transition entropy is 3 J/(K mol) [132], of which more than 80 % are gained at 295 K [133]. The transition entropy to the isotropic melt at 600 K is 5.7 J/(K mol) [3]. On the basis of NMR and X-ray data, Clark and Muus [134] proposed that torsional oscillations and hindered rotations can lead to an untwisting of the helix. The 1*15/7 helix of form IV is locked-in for a short temperature range because of its symmetry match with a trigonal 3/1 helix. Helix reversals were assumed to collect in the rigid crystal in domain boundaries normal to the backbone axis. Nucleation and movement of such domain boundaries were used to explain the transition kinetics II–IV [131]. Molecular motion was also studied by Raman spectroscopy [125]. A 581 cm^{-1} band, typical for the crystal form II at low temperature, decreases rapidly in the II to IV transition region, while a 601 cm^{-1} band appears. The latter band was attributed to the helix reversal conformational change. The regions of helix reversal are assumed to grow and to lead to equal trans(+) and trans(—) population above 400 K. Mobility in the transition range was also found by thermoluminescence [135].

Poly(vinylidene fluoride) can be considered an alternating copolymer of methylene and difluoromethylene. Usually during the synthesis the molecules acquire up to about 10 % head-to-head monomer reversals. Only because of extensive isomorphism between vinylidene fluoride and tetrafluoroethylene repeating units is the crystallinity of the higher defect content samples still appreciable (usually at about 50 %) [3]. There are 3 major crystal forms, and although many conformational defects have been described [136], and transformations between some polymorphs involve conformational mobility [137], no mesophases have been proposed for any of the crystal structures so far. A study of crystallization under elevated pressure [118c] revealed, however, that at 350 MPa, crystal form I can be grown as extended chain crystals. This is analogous to the crystallization of polytetrafluoroethylene. During this type of crystallization, large mobility within the crystal must exist. On cooling and removal of pressure, crystal form I remains as a metastable crystal and does not change to the stable crystal form II. Hasegawa et al. [138] showed that this metastable crystal form I has a close to planar zig-zag conformation which contains statistically disordered rotation angles of about $\pm 7°$, not far from the two trans conformational isomers of polytetrafluoroethylene. One might suggest that the metastable crystal form I is at low temperature perhaps a CD-glass. At elevated temperature and pressure it becomes a stable condis crystal similar to polytetrafluoroethylene I. The transformation to crystal form II which involves a change to trans-gauche(+)-trans-gauche(—) conformations seems not to be possible without fusion. The new conformation requires a $\pm 135°$ rotation angle. This is too big a change to occur in the crystal [3].

Poly(chloro trifluoroethylene) is another macromolecule which can, at pressures as little as 10 MPa, be crystallized into extended chain crystals [139]. Little is known about the detailed process, but a high mobility state was assumed to explain the special crystallization [19]. The only crystal structure reported is hexagonal with 17 monomers in the repeat distance along the chain axis. Conditions for conformational mobility seem favorable.

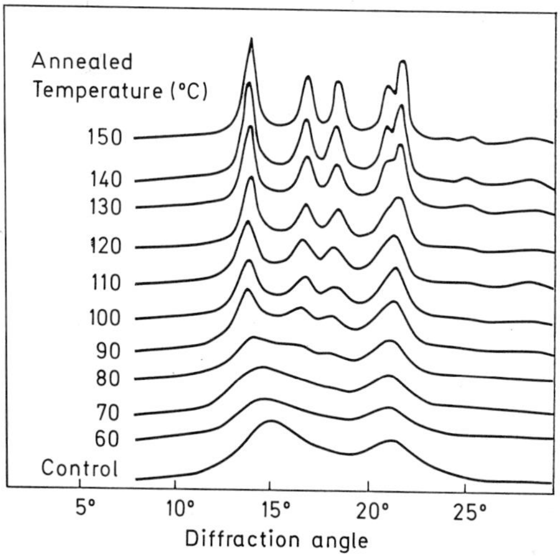

Fig. 21. X-ray diffraction pattern of the annealing effect on polypropylene quenched to the CD-glass (control trace). The changes may suggest a glass transition at 340 to 360 K. Curve courtesy of Dr. W. W. Cox of the Research Laboratory of Hercules Inc.

Isotactic *polypropylene* was shown by Natta et al. [140] to crystallize when cooled quickly from the melt into a crystal form which was called a smectic, mesomorphic form. The X-ray diffraction pattern resembled clearly that of a smectic material, as is shown in Fig. 21. The infrared spectrum seemed to indicate a helical, crystalline structure. Only the lateral packing was assumed to be different. Infrared absorption, density and the X-ray diffraction pattern indicated high crystallinity [140]. As with the other flexible, linear macromolecules, one can in the light of the definitions of Sect. 2 call this structure a condis structure. Although the isotactic polypropylene forms a helix, this helix is not as rigid as the polypeptide helices discussed in Sect. 5.1.1 and cannot serve as mesogen. Without a mesogen, a liquid crystal structure is unlikely. Later, this structure was also called para-crystalline to indicate the poor crystalline order [141]. It was found that this structure is metastable below 335 K. At room temperature it has been reported to persist for over 18 months [142], so that it should be called a CD-glass. The proposed transition mechanisms to the stable crystal form involve intramolecular helix perfection (removal of helix reversals) [141] and intermolecular alignment [143]. The transition is exothermic with about –0.7 kJ/mol enthalpy of transition [144]. The heat of fusion of the stable crystals is, in contrast, about 6.9 kJ/mol [3]. The question of interest remains whether there is at higher temperature and pressure a region of stability for the CD-glass. At atmospheric pressure no stable high temperature crystal form has been observed. Crystallization at elevated pressure (above 200 MPa) results in a new crystal form (gamma). Heating the crystal form stable at atmospheric pressure under elevated pressure shows no solid-solid transition [145], and X-ray diffraction shows no similarity of the CD-glass with the high pressure crystals (quenched to atmospheric pressure). The melting temperature at elevated pressure is lower for the high pressure crystal form, and on annealing at atmospheric pressure, it converts to the stable crystal form. Consequently, it seems that the high pressure crystal form is also not a stable crystal form under any of the studied conditions. The reason for the formation of the CD-glass (and the high pressure crystal form) must thus be looked for in the Ostwald Law of Successive States, which says that a new phase will occur step-by-step through successively more stable polymorphs. On crystallizing quickly, the condis crystal is frozen-in. This observation may have far-reaching implications for the understanding of crystallization of macromolecules.

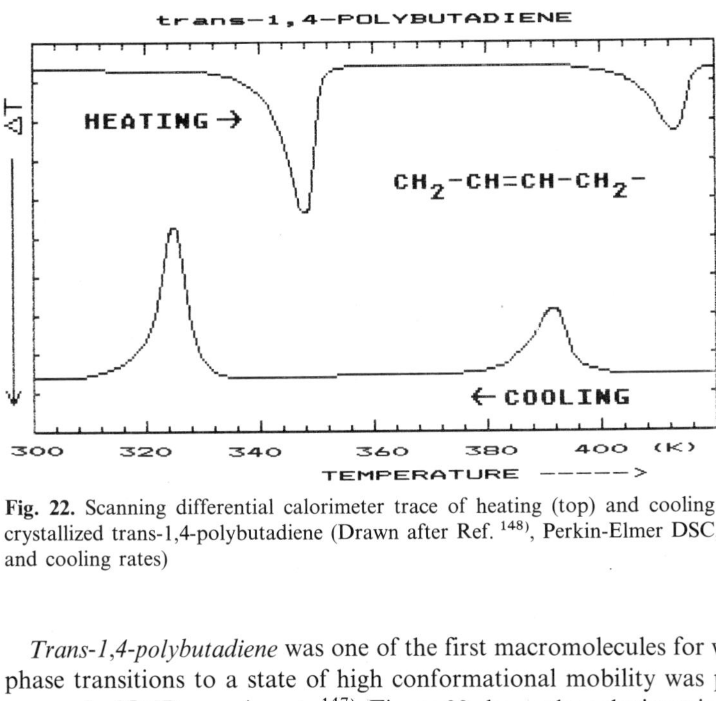

Fig. 22. Scanning differential calorimeter trace of heating (top) and cooling (bottom) of a melt crystallized trans-1,4-polybutadiene (Drawn after Ref. [148]), Perkin-Elmer DSC, unspecified heating and cooling rates)

Trans-1,4-polybutadiene was one of the first macromolecules for which a reversible phase transitions to a state of high conformational mobility was proposed [146] and proven by NMR experiments [147]. Figure 22 shows the calorimetric curves of cooling and heating. The condis crystals (crystal form II) is found between the two transitions. The heat of transition from the rigid crystal (form I) to the condis crystal is about twice as big as the transition of the condis crystal to the isotropic melt [148]. The supercooling of both transitions is similar. The two methylene groups have the normal trans-conformation in the monoclinic crystal form I, while the bonds next to the double bond are rotated +71° and −71° [18]. On transition to the pseudo-hexagonal condis crystals at about 350 K, the chains remain in their relative positions, but increase their chain separation by 7.5% and shorten the chain length by 4% and have an overall volume increase of 9% [146, 149, 150]. Estimations of the number of rotational isomers accessible in the condis crystal range from 9 per repeating units [149] to 4 per two repeating units [151] which would lead to conformational entropies of 18.3 to 5.8 J/(K mol), respectively. The total number of possible isomers is 27. The entropy change due to expansion has been estimated based on expansivity and compressibility data to be 3.2 J/(K mol) [151]. Transition temperatures of crystals I to condis crystals and then to the melt are 356 and 437 K. The corresponding entropies are 22 and 8.4 J/(K mol), respectively [152]. The total entropy of fusion of 30.4 J/(K mol) corresponds to 3 moles of bonds gaining conformational freedom, as expected empirically (see Sect. 2). It agrees also with the entropy of fusion of the cis-1,4-polybutadiene [32 J/(K mol)] which does not form stable condis crystals [3]. Although crystallization to condis crystals does not lead to extended chain crystals, considerably larger fold lengths (15–25 nm) are recorded on crystallization of condis crystals than for the crystal form I (8–10 nm) [152]. On heating the form I crystals, a doubling of the crystal thickness occurs at the transition to the condis crystal, with further thickening only close to the melting temperature of the condis crystals [19, 152]. No effort has been made as yet to observe a possible glass transition of the condis crystals. It is also interesting to note that the closely related trans-1,4-poly(2-methylbutadiene) (gutta percha) has also two polymorphs. The high temperature polymorph grows above 318 K, and is also the stable polymorph at low temperature [3]. A direct conversion between the two polymorphs seems, however, to be impossible or extremely slow, and both crystals have been assigned distinct chain conformations [112]. The entropies of fusion for both polymorphs show only a small difference, as is typical for different fully ordered crystal polymorphs [36.4 and 29.7 J/(K mol) [3]].

A different behavior which is not fully clarified is observed on the crystallization of *cis-1,4-poly(2-methylbutadiene)* (natural rubber). It has in its atmospheric pressure, monoclinic crystal polymorph an entropy of fusion of only 14.4 J/(K mol) compared to 32.0 J/(K mol) for the corresponding cis-1,4-polybutadiene [3]. Its crystal structure was found early to be statistically disordered [153], but there has been no report of mobility of the disordered chains in the monoclinic crystalline state. Increasing the pressure of crystallization, one finds an inhibition of nucleation of the monoclinic crystals; instead, a different, also disordered, hexagonal phase was discovered by Edwards and Phillips [154] at crystallization temperatures above 300 K and pressures above about 300 MPa. The crystal structure was found to resemble the monoclinic structure by having an identical chain axis repeat length, but a disordered packing normal to the chain. Great similarities exist between the high pressure phase of cis-1-4-poly(2-methylbutadiene) and the high pressure phase of polyethylene [155] so that the assignment as a condis crystal may be reasonable. As in polyethylene, large, tapered lamellae with secondary thickening were observed, which is an indication of increased mobility in the crystalline state. The ultimate thickness reached increased with pressure and temperature, but is much less than for polyethylene. Quenching to a CD-glass was possible in this case by cooling under pressure. A conversion of the condis crystal to the monoclinic crystal form was observed even at elevated pressure. It seems thus that the metastable, hexagonal condis crystal grows under most crystallization conditions because of nucleation difficulties of the more stable crystal form [155]. Much more detailed information is needed before the high pressure phase that has been characterized as "rotor phase solid" or as similar to a "smectic liquid crystal" can be fully characterized. The oldest known polymer crystal [156] is thus still not fully understood.

Polyoxybenzoate is a stiff chain, lyotropic liquid crystalline material, as was discussed on the basis of its copolymers with ethylene terephthalate (see Sect. 5.1.4). The crystal structure of the homopolymer polyoxybenzoate was shown by Lieser [157] to have a high temperature phase III, described as liquid crystalline. X-ray and electron diffraction data on single crystals suggested that reversible conformational disorder is introduced, i.e. a condis crystal exists. Phase III, which is stable above about 560 K, has hexagonal symmetry and shows an 11% lower density than the low temperature phases I and II. It is also possible to find sometimes the rotational disorder at low temperature in crystals grown during polymerization (CD-glass).

Nylon 6 was shown by Gogolewski and Pennings [158] to give extended chain crystals when grown under pressures of up to 800 MPa. No special crystal polymorph was discovered at elevated pressure. The transformation of the metastable gamma form to stable alpha, that occurs at atmospheric pressure close to melting [3], is visible as a separate exotherm at elevated pressure [159]. All extended chain crystals are of the stable polymorph after removal of pressure and seem to have undergone no phase transition after crystallization. The only indication that perhaps the alpha crystal form itself may permit higher mobility at elevated pressure (and temperature) is the fact that infrared analysis shows a high concentration of non-hydrogen-bonded NH-groups. [158]. Gogolewski [160] proposed that chemical reaction (transamidation) may also be operative in the chain extension on crystallization under elevated pressure. It may, however, be possible to have (in addition) conformational mobility in the gamma and alpha crystals. *Nylon 11* and *nylon 12* have similarly been crystallized under elevated pressure and yield extended chain crystals [161]. In both cases crystallinity and melting temperature were increased on crystallization under elevated pressure, and the alpha polymorph was the dominant crystal form. This applies even to nylon 12 which grows at atmospheric pressure as the gamma polymorph [162]. The question of the existence of a condis crystal in nylons must thus remain open. Although the quenched crystals of nylon 6 have been called mesomorphic [163], the detailed description and distinction between defect crystal and condis glass is not yet possible.

Poly(ethylene terephthalate) can be grown to extended chain crystals similar to the nylons [164]. No special high pressure polymorph has been suggested. But there may be, as in the nylons, the possibility that a continuous increase in mobility exists in the crystal phase at higher pressure and temperature. This mobility would have to be based on conformational changes, i.e. a condis crystal phase.

Poly(diethyl siloxane) was suggested by Beatty et al. [165] based on DSC, dielectric, NMR, and X-ray measurements to possess liquid crystalline type order between about 270 and 300 K. The macromolecule shows two large lower temperature first order transitions, one at about 200 K, the other at about 270 K [166, 167]. The transition of the possible mesophase to the isotropic liquid at 300 K is quite small and irreproducible, so that variable, partial crystallinity was proposed [165] [measured heat of transition about 150 J/mole [168]]. Very little can be said about this state which may even consist of residual crystals. It is of interest, however, to further analyze the high temperature crystal phase between 200 and 270 K. It is produced from the, most likely, fully ordered crystal with an estimated heat and entropy of transition of 5.62 kJ/mol and 28 J/(K mol), respectively [calculated from calorimetric data [166] assuming 60% crystallinity as determined by NMR [167]]. The transition to the melt (or possible mesophase) at 270 K has a similarly estimated heat and entropy change of 1.70 kJ/mol and 6.3 J/(K mol), respectively. The entropy gain at 200 K is so large, that a condis crystal seems likely. Some indication of higher mobility is also given in the NMR data [168] which were taken to support a possible liquid (viscous) crystalline state above 270 K. They showed, for example, that the spin-spin relaxation time T(2) in the amorphous and in the crystalline regions are equal above 200 K.

Alkoxy and aryloxy *polyphosphazenes* with the general structure

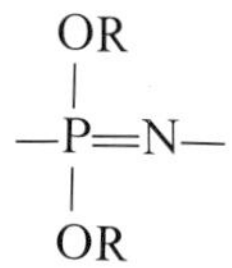

present a group of macromolecules which frequently have a wide temperature range of a stable mesophase. Schneider et al. [108] list 16 polymers of this type. In the light of the present discussion they should be classified as condis crystals. The polymers are flexible with low intrinsic barriers to rotation, as is indicated by their low glass transitions, which range from about room temperature down to 180 K, depending on substitution [108, 169]. In the fully ordered crystal, the chain conformation is cis-trans planar with a fiber repeat of 0.48 to 0.49 nm. The thermal behavior is dominated by the two first order transitions limiting the condis crystal state. Considerable data are available for poly[bis(trifluoroethoxy)phosphazene] and poly[bis(-p-chlorophenoxy)phosphazene] [170]. The disordering transition crystal to mesophase is strongly thermal history dependent. The heat of transition to the isotropic melt is only 1/10 or less of the initial disordering. Solution grown crystals of poly[bis(trifluoroethoxy)-phosphazene] have a spherulitic morphology and show at the disordering transition of 363 K a loss of X-ray diffraction to a single reflection. The morphology changes little up to melting at 513 K. Recrystallization from the melt leads to a needle-like morphology. The published transition data refer to 60–80% crystalline samples and one can estimate from these the following transition enthalpies, entropies and

Table 8. Transition Parameters of Macromolecular Condis Crystals

Macromolecule	T_d (K)	ΔH_d (kJ/mol)	ΔS_d [J/(K mol)]	T_i (K)	ΔH_i (kJ/mol)	ΔS_i [J/(K mol)]	Ref.
Polyethylene (500 MPa)	513	3.7	7.2	521	(1.4)	(2.7)	[121]
Polytetrafluoro-ethylene	300	0.9	3	600	3.4	5.7	[3]
trans-1,4-Poly-butadiene	356	7.8	22	437	3.7	8.4	[150]
Poly(diethyl siloxane)	200	5.6	28	270	1.7	6.3	[165]
Poly[bis(trifluoroeth-oxy)phosphazene]	365	12.5	34	513	1.2	2.3	[170]
Poly[bis(p-chlorophen-oxy)phosphazene]	442	11.8	27	629	~0	~0	[170]

volume changes for completely crystalline samples of poly[bis(trifluoroethoxy)-phosphazene]: $\Delta H_d = 12.5$ kJ/mol, $\Delta S_d = 34$ J/(K mol), $\Delta V_d = 7\%$ and $\Delta H_i =$ $= 1.2$ kJ/mol, $\Delta S_i = 2.3$ J/(K mol), $\Delta V_i = 8.6\%$; and for poly[bis(p-chlorophenoxy)-phosphazene]: $\Delta H_d = 11.8$ kJ/mol, $\Delta S_d = 27$ J/(K mol), $\Delta V_d = 5\%$ and almost zero heat effect, but 8% volume change for the transition to the isotropic melt [170]. The reported glass transitions seem to refer to amorphous glasses.

The thermal data on possible condis crystals are collected in Table 8. Substantial entropy gains are observed at the disordering transition, but variations are large, depending on the amount of conformational mobility gained. When compared to the total entropy of fusion, the listed entropies of disordering vary from 30%, for polytetrafluoroethylene, to close to 100% for the polyphosphazenes.

5.3.3 Small Molecules

Small molecules may also form condis crystals, provided they posses suitable conformational isomers. It is of interest to note that several of the organic molecules normally identified as plastic crystals are probably better described as condis crystals. Their motion was, as already shown in Sect. 5.2.2, not the complete reorientation of the presumed rigid molecule, but rather an exchange between a limited number of conformational isomers. The examples treated in Sect. 5.2.2 are 2,3-dimethylbutane, cyclohexanol and cyclohexane.

A series of other condis crystals are the larger cyclo-alkanes, analyzed by Grossmann [171]. Figure 23 illustrates the transition behavior of cyclotetracosane and cyclohexanonacontane. The melting transition of cyclotetracosane has only about 1/4 the heat of transition at the disordering to the condis phase. The condis phase has a much higher symmetry and fewer X-ray diffraction lines. Infrared and Raman spectroscopy indicate that practically no additional conformational isomers are introduced on

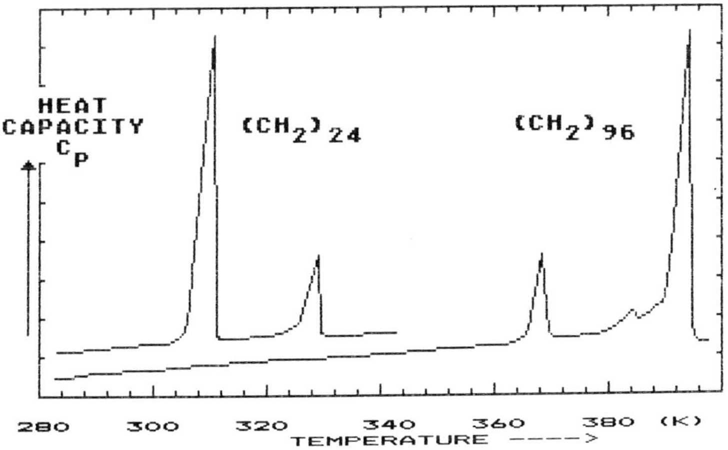

Fig. 23. Heat capacities of two cycloalkanes in the transition regions. The low temperature polymorphs are fully ordered. The high temperature transitions lead to the isotropic melts. The condis crystals exist between the two endotherms (Scanning calorimetry, heating rate 10 K/min, drawn after data of Ref. [171])

final melting. In rings with larger numbers of carbon atoms (cyclohexanonacontane, for example) the melting transition becomes the dominant transition, as seen in Fig. 23 [171].

Some paraffins show a high temperature crystal phase below the melting temperature which has been called the rotator phase [172]. It corresponds in size to the transition in the very large cycloalkanes (see Fig. 23). The details of their motion has been revealed recently through X-ray diffraction, Raman and infrared spectroscopy, dielectric relaxation studies and quasielastic neutron scattering [173]. A specially well studied example is tritriacontane. This paraffin shows two very small transitions (2.1 and 4.6 kJ/mol) at 328 and 339 K. Then follows the transition into the "rotor phase" at 341 K (29.3 kJ/mol) and the transition to the isotropic melt at 345 K (79.5 kJ/mol) [174]. Each transition corresponds to the gain of a specific defect and motion. First, the only change is a 180° jump of a full chain without translational motion, causing an occasional stacking fault. The crystal structure has changed from orthorhombic to monoclinic, but retained flat surfaces with aligned chain ends. At the next transition this rotation is coupled with a translational jump by one methylene repeat length (flip-flop-screw jump) leading to irregular end surfaces. These changes involve the whole molecule so that there is no conformational disorder involved in the motion in these two polymorphs. Their methylene chains are still fully extended, planar zig-zag chains. It was even suggested recently that these transition are partially irreversible and that motion occurs only at the transition [175]. Of interest to this discussion is the next higher temperature step at 341 K that does not involve full molecular rotation as originally proposed [172]. The crystals show after this transition shortened chains that posses a higher concentration of intrachain kinks. Estimates of 0.4 to 0.5 gauche-(+)-trans-gauche(—) kinks per chain have been made [176], suggesting a condis crystal. A major amount of the mobility involves intrachain defect diffusion and torsional oscillation [173].

Comparison between the various condis crystals shows that large variations in the amount of conformational disorder and motion is possible even in similar molecules. The tritriacontane in the condis state possesses about 3 gauche conformations per 100 carbon atoms. For cyclodocosane which is in its transition behavior similar to the tetracosane of Fig. 23, one estimates about 16 gauche conformations per 100 carbon atoms, and for the high pressure phase of polyethylene (see Sect. 5.3.2), one expects 37 gauche conformations per 100 carbon atoms [171]. The concentration of gauche conformations in cyclodocosane and polyethylene condis crystals are close to the equilibrium concentration in the melt, while the linear short chain paraffin condis crystals are still far from the conformational equilibrium of the melt.

6 Conclusions

The thermotropic mesophases are well enough understood to propose a subdivision into six types. Depending on the type of disorder, they are called liquid crystals, plastic crystals or condis crystals (positional and if applicable conformational disorder, orientational disorder, and conformational disorder, respectively). For the corresponding glasses, which represent the frozen-in mesophases, the names LC-, PC-, and CD-glasses are proposed (Fig. 2). For macromolecules not only equilibrium states, but multiphase, metastable equilibrium states must be considered (Figs. 3 and 4).

The kinetics of the transitions seem to be related to the corresponding limiting crystal-melt and glass-melt transitions, although only a limited amount of work has been done in this area.

Macromolecular liquid crystals and glasses can be produced by a variety of structures (Fig. 13). Closest to low molecular weight liquid crystals are macromolecules with mesogenic groups in the side or main chain, with the rest of the molecule providing a flexible spacer. The conformational motion of the flexible spacer permits the orientational alignment of the mesogens which leads to the liquid crystalline behavior. The transition to the isotropic phase is close to reversible. The transition parameters are only little different from the low molecular analogues. There is some inclusion of parts of the spacers into the orientational order. The orientation is, as in the low molecular weight analogues, rather imperfect (much lower entropies of transition to the isotropic state than expected for full orientational order, see Tables 2, 5, and 6). Of special interest are the changes with decreasing spacer length. The macromolecules become then more rigid and finally the whole macromolecule becomes one rigid rod which can only be brought into the liquid crystal state through the use of a solvent (lyotropic liquid crystals). Research on LC-glasses has been rather scarce. There seems to be a possibility of an increased cooperativeness of the glass transition. Another group of liquid crystals is based on supermolecular structure. It should be separated from the liquid crystals caused by identifiable mesogens. Examples of small molecules of this type are amphiphilic molecules. Macromolecules provide similar examples with their microphase separated block copolymers.

Plastic crystals seem to have been only observed for small molecules. Their positional order is rather perfect, as can be derived from their full positional entropy on transition to the isotropic phase (Table 7). PC-glasses have been observed. Some of the plastic crystals of more flexible organic molecules may better be described as condis crystals.

Condis crystals and glasses of macromolecules are a newly recognized type of mesophase. The mobility in this mesophase may lead to chain extension, and as a corollary, it may be possible that mechanical deformation can cause the stabilization of the condis state. Several examples of stable condis crystals are documented, but there seem to be also examples of metastable condis crystals which are produced as intermediates to crystallization. The size of the condis crystal transitions vary depending on the number of conformational isomers involved in the cooperative transitions.

A major effort should be undertaken to clarify the open questions. It is hoped that the classification as offered in this review can serve to focus future work in mesophases of macromolecules. Of particular interest is the question of the involvement of condis crystals as intermediates in polymer crystal deformation.

Acknowledgments: This work was supported by the Polymers Program of the National Science Foundation (Grant Number DMR 78-15279). Preliminary discussions were based on extensive literature reviews of Ms. D. Hornung and Dr. J. Menczel. Many of the new concepts were given trial discussions with several colleagues in the field, and all the students and coworkers at ATHAS were actively involved in the growth of the paper.

7 References

1. Dalton, J.: "A New System of Chemical Philosophy". Republication by Citadel Press (The Science Classic Library) of the 1808 first edition, New York, NY 1964, Chapter II, Section 4, Paragraph 2
2. Friedel, M. G.: "Les Etats Mésomorphes de la Matiére", Ann. Phys. (Paris) *18*, 273 (1922)
3. Wunderlich, B.: "Macromolecular Physics", Vol. 3, "Crystal Melting". Academic Press, New York, NY 1980
4. Staudinger, H.: "Organische Kolloidchemie". 3rd ed., Vieweg Verlag, Braunschweig 1950
5. Reinitzer, F.: "Beiträge zur Kenntnis des Cholesterins", Monatsh. *9*, 421, 1888. The term liquid crystals was first used by O. Lehmann in "Fluessige Kristalle". Engelmann, Leipzig, 1904. See also H. Kelker, History of liquid crystals. Mol. Cryst. Liq. Cryst. *21*, 1 (1973). For a history of the discovery and recognition of plastic crystals see J. Timmermanns, Plastic crystals, a historical review. J. Phys. Chem. Solids *18*, 1 (1961)
6. see for example White, J. L. and Fellers, J. F.: Macromolecular liquid crystals and their applications to high-modulus and tensile-strength fibers. J. Appl. Polymer. Sci., Appl. Polymer Symposium *33*, 137 (1978)
7. see Smith, G. W.: Plastic Crystals, Liquid Crystals, and the Melting Phenomenon. The Importance of Order. Adv. in Liquid Crystals, Vol. 1, G. H. Brown ed., Academic Press, 1975, New York, NY, see this review also for extensive listings of prior References
8. see for example: Ubbelohde, A. R.: "Melting and Crystal Structure". Oxford University Press (Clarendon), London and New York 1965; see also the recent update "The Molten State of Matter. Melting and Crystal Structure". Wiley, New York 1978
9. For a detailed discussion with many examples, see Chapter 8.2 of Reference 3
10. Richards, J. W.: Relations between the melting-points and the latent heats of fusion of the metals. Chem. News *75*, 278 (1897)
11. Walden, P.: Ueber die Schmelzwaerme, spezifische Kohaesion und Molekulargroesse bei der Schmelztemperatur. Z. Elektrochem. *14*, 713 (1908)
12. Chapter 8.4.7 of reference 3
13. Wunderlich, B.: Study of the change in specific heat of monomeric and polymeric glasses during the transition. J. Phys. Chem. *64*, 1052 (1960)
14. U. Gaur and B. Wunderlich, Additivity of the heat capacities of linear macromolecules in the molten state. Polymer Div. Am. Chem. Soc. Preprints *20*, 429 (1979)
15. Some general reviews of liquid crystals are:
    a. Gray, G. W.: "Molecular Structure and the Properties of Liquid Crystals". Academic Press, New York 1962
    b. Gray, G. W. and Winsor, P. A.: eds. "Liquid Crystals and Plastic Crystals". Wiley, Chichester 1974
    c. de Gennes, P. G.: "The Physics of Liquid Crystals". Clarendon Press, Oxford 1974
    d. Johnson, J. F. and Porter, R. S.: eds. "Liquid Crystals and Ordered Fluids". Vol. 1 and 2, Plenum Press, New York, NY, 1970, 1974
    e. Porter, R. S. and Johnson, J. F.: eds. "Ordered Fluids and Liquid Crystals". Am. Chem. Soc. Washington, DC 1967
    f. Symposium of the Faraday Soc. #5, "Liquid Crystals". Farad. Div. Chem. Soc. London 1972
    g. Proc. Int. Conf. Liq. Cryst., Gordon and Breach, London since 1965
    h. See also the journal Molecular Crystals Liquid Crystals, Gordon and Breach
    i. Ciferri, A., Krigbaum, W., and Meyer, R. B.: eds. "Polymer Liquid Crystals". (Materials Sci. and Tech. Ser.) Academic Press, New York, NY, 1983
16. Some general reviews of plastic crystals are:
    a. Sherwood, N.: ed. "The Plastically Crystalline State". (Orientationally-disordered crystals). J. Wiley and Sons, Chichester 1979
    b. Aston, J. G.: "Plastic Crystals" in D. Fox, M. M. Labes and A. Weissberger, eds. "Physics and Chemistry of the Organic Solid State". Interscience Publ., New York NY 1963, Vol. 1, Chpt. 9
    c. Staveley, L. A. K.: Phase transitions in plastic crystals. Annual Rev. of Phys. Chem." *13*, 351 (1962)
    d. DuPre, D. B., Samulski, E. T. and Tobolsky, A. V.: The mesomorphic state: liquid crystals

and plastic crystals, in Tobolsky, A. V. and Mark, H. F. eds. "Polymer Science and Materials". Chapter 7, Wiley-Interscience, New York, NY 1971
e. Proc. of the Symposium on Plastic Crystals and Rotation in the Solid State. April, 1960, Phys. Chem. Solids *18* (1) (1961)
f. See also Refs. 7 and 15b
17. Ehrenfest, P.: Phase changes classified according to the singularities of the thermodynamic potential. Proc. Acad. Sci., Amsterdam *36*, 153 (1933); Suppl. 75b, Mitt. Kammerlingh Onnes Inst., Leiden
18. Wunderlich, B.: "Macromolecular Physics", Vol. 1, "Crystal Structure, Morphology, Defects". Academic Press, New York, 1973
19. Wunderlich, B.: "Macromolecular Physics", Vol. 2, "Crystal Nucleation, Growth Annealing". Academic Press, New York 1976
20. Meesiri, W., Menczel, J., Gaur, U. and Wunderlich, B.: Phase transitions in mesophase macromolecules. III. The transitions in poly(ethylene terephthalate-co-oxybenzoate). J. Polymer. Sci., Polymer Phys. Ed., *20*, 719 (1982)
21. Menczel, J. and Wunderlich, B.: Phase transitions in mesophase macromolecules. II. The transitions of poly(p-acryloyloxybenzoic acid). Polymer *22*, 778 (1981)
22. Flory, P. J.: Phase equilibria in solutions of rod-like particles. Proc. Roy. Soc. London, Ser. A. 234, 73 (1956)
23. See Section 5.2.2 and refs. 104, 105 and 107
24. Wolpert, S. M., Weitz, A. and Wunderlich, B.: Time-dependend heat capacity in the glass transition region. J. Polymer. Sci., Part A-2, *9*, 1887 (1971)
25. Wunderlich, B., Bodily, D. M. and Kaplan, M. H.: Theory and measurement of the glass-transformation interval of polystyrene. J. Appl. Phys. *35*, 95 (1964)
26. Grebowicz, J. and Wunderlich, B.: The glass transition of p-alkyl-p′-alkoxy-azoxybenzene mesophases. Mol. Cryst. Liq. Cryst. *76*, 287 (1981)
27. Grebowicz, J. and Wunderlich, B.: Phase transitions in mesophase macromolecules. IV. The transitions in poly(oxy-2,2′-dimethylazoxybenzene-4,4′-diyloxydodecanedioyl). J. Polymer Sci., Polymer Phys. Ed. *21*, 141 (1983)
28. Menczel, J. and Wunderlich, B.: Heat capacity hysteresis of semicrystalline macromolecular glasses. J. Polymer Sci., Polymer Letters Ed. *19*, 261 (1981)
29. Poisson, S. D.: "Recherches sur la probabilité des jugements en matière criminelle et en matière civile", p. 206, Bachelier, Paris 1837
30. An equation of the type of eq. 7 for crystallization was first proposed by Kolmogoroff, A. N.: On the crystallization process in metals. Isvest. Akad. Nauk SSSR Ser. Math. *1*, 335 (1937) and than independently derived by Avrami and others
31. Price, F. P. and Wendorff, J. H.: Transitions in mesophase forming systems. I. Transformation kinetics and pretransition effects in cholesteryl myristate. J. Phys. Chem. *75*, 2839 (1971)
32. Jabarin, S. A. and Stein, R. S.: Light scattering and microscopic investigations of mesophase transitions of cholesteryl myristate. II. Kinetics of spherulite formation. J. Phys. Chem. *77*, 409 (1973)
33. Price, F. P. and Wendorff, J. H.: Transitions in mesophase forming systems. III. Transformation kinetics and textural changes in cholesteryl nonanoate. J. Phys. Chem. *76*, 276 (1972)
34. Price, F. P. and Wendorff, J. H.: Transitions in mesophase forming systems. II. Transformation kinetics and properties of cholesteryl acetate. J. Phys. Chem. *75*, 2849 (1971)
35. Price, F. P. and Fritzsche, A. K.: Kinetics of spherulite growth in cholesteryl esters. J. Phys. Chem. *77*, 396 (1973)
36. Adamski, P. and Klimczyk, S.: The crystallization rate constant and Avrami index for cholesterol pelargonate. Sov. Phys. Crystallogr. *23*, 82 (1978)
37. Adamski, P. and Czyzewski, R.: Activation energy and growth rate of spherulites of cholesterol liquid crystals. Soc. Phys. Crystallogr. *23*, 725 (1978)
38. Warner, S. B. and Jaffe, M.: Quiescent crystallization in thermotropic polyesters. J. Crystal Growth *48*, 184 (1980)
39. For a discussion see Ref. 19, Chapter 6.3.3
40. Hellmuth, E. and Wunderlich, B.: Superheating of linear high-polymer polyethylene crystals. J. Appl. Phys. *36*, 3039 (1965)
41. Wunderlich, B.: Molecular nucleation and segregation. Disc. Farad. Soc. *68*, 239 (1979)

42. Hellmuth, E., Wunderlich, B. and Rankin, J. M.: Superheating of linear high polymers. Polytetrafluoroethylene. Appl. Polymer Symposia, *2*, 101 (1966)
43. Wunderlich, B. and Shu, H. C.: The crystallization and melting of selenium. J. Crystal Growth 48, 227 (1980); and H.-C. Shu and B. Wunderlich, Crystallization of Selenium from the vapor phase. Polymer *21*, 521 (1980)
44. Landolt Boernstein "Zahlenwerte und Funktionen". Sixth Edition, K. Schaefer and E. Lax, eds., Berlin, 1960. Vol. II, Part 2a, Kast, W.: "Umwandlungstemperaturen kristalliner Fluessigkeiten", p. 266, Listing of 1500 small molecules. Vol. II, Part 6, Maier, W.: "Dielektrische Eigenschaften von kristallinen Fluessigkeiten", p. 607. Vol. II, Part 8, Maier, W.: "Optische und Magnetooptische Eigenschaften von kristallinen Fluessigkeiten", p. 553
45. First described by Chandrasekhar, S., Sadashiva, B. K. and Suresh, K. A. (Pramana *9*, 471, 1977). For a recent review see Billard, J. in W. Helfrich and G. Heppke, eds. "Liquid Crystals of One- and Two-Dimensional Order". Springer Verlag, Berlin 1980
46. Brooks, J. D. and Taylor, G. H.: The formation of graphitizing carbons from the liquid phase. Carbon *3*, 185 (1965) (see also the "Extended Abstracts of the 12th Biennial Conference on Carbon", Am. Carbon Soc., 1975).
47. See for example Benoit, H., Freund, L. and Spach, G.: in Fasman, G. ed. "Poly-alpha-amino acids". Vol. 1, p. 105, Dekker, 1967; and Samulski, E. T.: Liquid crystalline order in polypeptides, in A. Blumstein, ed., "Liquid Crystalline Order in Polymers". Academic Press, New York, NY 1978
48. Luzzati, V.: The structure of DNA as determined by X-ray scattering techniques. Progr. Nucleic Acid Res. *1*, 347 (1963). See also Ref. 52
49. See Ref. 15c p. 5
50. See for example:
    a. Gallot, B.: Liquid crystalline structure of block copolymers, p. 11 in A. Blumstein, ed., "Liquid Crystalline Order in Polymers". Academic Press, New York 1978
    b. J. Polymer Sci., Part C, Vol. 26 (1969)
    c. Aggarwal, S., ed.: "Block Copolymers". Plenum Press, New York, NY 1970
    d. Allport, D. C. and James, W. H. eds.: "Block Copolymers". Halstad Press, New York, NY 1973
51. See for example: Skoulios, A.: La structure des solutions aqueuses concentrées de savon. Adv. Colloid Interface Sci. *1*, 79 (1967)
52. See for example: Bouligand, Y.: Liquid crystalline order in biological materials, p. 261 in Blumstein, A. ed.: "Liquid Crystalline Order in Polymers". Academic Press, New York 1978. And Chapman, D.: "The Structure of Lipids". Methuen, London, 1965
53. Gaur, U. and Wunderlich, B.: Study of microphase separation in block copolymers of styrene and alpha-methylstyrene in the glass transition region using quantitative thermal analysis. Macromolecules *13*, 1618 (1980)
54. Charvolin, J. and Tardieu, A.: Lyotropic liquid crystals: Structures and Molecular Motions, in L. Liebert, ed. "Liquid Crystals". Solid State Physics, Supplement 14, Academic Press, New York, NY 1978, p. 209
55. Sackmann, H. and Demus, D.: The problem of polymorphism in liquid crystals. Mol. Cryst. Liq. Cryst. *21*, 239 (1973)
56. Tinh, N. H., Destrade, C. and Gasparoux, G.: Nematic disc-like liquid crystals. Phys. Lett. *72A*, 251 (1979)
57. Destrade, C., Tinh, N. H., Gasparoux, G., Malthete, J. and Levelut, A. M.: Disc-like mesogens: A classification. Mol. Cryst. Liq. Cryst. *71*, 111 (1981)
58. Petrie, S. E. B.: Smectic liquid crystals, in Saeva, F. D., ed. "Liquid Crystals, the Fourth State of Matter". Marcel Dekker, New York, NY 1979
59. Barrall II, E. M. and Johnson, J. F.: Thermal properties of liquid crystals in ref. 15b, p. 254
60. Sorai, M. and Suga, H.: Studies on mesogenic disc-like molecules. II. Heat capacity of benzene-hexa-n-heptanoate from 13 to 393 K. Mol. Cryst. Liq. Cryst. *73*, 47 (1981)
61. Vorlaender, D.: Remarks on Liquocrystalline Resins and Laquers. Trans. Farad. Soc. *29*, 207 (1933)
62. Sorai, M. and Seki, S.: Glassy liquid crystal of the the nematic phase of N-(o-Hydroxy-p-methoxybenzylidene)-p-butylaniline. Bull. Chem. Soc. Japan **44**, 2887 (1971)

63. Tsuji, K., Sorai, M. and Seki, S.: New finding of glassy liquid crystal — a non-equilibrium state of cholesteryl hydrogen phthalate. Bull. Chem. Soc., Japan *44*, 1452 (1971)
64. Sorai, M. and Seki, S.: Heat capacity of N-(o-hydroxy-p-methoxybenzylidene)-p-butylaniline: A glassy nematic liquid crystal. Mol. Cryst. Liq. Cryst. *23*, 299 (1973)
65. Petrie, S. E. B.: The effect of excess thermodynamic properties versus structure formation on the physical properties of glassy polymers. J. Macromol. Sci., Phys. *12*, 225 (1976)
66. Kessler, J. O. and Lydon, J. E.: Structure and thermal conductivity of supercooled MBBA. In Vol. 2 of of ref. 15d, p. 331
67. Cognard, J. and Gangguillet: Glassy transition in liquid crystal eutectic mixtures. Mol. Cryst. Liq. Cryst. Lett. *49*, 33 (1978)
68. Chistyakov, I. G., Schabischev, L. S., Jarenov, R. I. and Gusakova, L. A.: The polymorphism of the smectic liquid crystal. Mol. Cryst. Liq. Cryst. *7*, 279 (1969)
69. Deniz, K. U., Paranjpe, A. S., Mirza, E. B., Parvathanathan, P. S. and Patel, K. S.: DSC and X-ray diffraction investigations of phase transitions in HxBABA and NBABA. J. de Physique, C3, *40*, 136 (1979)
70. Blumstein, A. and Hsu, E. C.: Liquid crystalline order in polymers with mesogenic side groups; in Blumstein, A. ed.: "Liquid Crystalline Order in Polymers". Academic Press, New York, NY 1978, p. 105
71. See particularly the extensive work of Strzelecki, L. and Liebert, L.: published in the Bull. Soc. Fr. (1973) p. 597, 603, 605; (1975) p. 2073, 2750. See also Bouligand, Y., Cladis, P. E., Liebert, L. and Strzelecki, L.: Study of sections of polymerized liquid crystals. Mol. Cryst. Liq. Cryst. **25**, 233 (1974)
72. Ref. 3, Chapter 10.3.4
73. Plate, N. A. and Shibaev, V. P.: Comb-like polymers. Structure and properties. J. Polymer Sci., Macromol. Rev. *8*, 117 (1974)
74. Shibaev, V. P., Freidzon, Ya. S. and Plate, N. A.: Cholesterol-containing Liquid-crystalline polymers; Dokl. Akad. Nauk USSR *227*, 1412 (1976)
75. Shibaev, V. P. and Plate, N. A.: Liquid crystalline polymers, Vysokomol. Soedin. *A1***9**, 923 (1977), [Engl. translat. Polymer. Sci. USSR *19*, 1065 (1977)]
76. Finkelmann, H., Ringsdorf, H. and Wendorff, J. H.: Model considerations and examples of enantiotropic liquid crystalline polymers. Makromol. Chem. *179*, 273 (1978)
77. Maganini, P. L.: Structure and properties of polymers with strongly anisometric side groups. Makromol. Chem. Suppl. *4*, 223 (1981)
78. Frosini, V.: Mechanical relaxation in polymer mesophases. Proc. 28th Macromol. Symp. IUPAC, U. Mass., Amherst, MA, p. 806, 1982
79. Shibaev, V. P., Plate, N. A. and Freidzon, Y. S.: Thermotropic cholesterol-containing liquid crystalline polymers, in A. Blumstein, ed. "Mesomorphic Order in Polymers". ACS Symposium Series 74, Am. Chem. Soc. Washington, D.C. 1978. See also J. Poly. Sci., Polymer Chem. Ed. *17*, 1655 (1979)
80. Lupinacci, D., Frosini, V. and Magagnini, P. L.: Mesomorphic structure of a homopolymer of N-(4-biphenyl)acrylamide and of copolymers with 4-biphenylacrylate. Makromol. Chem. Rapid Commun. *1*, 671 (1980)
81. Finkelmann, H., Happ, M., Portugal, M. and Ringsdorf, H.: Liquid crystalline polymers with biphenyl-moieties as mesogenic group. Makromol. Chem. *179*, 2541 (1978)
82. Frenzel, J. and Rehage, G.: PVT-measurements on liquid crystalline polymers. Makromol. Chem., Rapid Commun. *1*, 129 (1980)
83. Finkelmann, F., Ringsdorf, H., Siol, W. and Wendorff, H.: Synthesis of cholesteric liquid crystalline polymers. Makromol. Chem. *179*, 829 (1978)
84. Shibaev, V. P., Moiseenko, V. M., Plate, N. A.: Thermotropic liquid crystalline polymers, 3, Comb-like polymers with side chains simulating the smectic type of liquid crystals. Makromol. Chem. *181*, 1381 (1980)
85. Finkelmann, H. and Rehage, G.: Investigations on liquid crystalline polysiloxanes, 1, Synthesis and characterization of linear polymers. Makromol. Chem. Rapid Commun. *1*, 31 (1980)
86. Wendorff, J. H.: Scattering in liquid crystalline polymer systems, in A. Blumstein, ed. "Liquid Crystalline Order in Polymers". Academic Press, New York, NY 1978, p. 1
87. Hardy, Gy., Cser, F., Nyitrai, K., Samay, G. and Kallo, A.: Investigation of the mesomorphic structure of p-alkoxy-phenyl-p-acryloyloxybenzoate polymers. J. Cryst. Growth, *48*, 191 (1980)

88. Shibaev, V. P., Tal'rose, R. V., Karakhanova, F. I. and Plate, N. A.: Thermotropic liquid crystals. II. Polymers with aminoacid fragments in the side chains. J. Polymer. Sci. Polymer Chem. Ed. *17*, 1671 (1979)
89. Finkelmann, H., Lehmann, B. and Rehage, G.: Phase behaviour of lyotropic liquid crystalline side chain polymers in aqueous solutions. Colloid Polymer Sci. *260*, 56 (1982)
90. Asrar, J., Thomas, O., Zhou, Q. and Blumstein, A.: Thermotropic liquid crystalline polyesters: Structure property relationship. Proc. 28th Macromol. Symp. IUPAC, U. Mass., Amherst, MA, p. 797, 1982
91. Jannelli, P., Roviello, A. and Sirigu, A.: Mesophasic properties of linear copolymers. Proc. 28th Macromol. Symp. IUPAC, U. Mass., Amherst, MA, p. 803, 1982. See also J. Polymer Sci., Polymer Lett. Ed. *13*, 455 (1975), Makromol. Chem. *181*, 1799 (1980) and Europ. Polymer J. *15*, 61, (1979)
92. Jo, B. W. and Lenz, R. W.: Liquid crystalline polymers, 7, thermotropic polyesters with main chain phenyl-1,4-phenylene, 4,4'-biphenylene, and 1,1'-binaphthyl-4,4'-ylene units. Makromol. Chem. Rapid Commun. *3*, 23 (1982)
93. Griffin, A. C. and Havens, S. J.: Mesogenic polymers. III. Thermal properties and synthesis of three homologous series of thermotropic liquid crystalline "backbone" polyesters. J. Polymer Sci., Polymer Phys. Ed. *19*, 951 (1981)
94. Iimura, K., Koide, N., Ohta, R. and Takeda, M.: Synthesis of thermotropic liquid crystalline polymers, 1, azoxy and azo-type polyesters. Makromol. Chemie *182*, 2563 (1981)
95. Iimura, K., Koide, N., Tanabe, H. and Takeda, M.: Synthesis of thermotropic liquid crystalline polymers, 2, polyurethanes. Makromol. Chemie *182*, 2569 (1981)
96. Roviella, A. and Sirgu, A.: Odd-even effects in polymeric liquid crystals. Makromol. Chem. *183*, 895 (1982)
97. Takeda, M.: Recent developments in the area of thermotropic liquid crystalline polymers and their thermal analysis, in B. Miller, ed. "Thermal Analysis", Vol. II, Proceedings of the 7th ICTA, Wiley and Sons 1982, p. 927
98. Blumstein, R. B. and Stickles, E.: Influence of molecular weight on some properties of polymeric liquid crystals. Proc. 28th Macromol. Symp. IUPAC, U. Mass., Amherst, MA, p. 799, 1982
99. Preston, J.: Synthesis and properties of rodlike condensation polymers. A. Blumstein, ed. "Liquid Crystalline Order in Polymers". Academic Press, New York 1978, pg. 141
100. Winsor, P. A.: Non-amphiphilic cubic mesophases "plastic crystals". Chapter 2.2 in Ref. 15b
101. Dunning, W. J.: The crystal structure of some plastic and related crystals in Ref. 16a, pg. 1
102. Westrum, Jr., E. F. and McCullough, J. P.: Thermodynamics of crystals, in Fox, D., Labes, M. M. and Weissberger, A.: "Physics and Chemistry of the Organic Solid State". Interscience Publ., New York, NY 1963, p. 1–178
103. Boden, N.: NMR studies of plastic crystals. Chapter 5 in Ref. 16a
104. Adachi, K., Suga, H. and Seki, S.: Calorimetric study of the glassy state. VI. Phase changes in crystalline and glassy-crystalline 2,3-dimethylbutane. Bull. Chem. Soc. Japan **44**, 78 (1971)
105. Adachi, K., Suga, H. and Seki, S.: Phase changes in crystalline and glassy-crystalline cyclohexanol. Bull. Chem. Soc. Japan *41*, 1073 (1968)
106. Huffman, H. M., Todd, S. S. and Oliver, G. D.: Low temperature data on eight alkylcyclohexanes. J. Am. Chem. Soc. *71*, 584 (1949)
107. Adachi, K., Suga, H. and Seki, S.: The glassy crystalline state — a non-equilibrium state of plastic crystals. Bull. Chem. Soc. Japan *43*, 1916 (1970)
108. Schneider, N. S., Desper, C. R. and Beres, J. J.: Mesomorphic structure in polyphosphazenes in A. Blumstein, ed. "Liquid Crystalline Order in Polymers". Academic Press, New York, NY 1978
109. Pechhold, W. and Blasenbrey, S.: Phase transitions, relaxations, and properties of high polymers. Angew. Makromol. Chem. *22*, 3 (1972); see also Rheol. Acta *6*, 174 (1967); Kolloid Z. Z. Polymere *216/217*, 235 (1967); Ber. Bunsenges. *74*, 784 (1970); and Kolloid Z. Z. Polymere *241*, 955 (1970)
110. Baur, H.: Bemerkungen zur Defekttheorie von n-Paraffinen und Polymeren. Colloid and Polymer Sci. *252*, 641 (1974), and: Zur Theorie der Umwandlungserscheinungen in n-Alkyl-Lamellen. Habilitationsschrift DB 1992+a, Technical University Hannover 1977
111. Statton, W. O.: Directional crystallization of polymers. Ann. N.Y. Acad. Sci. *83*, 27 (1959)
112. Tadokoro, H.: "Structure of Crystalline Polymers". Wiley-Interscience, New York, NY 1979

113. Wunderlich, B. and Arakawa, T.: Polyethylene crystallized from the melt under elevated pressure, J. Polymer Sci., Part A, *2*, 3697 (1964)
114. Geil, P. H., Anderson, F. R., Wunderlich, B. and Arakawa, T.: Morphology of polyethylene crystallized from the melt under pressure. J. Polymer Sci., Part *A*, *2*, 3703 (1964)
115. Arakawa, T. and Wunderlich, B.: Thermodynamic properties of extended chain polymethylene single crystals. J. Polymer Sci., Part C, *16*, 653 (1967)
116. Wunderlich, B. and Melillo, L.: Morphology and growth of extended chain crystals of polyethylene. Makromol. Chemie *118*, 250 (1968)
117. Bassett, D. C. and Turner, B.: Chain extended crystallization of polyethylene: Evidence of a new high-pressure phase. Nature (London) Phys. Sci. *240*, 146 (1972); Bassett, D. C., Block, S. and Piermarini, G. J.: A high-pressure phase of polyethylene and chain-extended growth, J. Appl. Phys. *45*, 4146 (1974); Yasuniwa, M., Nakafuku, C. and Takemura, T.: Melting and crystallization process of polyethylene under high pressure, Polymer J. *4*, 526 (1973)
118. For general reviews of the topic of extended chain crystals see:
a. ref. 3, Chapter 8.5.2; ref. 18, Chapter 3.3.1; ref. 19, Chapter 6.3.3
b. Bassett, D. C., Chain-extended polyethylene in context: a review. Polymer *17*, 460 (1976)
c. Matsushige, K. and Takemura, T.: Crystallization of macromolecules under high pressure. J. Crystal Growth *48*, 343 (1980)
d. Maeda, Y., Kanetsuna, H., Nagata, K., Matsushige, K. and Takemura, T.: Direct observation of phase transitions of polyethylene under high pressure by a PSPC X-ray system. J. Polymer Sci., Polymer Phys. Ed. *19*, 1313 (1981)
119. Hikosaka, M., Minomura, S., Seto, T.: Melting and solid-solid transition of polyethylene under pressure. Japan. J. Applied Phys. *19*, 1763, 1980
120. Yamamoto, T., Miyagi, H. and Asai, K.: Structure and properties of high pressure phase of polyethylene. Japan. J. Appl. Phys. *16*, 1891 (1977)
121. Yasuniva, M., Enoshita, R. and Takemura, T.: X-ray studies of polyethylene under high pressure. Japan. J. Appl. Phys. *15*, 1421 (1976)
122. Ide, T., Taki, S. and Takemura, T.: The high pressure and high temperature dilatometer. Japan. J. Appl. Phys. *16*, 647 (1977)
123. Yasuniwa, M. and Takemura, T.: Microscopic observation of the crystallization process of polyethylene under high pressure. Polymer *15*, 661 (1974)
124. Yamamoto, T.: Nature of disorder in the high pressure phase of polyethylene. J. Macromol. Sci.-Phys. *B16*, 487 (1979)
125. Tanaka, H. and Takemura, T.: Studies on the high-pressure phases of polyethylene and polytetrafluoroethylene by Raman spectroscopy. Polymer J. *12*, 355 (1980)
126. Nagata, K., Tagashiva, K., Taki, S. and Takemura, T.: Ultrasonic study of high pressure phase in polyethylene. Japan. J. Appl. Physics *19*, 985 (1981)
127. Starkweather, H. W.: A comparison of the rheological properties of polytetrafluoroethylene below its melting point with certain low-molecular weight smectic states. J. Polymer Sci., Polymer Phys. Ed. *17*, 73 (1979)
128. Melillo, L. and Wunderlich, B.: Extended chain crystals VIII. Morphology of polytetrafluoroethylene. Kolloid Z. Z. Polymere *250*, 417 (1972)
129. Bassett, D. and Davitt, R.: On the crystallization phenomena in polytetrafluoroethylene. Polymer *15*, 721 (1974)
130. Bates, T. W. and Stockmayer, W. H.: Conformational energies of perfluoroalkanes. III. Properties of polytetrafluoroethylene. Macromolecules *1*, 17 (1968)
131. Natarajan, R. T. and Davidson, T.: Kinetics of the 20 °C phase transformation in polytetrafluoroethylene. J. Polymer Sci., Polymer Phys. Ed. *10*, 2209 (1972)
132. Starkweather, Jr., H. W., Zoller, P., Jones, G. A. and Vega, A. J.: Heat of fusion of polytetrafluoroethylene. Proc. of the 11th NATAS Conference, New Orleans, LA (1981) pg. 361; see also J. Polymer Sci. Polymer Phys. Ed. *20*, 751 (1982)
133. Marx, P. and Dole, M.: Specific heat of synthetic high polymers. V. A. study of the order-disorder transition in polytetrafluoroethylene. J. Am. Chem. Soc. *17*, 4771 (1955)
134. Clark, E. S. and Muus, L. T.: Partial disordering and crystal transitions in polytetrafluoroethylene. Z. Krist. *117*, 119 (1962); see also E. S. Clark, J. Makromol. Sci. Phys., *B1*, 795 (1967)
135. Mele, A., Site, A. D., Bettiniali, C. and DiDominico, A.: Thermoluminescence and phase transitions of irradiated fluorinated polymers. J. Chem. Phys. *49*, 3297 (1968)

136. Gohil, R. M. and Petermann, J.: Chain conformational defects in polyvinylidene fluoride. Polymer *22*, 1612 (1981); Takahashi, Y. and Tadokoro, H.: Formation mechanism of kink bands in modification II of poly(vinylidene fluoride). Evidence for flip-flop motion between TGT$\bar{G}$ and T$\bar{G}$TG conformations. Macromolecules *13*, 1316 (1980); Takahashi, Y., Tadokoro, H. and Odajima, A.: Kink bands in form I of poly(vinylidene fluoride). Macromolecules *13*, 1318 (1980)
137. See for example Lovinger, A. J.: Annealing of poly(vinylidene fluoride) and formation of a fifth phase. Macromolecules *15*, 40 (1982)
138. Hasegawa, R., Kobayashi, M. and Tadokoro, H.: Molecular conformation and packing of poly(vinylidene fluoride). Stability of three crystalline forms and the effect of high pressure, Polymer J. *3*, 591 (1972); Hasegawa, R., Takahashi, Y., Chatani, Y. and Tadokoro, H.: Crystal structures of three crystalline forms of poly(vinylidene fluoride). Polymer J. *3*, 600 (1972)
139. Miyamoto, Y., Nakafuku, C. and Takemura, T.: Crystallization of polychlorotrifluoroethylene. Polymer J. *3*, 120 (1972)
140. Natta, G., Peraldo, M. and Corradini, P.: Modificazione mesomorfa smettica del polipropilene isotattico. Rend. Accad. Naz. Lincei, Vol. *24*, 14 (1959)
141. Zannetti, R., Celotti, G. C., Fichera, A. and Francesconi, R.: The structural effects of annealing time and temperature on the paracrystal-crystal transition in isotactic polypropylene. Makromol. Chemie *128*, 137 (1969)
142. Miller, R. L.: On the existence of near-range order in isotactic polypropylenes. Polymer *1*, 135 (1960)
143. Corradini, P.: The stereochemistry of Macromolecules. Dekker, New York, NY Vol. 3, 1968
144. Fichera, A. and Zannetti, R.: Thermal properties of isotactic polypropylene quenched from the melt and annealed. Makromol. Chemie *176*, 1885 (1975)
145. Nakafuku, C.: High pressure d.t.a. study on the melting and crystallization of isotactic polypropylene. Polymer *22*, 1673 (1981)
146. Natta, G. and Corradini, P.: Conformation of linear chains and their mode of packing in the crystal state. J. Polymer Sci. *39*, 29 (1959); see also Natta, G., Corradini, P. and Porri, D.: Rend. Accad. Nazl. Lincei *20*, 728 (1956)
147. Iwayanagi, S. and Miura, J.: Nuclear magnetic resonance study of solid phase transition of trans-1,4-polybutadiene, Rept. Progr. Polymer Phys. Japan *8*, 303 (1965)
148. Moraglio, G., Polizzotti, G. and Danusso, F.: Enantiotropic polymorphism of transtactic poly-1,3-butadiene, Europ. Polymer J. *1*, 183 (1965)
149. Corradini, P.: On the chain conformation of the high temperature polymorph of trans-1,4-polybutadiene. Polymer Letters *7*, 211 (1969); see also J. Polymer Sci., Symposia *50*, 327 (1975)
150. Suehiro, K. and Takayanagi, N.: Structural studies of the high temperature form of trans-1,4-polybutadiene crystal. J. Macromol. Sci. Phys. *B4*, 39 (1970)
151. Bautz, G., Leute, U., Dollhopf, W. and Haegele, P. C.: On the solid state phases of poly(trans-1,4-butadiene). Colloid and Polymer Sci. *259*, 714 (1981)
152. Finter, J. and Wegner, G.: The relation between phase transition and crystallization behavior of 1,4-trans-poly(butadiene). Makromol. Chemie *182*, 1859 (1981) (see here and Ref. 3 for older data)
153. Natta, G. and Corradini, P.: The crystal structure of cis-1,4-polybutadiene. Nuovo Cimento Suppl. *15*, 111 (1960); see also Angew. Chemie *68*, 615 (1956) and Nyburg, S. C.: Acta Cryst. *7*, 385 (1954)
154. Edwards, B. C. and Phillips, P. J.: The structure of the high pressure phase of cis-polyisoprene. J. Mat. Sci. *10*, 1233 (1975); see also J. Polymer Sci. *B10*, 321 (1972) and Polymer *15*, 491 (1974)
155. Phillips, P. J. and Edwards, B. C.: High pressure phases in polymers. III. The nature of the high-pressure phase in cis-polyisoprene. J. Polymer Sci., Polymer Phys. Ed. *14*, 377 (1976); see also ibid. *13*, 1819, 2117 (1975) and *14*, 391 (1976)
156. Rossem, A. van and Lotichius, J.: Das Einfrieren des Rohkautschuks. Kautschuk *5*, 2 (1929); N. Bekkedahl, Forms of rubber as indicated by temperature-volume relationship, J. Res. Natl. Bur. Stand. *13*, 411 (1934)
157. Lieser, G.: Polymer single crystals of poly(4-hydroxybenzoate). J. Polymer Sci., Polymer Phys. Ed., *21*, 1611 (1983)
158. Gogolewski, S. and Pennings, A. J.: Crystallization of polyamides under elevated pressure. Nylon 6. Polymer *14*, 463 (1973); see also *18*, 647, 654 (1977)

159. Hiramatsu, N. and Hirakawa, S.: Melting and transformation behavior of gamma form Nylon 6 under high pressure. Polymer J. *14*, 165 (1982)
160. Gogolewski, S.: A possible mechanism of chain extension in nylon-6 during crystallization under pressure. Polymer *18*, 63 (1977)
161. Gogolewski, S. and Pennings, A. J.: Crystallization of polyamides under elevated pressure: 5 Pressure-induced crystallization from the melt and annealing of folded chain crystals of nylon-11, poly(aminoundecaneamide) under pressure, Polymer *18*, 660 (1977); Stamhuis, J. E. and Pennings, A. J.: Crystallization of polyamides under elevated pressure: 6. Pressure-induced crystallization from the melt and annealing of folded chain crystals of nylon-12, polylaurolactam under pressure. Polymer *18*, 667 (1977)
162. Hiramatsu, N., Hashida, S. and Hirakawa, S.: Formation of alpha form nylon 12 under high pressure. Japan. J. Appl. Phys. *21*, 651 (1982)
163. Kast, W.: Die Molekel-Struktur der Verbindungen mit kristallin-fluessigen (mesomorphen) Schmelzen. Angew. Chemie *67*, 592 (1955)
164. Siegmann, A. and Harget, P. J.: Melting and crystallization of poly(ethylene terephthalate) under pressure. J. Polymer Sci., Polymer Phys. Ed. *18*, 2181 (1980)
165. Beatty, C. L., Pochnan, J. M., Froix, M. F. and Hinman, D. D.: Liquid crystalline type order in polydiethylsiloxane. Macromolecules *8*, 547 (1975)
166. Pochan, J. M., Hinman, D. F. and Froix, M. F.: Morphological studies on the viscous crystalline phase of poly(diethylsiloxane) including the dynamics of phase formation and the relationship of viscous crystalline structure and crystalline structure. Macromolecules *9*, 611 (1976)
167. Beatty, C. L. and Karasz, F. E.: Transitions in poly(diethyl siloxane). J. Polymer Sci., Polymer Phys. Ed. *13*, 971 (1975); see also Pochnan, J. M., Beatty, C. L., Hinman, D. D. and Karasz, F. E.: ibid. 977
168. Froix, M. F., Beatty, C. L., Pochnan, J. M. and Hinman, D. D.: Nuclear spin relaxation in poly(diethylsiloxane). J. Polymer Sci., Polymer Phys. Ed. *13*, 1269 (1975)
169. Singler, R. E., Schneider, N. S., Hagnamer, G. L.: Polyphosphazenes: Synthesis-properties-applications. Polymer Eng. Sci. *15*, 321 (1975)
170. Schneider, N. S., Desper, C. R., Singler, R. E.: The Thermal transition behavior of polyorganophosphazenes. J. Appl. Polymer Sci. *20*, 3087 (1976)
171. Grossmann, H.-P.: Investigation of conformational transitions in cycloalkanes. Polymer Bulletin *5*, 137 (1981)
172. Mueller, A.: An X-ray investigation of normal paraffins near their melting points. Proc. Roy. Soc. *A*. *138*, 514 (1932)
173. For a summary see Ewen, B., Strobl, G. R. and Richter, D.: Phase transitions in crystals of chain molecules. Disc. Farad. Soc. *69*, 19 (1980)
174. Strobl, G. R.: Molecular motion, thermal expansion, and phase transitions in paraffins: A model for polymers. J. Polymer Sci. Polymer Symposium *59*, 121 (1977); see also Colloid Polymer Sci. *254*, 170 (1976)
175. Takamizawa, K., Ogawa, Y. and Oyama, T.: Thermal behavior of n-alkanes synthesized with attention paid to high purity. Polymer J. *14*, 441 (1982)
176. Strobl, G., Ewen, B., Fischer, E. W. and Piesczek, W.: Defect structure and molecular motion in the four modifications of n-tritriacontane. J. Chem. Phys. *61*, 5257, 5265 (1974)

M. Gordon/H.-J. Cantow (Editors)
Received February 7, 1983

Properties and Applications of Liquid-Crystalline Main-Chain Polymers

M. G. Dobb and J. E. McIntyre
Department of Textile Industries, The University of Leeds,
Leeds LS2 9JT, UK

The development is reviewed of liquid-crystalline polymers whose mesophase formation derives from the nature of the chemical units in the main chain. The emphasis lies primarily on highly aromatic condensation polymers and their applications. The general properties of nematic phases formed by such polymers are surveyed and some chemical structures capable of producing nematic phases are classified in relation to their ability to form lyotropic and thermotropic systems. The synthesis, properties, physical structure and applications of two of the most important lyotropic systems and of a range of potentially important thermotropic polymers are discussed with particular reference to the production and use of fibres, films and anisotropic mouldings.

Advances in Polymer Science 60/61

1 Historical Introduction

Simple liquid-crystalline compounds of low molecular weight have been recognised and widely studied for nearly 100 years. As far as polymers are concerned, the occurrence of liquid-crystalline phases has long been reported for various natural polymers, usually in water, over a number of years but little characterisation of this property has been carried out. A number of synthetic polymers were intermittently reported to exhibit smectic, ψ-hexagonal or 2-dimensional order under certain conditions. In the case of crystallisable polymers such as nylon 6 or polypropylene this phase appears to be monotropic in character, whereas in non-crystallisable polymers such as atactic polyacrylonitrile, which possess highly dipolar groups poorly in register, the phase is enantiotropic. In all these synthetic polymers the system seems to be heterogeneous, in that a smectic phase and an isotropic or oriented amorphous phase are simultaneously present.

Theoretical studies of the effect of segmental rigidity in polymers led in 1956 to the formulation by Flory [1,2] of criteria for the formation of a single anisotropic phase in solutions of rigid and semi-flexible polymers. These criteria can also be interpreted as applying to polymers where the solvent concentration is zero, and therefore to thermotropic as well as lyotropic systems. At about the same time it was shown by Conmar Robinson [3,4] that poly(γ-methyl-L-glutamate) (PMLG) and poly(γ-benzyl-L-glutamate) (PBLG), synthesised by Elliott and Ambrose [5] at Courtaulds Ltd., as part of a programme of evaluation of fibre formation in synthetic polypeptides, gave liquid crystalline solutions in relatively non-protonating solvents such as dioxan and methylene chloride. Fibres were wet-spun from both by Ballard [6,7] but were not commercialised. In 1967, Hermans [8] reported that anisotropic solutions with very low viscosities at low shear rates were formed by PBLG provided that (a) the solvent was not too strongly protonating in character, (b) the molecular weight was high enough, and (c) the concentration of the polymer in solution was high enough. It has since been shown [9] that the phase diagram for PBLG solution in terms of polymer concentration and temperature conforms remarkably closely to that previously predicted on theoretical grounds for semi-flexible polymers by Flory [2].

A major impetus was given to work, both academic and industrial, in the field of lyotropic systems by the development by duPont of commercial fibres having exceptionally high tensile strength and modulus through use of nematic anisotropic solutions of relatively rigid-chain aromatic polyamides. The earliest product to appear, Fibre B, was based upon poly (*p*-benzamide) (I) [10], but was replaced by the fully commercial product, Kevlar, based upon poly (*p*-phenylene terephthalamide) (II) [11]. Arenka, from Akzo, also has the latter chemical repeating unit.

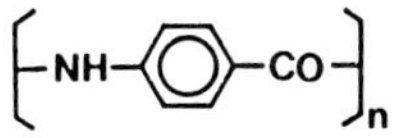

I

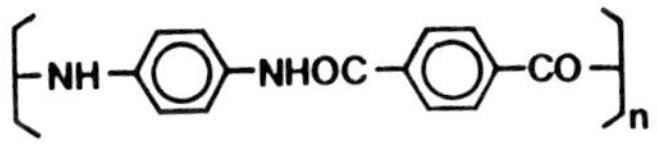

II

Although some of the polyesters described in earlier patents issued to ICI [12,13] and to the Carborundum Co. [14] are thermotropic, no statement to this effect was made and the first public ascriptions to a polymer of thermotropic character arising from main-chain conformation appeared in 1975. At least three independent publications can be cited: a paper by Roviello and Sirigu [15] that describes a polyester containing regularly alternating rigid and flexible segments where the rigid segment is derived from a nematic compound of low molecular weight; a patent to duPont [16] laid open for inspection in November 1975, but filed in May 1974, that describes a range of highly aromatic polyesters with a substantial proportion of extended-chain conformation; and a paper by Jackson and Kuhfuss [17] of the Tennessee Eastman Co. that describes random copolyesters of poly(ethylene terephthalate) with poly(*p*-oxybenzoyl), previously patented [18] with a filing date in 1972. This last patent, however, does not include a statement relating to thermotropic behaviour. Copolyesters of the type described in it were made freely available for commercial and industrial evaluation by the Tennessee Eastman Co. under codes such as X7G and X7H, and have consequently been widely studied.

The possibility of obtaining thermotropic polymers was considered theoretically by Papkov and his collaborators in 1973 [19] and by Ciferri in 1975 [20] and is implicit in Flory's treatment [2] of the rigid-rod situation.

It was also shown in 1975 that incorporation of rigid-rod and mesogenic units into the side-chain of acrylate or methacrylate polymers could lead to thermotropic products [21]. The properties of these polymers, which are not further considered in this review, differ in a number of respects from those of polymers whose thermotropic behaviour depends upon the constitution of their main chain.

2 Special Properties of Main-Chain Liquid-Crystalline Polymers

Academic and industrial interest in liquid-crystalline polymers of the main-chain type has been stimulated by certain special properties shared by lyotropic and thermotropic systems that exhibit a nematic phase. Although these special properties affect both the processing into fibres and other shaped articles and the physical behaviour of the products, the product behaviour is at least partly attributable to the novel processing behaviour.

Outstanding among the properties relevant to the processing of the polymers is the response of the nematic phase to shearing. Even quite low shear rates are sufficient to cause alignment of the nematic domains parallel to the direction of flow, with the result that apparent viscosities are remarkably low in the anisotropic phase. Where the composition under examination forms an isotropic phase at a higher temperature, the apparent viscosity in that isotropic phase is higher, and sometimes much higher, than in the anisotropic phase at the lower temperature. Such behaviour also occurs in simple compounds that exhibit both nematic and isotropic liquid phases [22], but is much more pronounced in polymers. Under conditions of extensional flow, such as those encountered during fibre or film extrusion at quite modest haul-off speeds, the degree of parallelisation of nematic domains becomes so high that the product is already oriented to an extent similar to or greater than that attainable by the conventional process of drawing fibres or films initially of low orientation. The undrawn products therefore have mechanical properties that in terms of tensile strength and modulus greatly surpass the conventional undrawn fibres and films and even the majority of drawn fibres and films. Further substantial improvements in tensile strength or modulus can be obtained by annealing these products at high temperatures, but it is not clear to what extent, if at all, the nematic phase is involved in this stage.

It should be noted that extrusion of a nematic phase is not the only means of attaining high strength and high modulus. Polyethylene drawn to very high draw ratios has been shown to produce extended-chain structures with this type of mechanical property [23], and an aromatic copolyamide, poly(*p*-phenylene *co* 3,4^1-oxydiphenylene terephthalamide), that is spun from an isotropic solution then drawn at a high temperature, gives a fibre with tensile properties similar to those of Kevlar 29 and Arenka 900 [24,25].

In addition to their unusual rheological properties, the nematic phases of polymers, like those of simple compounds, can be oriented by the application of magnetic or electrical fields. These properties have been more fully examined for comb-type polymers with mesogenic side-chains than for polymers with the mesogenic groups in the main chain, since in the comb polymers it is possible to influence the side-chain orientation independently of the main-chain orientation.

3 Survey of the Chemical Structures Involved

At least two types of polymeric chain structure capable of producing linearly rigid polymers that possess liquid-crystalline character can be distinguished. One type consists of a helix in which successive turns in the helix are maintained in register through steric hindrance or by intramolecular bonding through, for example, hydrogen bonds. The individual bonds within the main chain of the helix may be, but need not be, inherently quite flexible, as in the α-helical form of polypeptides. Most polymers of this type contain chiral centres and therefore form cholesteric rather than nematic phases. Much the most widely studied examples are esters of poly(L-glutamic acid), where the side chain R in the peptide repeating unit —HNCHRCO— is $—CH_2CH_2COOR^1$ and R^1 is benzyl, cyclohexyl or 1-6C alkyl [3,26]. This class is reviewed separately in the present series by Uematsu [27].

The majority of synthetic liquid-crystalline polymers, however, fall into the second class. These polymers consist of a succession of rigid units that are individually characterised by possessing chain-continuing bonds that are either collinear or parallel and oppositely directed. A simple example of the former unit is *para*phenylene, and of the latter unit is 1,5-naphthylene. Figure 1 illustrates a wider range of such groups based upon cyclic structures. The inclusion of 1,4-*trans*cyclohexylene illustrates

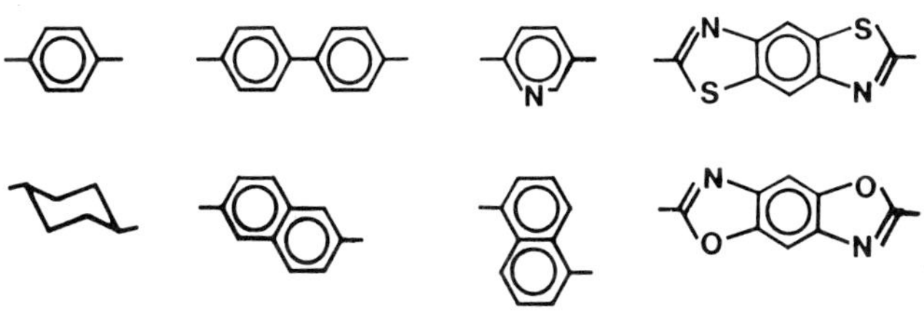

Fig. 1. Cyclic units present in rigid-rod segments of liquid-crystalline polymers

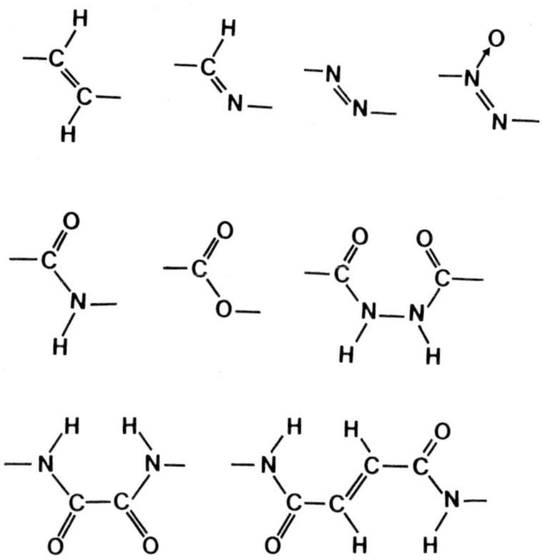

Fig. 2. Noncyclic units present in rigid-rod segments of liquid-crystalline polymers

that the rings involved need not be aromatic in character. These ring structures are inherently rigid, and are main-chain components of nearly all the liquid-crystalline synthetic polymers of this type known. They may, however, be connected by noncyclic units such as those illustrated in Fig. 2, the common features of which are again that their chain-continuing bonds are either parallel or approximately parallel, and also oppositely directed. The *trans* conformations required to meet this criterion are maintained either by the presence of double bonds or, as in the case of the amide and ester groups, by a substantial energetic preference for that conformation. A deviation of at least 9° from parallelism in units containing two main-chain atoms (amide) [28] (cf. 1° for ester) is tolerable.

It is evident that these structural guidelines for the second type of polymer are closely akin to those that define mesogenic segments in simple liquid-crystalline compounds.

3.1 Lyotropic Systems

Both lyotropic and thermotropic liquid-crystalline synthetic polymers have been widely studied. Aromatic polyamides constitute the most important class forming liquid-crystalline solutions; the solvents are either powerfully protonating acids such as 100% sulphuric acid, chloro-, fluoro- or methane-sulphonic acid, and anhydrous hydrogen fluoride, or aprotic dipolar solvents such as dimethyl acetamide containing a small percentage, usually 2–5%, of a salt such as lithium chloride or calcium chloride. Such solutions constitute a nematic phase within certain limits. Some criteria for formation of a nematic instead of an isotropic phase are:

i polymer concentration above a critical value;
ii polymer molecular weight above a critical, high value;
iii temperature below a critical value.

These critical values differ from solvent to solvent. With the solvents most fully investigated, sulphuric acid for poly-(p-phenylene terephthalamide) and dimethyl acetamide/lithium chloride for poly (p-benzamide), there are also critical values of solvent composition: the sulphuric acid must exceed a critical strength, the lithium chloride in the dimethyl acetamide must exceed a critical concentration. The critical values are, of course, interdependent rather than absolute. Diagrams that display some of the critical values for the two systems cited have been published in patents [10,11]. Figures 3 and 4 illustrate the type of information available.

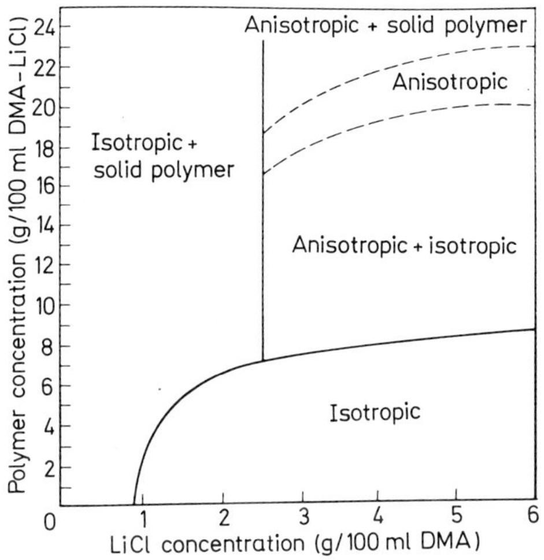

Fig. 3. Phase diagram for solutions of poly-*p*-benzamide in dimethyl acetamide/lithium chloride [10]. Polymer inherent viscosity 1.18

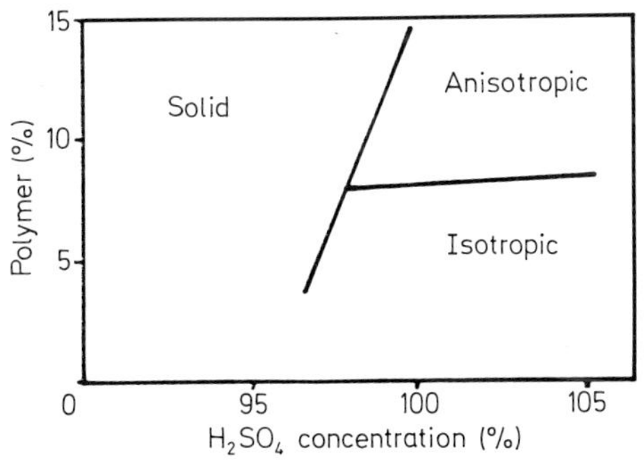

Fig. 4. Phase diagram for solutions of poly (*p*-phenylene terephthalamide) in sulphuric acid [74]. Polymer inherent viscosity 3.32

It has been found for some systems, and may be true for all, that there is no transition directly from the isotropic to the nematic phase as the critical condition is attained. Instead, a narrow biphasic region is found in which isotropic and nematic phases co-exist. This behaviour was predicted by Flory [2], even although his initial calculations related to monodisperse polymers. It is accentuated by polydispersity (see Flory's review in Vol. 59 of Advances in Polymer Science), and indeed for a polydisperse polymer the nematic phase is found to contain polymer at a higher concentration and of a higher average molecular weight than the isotropic phase with which it is in equilibrium.

For the formation of the lyotropic systems studied so far it has been essential to use a polymer of high molecular weight. The higher the molecular weight, the lower the critical concentration for nematic solution formation. At the low molecular weight end of the range, Papkov [29] found (Fig. 5) that a weak maximum in solution viscosity as a function of polymer concentration can still be observed for PBA at a limiting viscosity number of 0.47 in dimethyl acetamide/3% LiCl at 20 °C. He derived a molecular weight, $\overline{M}$, of 11,000 for this polymer from the Flory relationship between the critical volume concentration of polymer, v*, and the axial ratio of rigid and rod-like particles, x,

$$v^* = \frac{8}{x}\left(1 - \frac{2}{x}\right)$$

In principle the possibility exists of obtaining lyotropic systems from polymers of low molecular weight or of low average rigid-rod segment length at much higher volume fractions of polymer, such that the liquid-crystalline phase might also be considered to be a plasticised melt.

In isotropic solutions of PPT in solvents of the N,N-disubstituted amide/LiCl type, a maximum in limiting viscosity number, [η], has been observed when the lithium chloride content of the solution is chemically equivalent to the secondary amide group content provided by the dissolved polymer [30]. Titration of model secondary amides with lithium chloride in tertiary amide solvents confirms that there is a 1:1

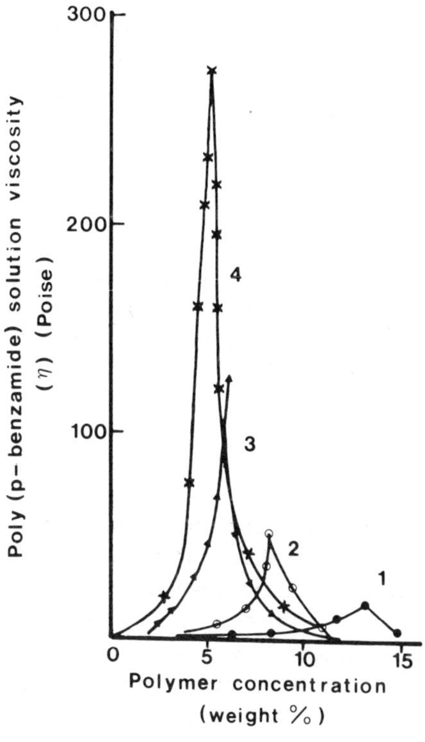

Fig. 5. Viscosity of poly-*p*-benzamide solutions in dimethyl acetamide/lithium chloride as a function of concentration and intrinsic viscosity [29]. Values of intrinsic viscosity are: (1) 0.47, (2) 1.12, (3) 1.68, (4) 2.96

association, and electrophoretic data on other aromatic polyamides show that the chloride anion associates with the secondary amide group[31]. The lithium cation associates with the tertiary amide solvent, thus

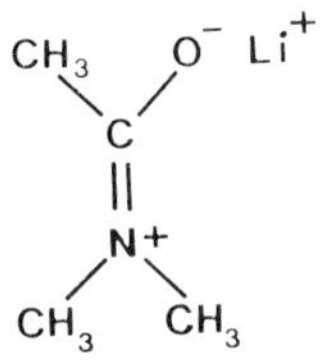

III

Substituents in the rings, if not too large, still permit the formation of a nematic phase but make its attainment, in terms of the critical parameters, more difficult. Poly(chloro-*p*-phenylene terephthalamide), poly(methyl-*p*-phenylene terephthalamide) and poly(*p*-phenylene chloroterephthalamide) are examples of such substituted polymers that form anisotropic solutions[32]. Copolymerisation of units conforming to the general rules stated earlier does not destroy the capacity for nematic phase formation, so that for example PBA/PPT copolymers, both random and ordered, form anisotropic solutions. However, copolymerisation with quite small proportions of units that do not conform to the requirements for rigid rod formation, for example replacement of more than about 10% of the terephthaloyl groups in PPT by isophthaloyl groups, leads to loss of nematogenic ability.

Lyotropic polymers consisting solely of ring structures are also known, and are exemplified by poly(p-phenylene benzobisthiazole) (IV), which forms a nematic phase in several strongly protonating acids including polyphosphoric acid. Its synthesis and application, like those of the aromatic amide polymers, are discussed later.

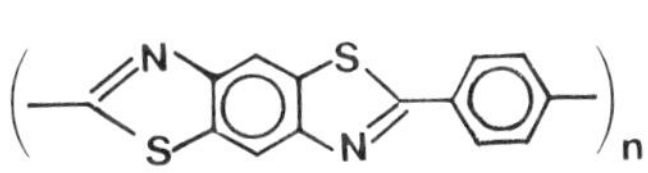

IV

Poly(terephthaloyl hydrazide) forms anisotropic solutions in concentrated solutions of some quarternary ammonium hydroxides such as tetramethyl ammonium hydroxide[33]. These solutions contain typically about 10% each of the organic base and the polymer. Aromatic polyhydrazides such as poly(chloroterephthaloyl hydrazide) and various co-polyhydrazides also form anisotropic solutions in some of the solvents used for PPT, such as 100% sulphuric acid and fluorosulphonic acid[34].

3.2 Thermotropic Systems

Thermotropic polymers require no solvent for formation of a liquid-crystalline phase, which occurs instead within a defined temperature range. The chemical units useful for thermotropic polymer formation are generally those already exemplified in Figs. 1 and 2, but these units in homopolymer form give rise to crystalline polymers with melting points above their decomposition temperatures. The problem of polymer design is to reduce the melting temperature in order to obtain a liquid-crystalline phase at a temperature below that of decomposition. Whereas the lyotropic systems are commonly based upon polymers that contain amide groups in the main chain, thermotropic systems are more usually based upon polymers that contain ester groups or other linking groups free from hydrogen bonds. Even without the inter-chain association caused by amide groups, considerable disruption of main-chain order is necessary to depress crystallinity. Thus the homopolyesters poly(*p*-oxybenzoyl), poly(*p*-phenylene terephthalate) and poly(*p*-phenylene naphthalene-2,6-dicarboxylate) melt at temperatures of 610° [35], 596° and 577 °C [36] respectively, too high for formation of a stable nematic melt. Methods of disruption of order will be considered in two main sections each containing two sub-sections: polymers that contain only rigid-rod segments, which are divisible into those free from and those including contributions to disorder from unsymmetrically substituted rings, and polymers that contain

Table 1. Some thermotropic copolymers containing rigid-rod units only

Polymer No.	AB units	AA units	BB units	Ref.
V	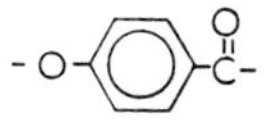	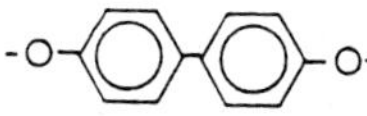	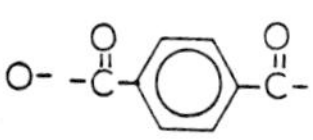	37–39)
VI	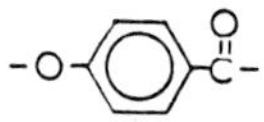	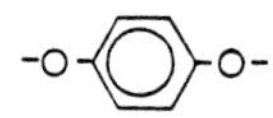	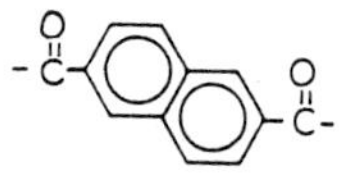	35, 40)
VII	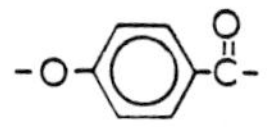	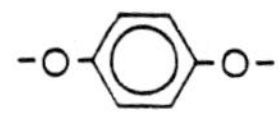	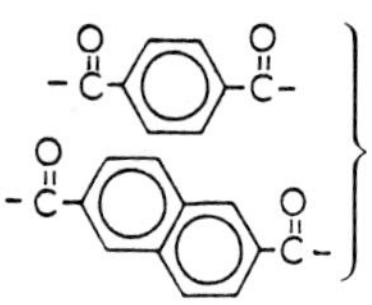	36)
VIII	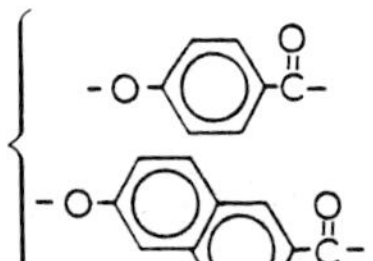	—	—	41)
—	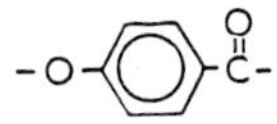	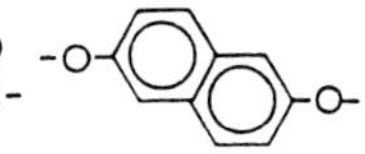	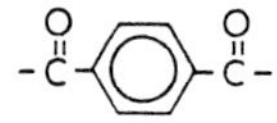	42)
—	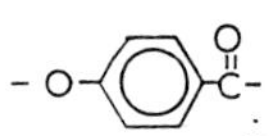	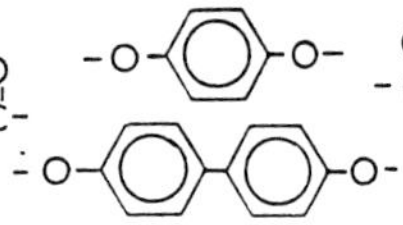	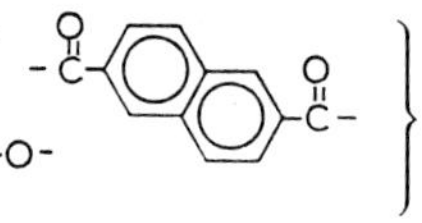	40)
IX	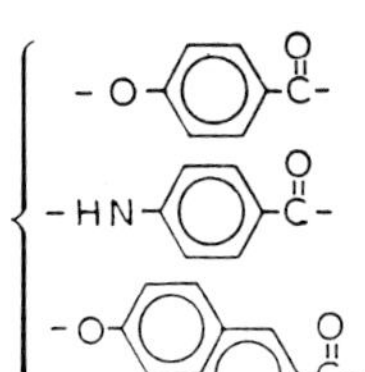	—	—	43)

both rigid-rod segments and flexible or angular spacer units, which are divisible into those where the rigid-rod units are regularly disposed in the structure and those where they are randomly disposed and therefore of variable length.

3.2.1 Polymers Containing Rigid-Rod Segments Only

3.2.1.1 Disorder not due to Unsymmetrical Ring Substitution

Copolymers that contain only rigid-rod segments are of particular interest because highly oriented samples are expected to exhibit higher moduli than those derived from copolymers that contain flexible or angular segments. Some products of this type, free from nuclear substitution, are listed in Table 1.

The polymer V containing 67 moles % of *p*-oxybenzoyl units was marketed by the Carborundum Co. as Ekkcel 12000. According to Jackson [35], the properties of this copolymer, which melts at 380 °C, are highly anisotropic when it is injection-moulded at 400 °C but the melting point proves too high for fibre extrusion.

Polymer VI exhibits a minimum in melting point at about 325 °C at a composition containing about 70 moles % of *p*-oxybenzoyl repeating units and 30 moles % of *p*-phenylene naphthalene-2,6-dicarboxylate repeating units. Such products are readily melt-processable, in contrast to the corresponding copolymers containing terephthaloyl instead of naphthalene-2,6-dicarbonyl groups which melt above 500 °C. However, terephthaloyl groups can be substituted for at least 60 % of the naphthalene-2,6-dicarbonyl groups without significantly enhancing the melting point (Polymer VII). The 2,6-naphthylene group appears particularly effective in lowering the melting point of copolymers to reveal nematic behaviour. Noteworthy are the copolymers of *p*-oxybenzoyl and 2-oxy-6-naphthoyl (Polymer VIII), which are anisotropic and melt-processable at proportions from 75/25 to 40/60 molar and possibly over a wider range and which give a minimum melting point of about 265 °C at 60/40 molar proportion.

Polymer IX illustrates that it is possible to include a limited proportion, for example 20 moles %, of amide linking groups into this type of copolymer without destroying the thermotropic property. In this case the melting point of the copolymer containing 60 moles % of 2,6-oxynaphthoyl units and 20 moles % each of *p*-oxybenzoyl and *p*-iminobenzoyl units is only 277 °C [43], similar to that just noted for the copolymer with ester instead of amide groups.

3.2.1.2 Disorder Arising from Unsymmetrical Ring Substitution

In a rigid-rod polymer of the AA–BB type based upon p-phenylenedioxy and terephthaloyl segments, the para-disubstituted phenylene rings possess a plane of symmetry normal to the chain axis. Introduction of a single substituent into the p-phenylene ring of either segment destroys the plane of symmetry and leads effectively to copolymerisation, since the random occurrence of head-to-head, head-to-tail isomerism disrupts the ability of the chain segments to pack into crystallites. This principle was exploited by Goodman [12, 13] in a series of patents in which several of the examples represent thermotropic polyesters, although no statement to this effect was made. The principle was applied by Schaefgen and his colleagues [44] to reduce the crystalline melting point of rigid-rod polymers and to make it possible not only to observe the nematic phase at a temperature below the decomposition temperature but also to process the nematic melt. In these polyesters, substitution in the p-phenylenedioxy ring proves more effective than substitution in the terephthaloyl ring. The size of the substituent is important. Methyl and chloro-substituents have proved particularly effective. Fluoro-substituents can be too small to prevent crystalline packing, and progressively larger substituents such as methoxy, bromo, etc. reduce the nematic phase stability. Nevertheless substituents as large as phenyl [45] and *p*-tolyl [46] are effective in revealing nematic melts.

Extensions of this principle, for example to substitution in one of the rings of a 4,4′-biphenylene segment [47], are readily visualised, and the principle has also been applied to polyazomethines [48]. There is a further possibility that does not involve an additional nuclear substituent, and this is to use a reactant such as a *p*-mercaptophenol derivative [49] or a *p*-aminophenol derivative [50, 51] that also leads to loss of the plane of symmetry in the *p*-phenylene ring. In neither case has this variant been demonstrated as being effective on its own, although the latter in combination with other methods of reducing crystallisability has made it possible to produce thermotropic polyesteramides in spite of the strong inter-chain ordering forces due to hydrogen bonding of the amide groups.

3.2.2 Polymers Containing Spacer Units

3.2.2.1 Regularly Disposed

Thermotropic behaviour is observed in many polymers that contain rigid-rod units alternating regularly with relatively flexible spacer units. A first requirement for nematic melt formation is that the rigid-rod unit should exceed a certain length and length: diameter ratio. The shortest rigid-rod unit so far reported [52] to permit nematic melt formation is

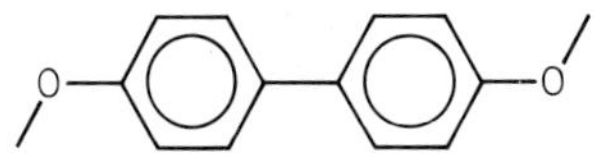

with a rigid-rod length of only about 1.1 nm. A second requirements is that the flexible spacer unit should not exceed a certain critical length, that length depending upon the structure of the rigid-rod unit and the structure of the flexible spacer unit. In the majority of cases the smectic or crystalline phase stability becomes higher than the nematic phase stability beyond that critical length.

The nematic phase stability is higher the greater the length and length: diameter ratio of the rigid-rod unit, and the shorter the flexible spacer unit, and rises ultimately to values above the temperature of decomposition of the polymer.

Copolymerisation of a single rigid-rod unit randomly with two different spacer units leads to a nematic stability that is a weighted average of the two homopolymeric values, but expands the thermal range over which the nematic phase is observed because it lowers crystalline and smectic phase stabilities [53]. Particularly wide nematic ranges have been observed with oligosiloxane spacer units and various rigid rod units containing three or four rings and two ester groups, where the glass transition temperatures are as low as −111 °C. Here the polymers with a uniform length of oligosiloxane segment generally give smectic mesophases only, but use of two randomly distributed oligosiloxane segments of significantly different length leads to nematic phase formation and in one case [54] to a nematic range of 311 °C.

The spacer unit may alternatively be inflexible and relatively rigid and angular. The oxygen atoms of 4,4′-disubstituted diphenyl ether groups introduce an angle of approximately 120° into the rigid chain, and such units appear to be particularly effective in reducing or suppressing the tendency to crystallise. Polymers from mono-substituted *p*-phenylenedioxy groups and diphenyl ether 4,4′-dicarbonyl groups contain these angular units regularly disposed along the chain and have been found to give nematic melts sufficiently stable for extrusion [55].

Shorter spacer length does not necessarily assist the developement of thermotropic behaviour. The polymer X does not appear to be thermotropic [56] yet the same rigid-rod unit with a flexible alkylenedioxy spacer forms a nematic melt [57].

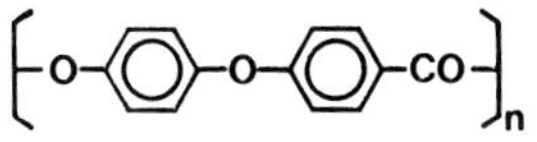

X

3.2.2.2 *Randomly Disposed*

Flexible or rigid but angular units can also be used as non-mesogenic components of thermotropic random copolymers. In such cases the rigid-rod length and/or spacer length are no longer constant throughout the chain. Where there is a distribution of rigid-rod lengths, the effect of this distribution upon the upper limit of nematic phase stability has not been established. Since however the crystal → nematic transition temperature is lowered relative to the corresponding products with regularly disposed spacers, the temperature range of nematic stability is wider. This method of producing thermotropic polymers has therefore been widely applied, particularly in patent examples. In addition to the angular rigid groups already discussed, isophthaloyl, *m*-phenylenedioxy, *m*-oxybenzoyl and other groups containing *m*-phenylene units also introduce a relatively rigid angle of 120° into the chain and have frequently been used.

Much the most widely studied liquid-crystalline polyesters are of this type. They are the copolymers of poly(ethylene terephthalate) with poly-*p*-oxybenzoyl, produced by reacting pre-formed poly(ethylene terephthalate) with *p*-acetoxybenzoic acid then repolymerising the reaction product [17, 18]. At less than about 30 moles % of oxybenzoyl units the melts remain isotropic, but above this value thermotropic behaviour occurs. Table 2 lists some of the phenomena that appear in the thermotropic range. The anisotropy of mechanical properties is highest at a composition close to 60 moles % of oxybenzoyl units, and in all cases depends greatly on the conditions of moulding and the dimensions of the moulded sample. Higher values of strength and modulus are obtained at the same molecular weight if the moulding temperature is increased, and at the same moulding temperature if the molecular weight is increased. In the range 60–80 mole % *p*-oxybenzoyl units, tensile strengths, elongations at break, flexural moduli and impact strengths are as good as, or better than, those of glass-reinforced polyesters.

Numerous other copolyester systems incorporating ethylene terephthalate units and rigid rod polyester units give similar behaviour. With *p*-phenylene terephthalate

Table 2. Some phenomena arising from thermotropic behaviour in copolyesters (Comparisons are with analogous polymers giving isotropic melts)

Melt behaviour	
i	Melt turbid and stir opalescent
ii	Melt orientable in a magnetic field
iii	Melt orientable in an electric field
iv	Broad-line NMR structure persists in melt
v	Lower melt viscosity at similar dilute solution viscosity
Solid behaviour	
i	Higher quenched-polymer density
ii	Anisotropic tensile strength (injection-moulded)
iii	Anisotropic flexural strength (injection-moulded)
iv	Anisotropic flexural modulus (injection-moulded)
v	Anisotropic impact strength (injection-moulded)
vi	Lower coefficient of thermal expansion
vii	Lower thermally-induced shrinkage

rigid-rod units, turbid melts are observed above about 15 mole %; then high-melting products are formed above about 35 mole %, but substitution of a methyl or chloro-group in the *p*-phenylene ring permits formation of products analogous to those from *p*-oxybenzoyl units over a range from about 40 to 70 mole % [58]. The copolyester containing 40 mole % ethylene terephthalate, 30 mole % *p*-oxybenzoyl and 30 mole % *p*-phenylene terephthalate units is anisotropic, as are a wide range of copolyesters containing other rigid rod groups in addition to *p*-oxybenzoyl or *p*-phenylene terephthalate [58]. The flexible component in anisotropic copolymers of these basic types may alternatively be ethylene naphthalene-2,6-dicarboxylate [58] or ethylene diphenoxyethane-4,4′-dicarboxylate [59].

According to Lenz and Feichtinger [60], copolymers of these types containing *p*-oxybenzoyl or 3-mono- or 3,5-di-substituted p-oxybenzoyl units tend to reorganise to form block copolymers when held in the molten state, and sufficient sequences of the rigid-rod units are built up for the polymers to become semi-crystalline, infusible and insoluble. Such behaviour can clearly create problems of irreproducible processing behaviour in samples subjected to prolonged periods in the molten state.

Analogous polyesteramides have been made from poly(ethylene terephthalate) and *p*-acetamidobenzoic acid (PAB) [61]. The products exhibited similar anisotropic strength and stiffness to liquid-crystalline oxybenzoate/ethylene terephthalate copolymers, but at somewhat higher ethylene terephthalate content. Only about 30 moles % of PAB units could be incorporated before the melting point rose to a temperature too high for melt processing. Although the polyesteramide melts were turbid, they exhibited only one line when examined by proton NMR at 90 MHz and did not exhibit general birefringence when observed between crossed polars. Consequently they do not appear to form nematic mesophases. Nevertheless the melts do have very long relaxation times, and this seems sufficient to account for the anisotropic nature of the physical properties of thin moulded specimens.

3.2.3 Effect of Molecular Weight on Thermotropic Properties

Polymers that consist entirely of rigid-rod units form anisotropic melts at a very early stage of melt polymerisation if they form them at all. This observation suggests that a very low degree of polymerisation suffices for the development of a liquid-crystalline phase and is consistent with the fact that very short rigid-rod segments are sufficient in low molecular weight compounds and with the common use of compounds that are already liquid-crystalline as intermediates for the synthesis of liquid-crystalline polymers. However, the nematic-isotropic transition temperature appears to be a function of molecular weight for polymers of high rigid-rod content, as expected since higher molecular weight will lead to greater average rigid-rod length. With copolymers of a disordered type that contain angular or flexible spacer units, liquid-crystalline behaviour may not develop until a somewhat higher critical molecular weight has been attained. This too is expected, since at low molecular weights the average rigid-rod length present in the polymer will be low and further polymerisation is required to reach a critical average length for mesophase formation.

Rather more surprisingly, there is evidence that even with a fixed rigid-rod length the nematic-isotropic transition temperature can depend upon molecular weight, although to a much lesser extent than in the cases already considered. Thus in the polymer XI the transition temperature rose from 135.5 °C to 165 °C as $\overline{M}_n$ was increased from 4,000 to 20,000 [62].

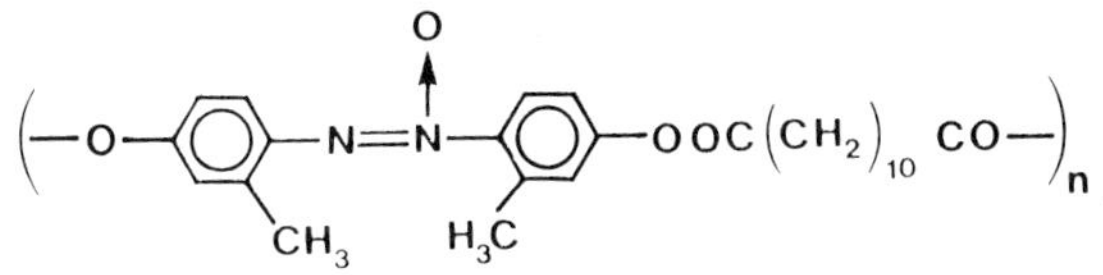

XI

3.2.4 Thermal Transitions

In many cases the disorder introduced into rigid-rod structures in order to depress the crystalline melting point to a temperature below the degradation temperature is such that the polymer can crystallise only to a small extent or not at all. In such cases, the polymer exhibits a frozen nematic structure at temperatures below the glass transition, with a typical micaceous sheen.

The glass transition temperature observed in copolyesters of this type is, however, often much lower than would be predicted for a polymer of this structure, using established correlations such as that of van Krevelen [63]. For instance. Warner and Jaffe have shown that Tg measured by D.S.C. on the copolyester poly(oxybenzoate *co p*-phenylene naphthalene-2,6-dicarboxylate *co* isophthalate) (75/12.5/12.5)[1] is only 75 °C [64], whereas a predicted value would be 159 °C. The transition is presumably connected not with segmental motion of the type observed in isotropic polymers, but with the onset of some other phenomenon peculiar to rigid-rod polymers. A possible source of such a transition is displacement of parallelised rigid-rod segments relative to each other along the main chain axis. If Tg is regarded as a temperature associated with attainment of a particular viscosity, the low value might be correlated with the low viscosity of the nematic melt, which can be several orders of magnitude below that predicted from extrapolation of isotropic viscosities to the temperature

[1] Molar ratios have been normalised throughout this review such that the sum of the simplest possible set of *repeating* units is 100.

of measurement, although there is a difficulty here in that this low viscosity is not strictly associated with an unstirred melt.

This low temperature transition is potentially of some technological importance, since it can lead to a fall in modulus in semi-crystalline shaped articles at unexpectedly low temperatures. Further study of this phenomenon seems desirable, since rigid-rod polymers with Tg values approximating to those predicted have also been reported.

Whereas slow cooling of crystallisable rigid-rod copolyesters can lead to well-developed crystallites whose crystal structure can be examined by wide-angle x-ray techniques, quenching normally leads to samples that give almost 'amorphous' x-ray patterns. Nevertheless such samples may exhibit endotherms that are attributable to crystallite fusion, and the heats of fusion derived from these endotherms are identical for the two types of sample [65]. This result, observed for both an unsubstituted rigid rod copolymer, poly(*p*-oxybenzoate-*co*-2,6-oxynaphthoate) (40/60) and a substituted copolymer with a spacer, poly(chloro-*p*-phenylene terephthalate-*co*-*p*-phenylene diphenoxyethane-4,4′-dicarboxylate) (50/50), has been explained in terms of the formation in quenched samples of microcrystals smaller than those in conventional polymers. These are stabilised by their very low surface energy, which is a consequence of the nematic morphology and the absence of chain folds at crystallite surfaces.

3.2.5 Thermotropic Derivates of Natural Polymers

A few derivatives of cellulose have been shown to be thermotropic. Benzoyl [66], propionyl [67] and acetyl [68] hydroxypropyl cellulose are reported to give cholesteric melts that clear at 164 °C, 167 °C and 160 °C respectively. Trifluoroacetoxypropyl cellulose [69] clears at 155 °C but cellulose acetate butyrate [70] is not thermotropic. Commercial ethyl cellulose [70] and hydroxypropyl cellulose [70,71] are also liquid-crystalline and the latter has been melt-spun [71]. Most of these polymers (including cellulose acetate butyrate) have been shown to give cholesteric solutions at high concentrations in various solvents.

Cholesteric melts have even been reported for several polypeptides, each copolymers of a relatively small (methyl, propyl) and a relatively large (hexyl, octyl, benzyl) ester of glumatic acid [72,73] of a very low crystallinity. In these cases the products were probably insufficiently stable for melt-processing.

4 Development of Lyotropic Polymers

This section concentrates upon the development of materials based on aromatic polyamides, particularly poly(*p*-phenylene terephthalamide) (PPT) (II) [11,74] and on poly(*p*-phenylenebenzobisthiazole) (PBT) (IV) [75]. These materials are primarily intended to satisfy the demands of industry for lightweight, high-performance materials suitable for reinforcement purposes, especially in composites. The polymers have therefore been fabricated mainly into fibrous products.

The former polymer has been available commercially for some years in the form of high-modulus fibres (Kevlar, Arenka) and current work on the latter points towards its possible use as a material exhibiting even higher mechanical performance.

4.1 Poly(*p*-Phenylene Terephthalamide)

4.1.1 Synthesis

Poly(*p*-phenylene terephthalamide) (PPT) is usually synthesised [74] via a condensation reaction based on *p*-phenylene diamine and terephthaloyl chloride according to the equation shown in Fig. 6.

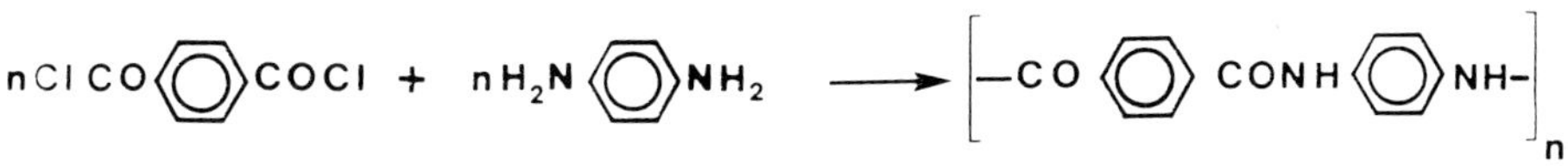

Fig. 6. Synthesis of poly(*p*-phenylene terephthalamide)

One method involves dissolving appropriate amounts of *p*-phenylene diamine in a mixture of hexamethylphosphoramide and N-methylpyrrolidone, cooling to 15 °C in a nitrogen atmosphere and adding terephthaloyl chloride. The resulting product is ground with water, filtered and subsequently washed to remove solvent and hydrogen chloride.

Alternative routes designed to eliminate the use of hexamethylphosphoramide (considered to be a carcinogenic agent) have been devised [76]. It is claimed that these methods, which involve the use as solvent of N-methylpyrrolidone containing about 10% of calcium chloride, produce products of adequate molecular weight suitable for fibre production. In all cases the presence of impurities, including traces of water in the monomers and solvent, tends to impair the formation of high molecular weight polymer.

Para aromatic polyamides under certain conditions of concentration, type of solvent and temperature form liquid-crystalline solutions. A typical phase diagram is shown in Fig. 4. To prepare anisotropic dopes suitable for spinning fibres (see Sect. 5.5), 99.8% sulphuric acid has been used as a solvent together with a high solids concentration (~20% polymer). The dopes are solid at room temperature but melt at higher temperatures, become less viscous and show optical anisotropy under the polarising microscope. The liquid-crystal type involved is thought to be nematic in character. Since excessive heating promotes a decrease in anisotropy the dopes are usually extruded at temperatures in the range 70–90 °C.

4.1.2 Fibre Formation

Production of fibres from the predecessor of PPT, poly(*p*-benzamide), involved use of an anisotropic solution in dimethyl acetamide or tetramethyl urea containing lithium chloride, which could be either wet- or dry-spun [10]. The filaments obtained form the spinning process, particularly from the preferred dryspinning process, were already quite highly oriented and had very useful tensile properties. A tensile strength of 1.05 GPa and Young's modulus of 65 GPa at an extension to break of 3.1% could be achieved by dry-spinning at a wind-up speed of 123.5 m · min^{-1} and a spin-stretch factor of 3.2. Brief annealing, for a duration of only seconds at a high temperature at constant length under nitrogen, increased the tensile strength and modulus and reduced the extension at break; use of a maximum temperature of 525 °C gave values of 2.2 GPa, 137 GPa and 1.9% respectively.

Production of fibres from PPT using sulphuric acid as solvent, on the other hand, necessitates use of a wet-spinning process [11]. Although fibres with excellent tensile properties can be made by conventional wet-spinning followed by annealing at about 500 °C, still better properties can be achieved by introducing an inert fluid layer of gas or a non-coagulating liquid between the face of the spinneret and the coagulant [77]. In practice the inert fluid is normally air and the gap is of the order 0.5 to 2 cm. The process [78] is known as air-gap spinning or dry-jet wetspinning. One of the functions of the air gap is to permit operation with very different dope and coagulant temperatures. The dope temperature is about 80 °C, a temperature which is above the melting point of a solution containing about 20 weight % of polymer and thus permits use of a concentrated anisotropic solution, whereas the coagulant temperature is preferably below 5 °C for the highest tenacity fibres. The coagulant is normally water or dilute sulphuric acid, although use of water-miscible organic solvents has also been described.

Use of a vertical spin tube with co-current flow of coagulant improves the mechanical properties of the as-spun fibres [77]. Tapering tubes are preferred to those possessing parallel sides for producing as-spun fibres exhibiting higher modulus and lower extensibility. The presence of the air gap and use of a spin tube both permit the use of higher spin-stretch factor (SSF), a parameter defined as the ratio of the velocity of the fibre leaving the coagulating bath to the average velocity of the dope emerging from the spinneret capillary. Generally with increasing SSF the tenacity and modulus both increase but at the expense of the breaking elongation.

Thorough washing of the fibres is mandatory to remove traces of acid which otherwise have adverse effects on the tenacity. It is usual for the thread line to be sprayed with a weakly alkaline solution before final washing in water and subsequent drying.

Perhaps the most important stage in producing the highest modulus variants is the thermal treatment of as-spun fibres [79]. Essentially, tensioned fibres are passed continuously through a heated tube in an inert atmosphere at zone temperatures between 450 and 550 °C for 0.5 to 5 seconds. When heated above 450 °C for periods of time greater than a few minutes excessive degradation occurs resulting in an undesirable drop in strength.

It is probable that suitable thermal treatment of the tensioned fibres promotes molecular chain displacements, thus permitting the formation of regular bonding between adjacent chains leading to highly oriented regions of crystalline order.

4.1.3 Fibre Properties

4.1.3.1 Mechanical Behaviour

Depending on the choice of processing conditions PPT fibres can be produced with significantly different mechanical performance. Perhaps the best known commercially available variants are Kevlar 29 and Kevlar 49 [80]. Although production details are confidential, it is probable that the latter has been subjected to a thermal treatment.

The stress-strain characteristics of the two types are shown in Fig. 7 which indicates that the curves to failure are almost linear. It has been suggested [81] that for a large part of the curve, the strain is mainly brought about by elongation of the crystal lattice

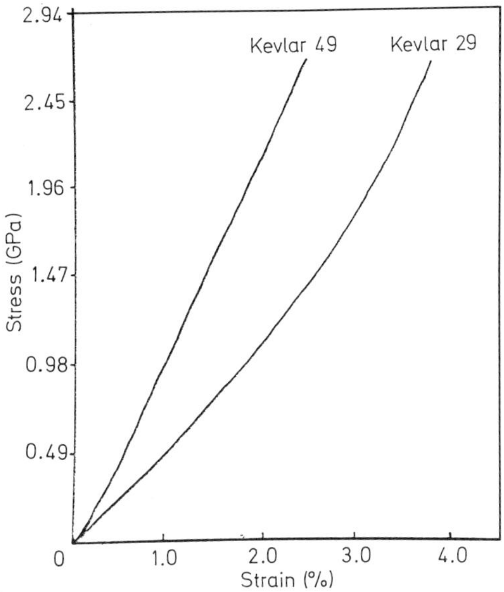

Fig. 7. Stress-strain diagram for two types of Kevlar (PPT) aramid fibre

through valence angle deformation and bond stretching of the polymer chain. Deviations from linearity are attributed to chain breakage and other irreversible processes. Tensile failure in air is characterised by stepped-fibrillar type fracture faces extending over considerable lengths. However when fractured in glycerol the fracture faces are often wedge-shaped and smooth, often revealing an internal periodicity (pleat structure) as shown in Fig. 8. The fibrous polymers are not notch sensitive which is compatible with the notion of relatively weak interchain bonding which

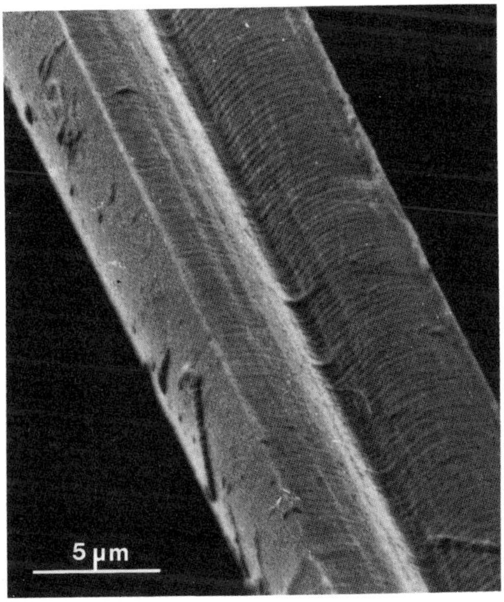

Fig. 8. Scanning electron micrograph of Kevlar 49 fractured in glycerol showing details of internal axial pleating

TENSILE PROPERTIES							
	KEVLAR 29	KEVLAR 49	NOMEX	CARBON		PBT/PPA	
				TYPE I	TYPE II	heat set	as spun
DENSITY (kg m^{-3})	1440	1450	1400	1950	1750	1500	
TENSILE STRENGTH (GPa)	2.64	2.64	0.7	2.0	2.6	0.8–2.5	0.5–2.2
MODULUS (GPa)	58.9	127.5	17.3	400	260	76–265	50–76
ELONGATION (%) AT BREAK	4.0	2.4	22.0	0.5	1.0	1.0–1.5	≈ 4.5

Fig. 9. Tensile properties of various high-strength, high-modulus fibres

allows the deflection of an originally transverse crack along a path parallel to the chain direction.

Typical values of the tensile parameters are summarised in Fig. 9 and illustrate the high mechanical performance of such polymers. Kevlar 49 has approximately double the modulus of Kevlar 29 with almost half the elongation. The combination of low density with high strength and modulus gives Kevlar the highest specific tensile strength of any commercially available material and a reasonably high specific modulus even when compared with carbon fibres. It is also reported that PPT fibres exhibit high dimensional stability under a static load (low creep) which is a particularly desirable characteristic in use as re-inforcing elements in composites.

Although structurally Kevlar fibres are highly crystalline or ordered it is interesting to note that appreciable moisture is absorbed at equilibrium. For example in variant T950 the moisture uptake is about 5% at 22 °C and 55% relative humidity. As discussed in Sect. 4.1.4.1 it is likely that the water is retained in microvoids distributed close to the surface of the fibres. Certainly in the short term there appears to be little effect of moisture on the tensile properties.

It is clear that under selected conditions PPT has a capacity for forming fibres of outstanding tensile properties much superior to conventional organic polymers. Unfortunately the compressional behaviour is somewhat disappointing [82,83]. As indicated in Fig. 10 for a unidirectional Kevlar composite, the deformation is elastic at low strain values but at higher strains is almost perfectly plastic. The onset of plastic deformation arises from shearing of the molecular chains which is aided by the relatively weak lateral interchain bonds and leads subsequently to the formation of oblique kink bands within the fibre. Examination of fibres extracted from highly compressed composites shows lateral translation of whole segments of the fibres. Although PPT variants differ somewhat in their abilities to accommodate compression, generally they are unsuitable for applications involving high compressive loading. Nevertheless, in contrast to carbon fibres which exhibit good compressional characteristics, the aramids can survive bending intact by yielding in axial compression. This mode of behaviour is particularly important technologically in terms of translating fibres into complex textile structures by the weaving process.

4.1.3.2 Thermal Behaviour

PPT fibres, unlike most organic fibrous polymers, exhibit high thermal stability, do not melt and only decompose in air at temperatures in excess of about 450 °C. For example Kevlar 49 when heated at 300 °C for 30 minutes retains 60% of its room temperature strength, 80% of its tensile modulus and 70% of its breaking extension. A high dimensional stability at elevated temperatures is a desirable feature in many engineering applications. It is reported that shrinkage at 160 °C is of the order of 0 to 0.2% and the tensile creep less than 0.03%. As indicated previously, heat treatment of the as-spun fibres produces a dramatic improvement in modulus yet increasing temperature on commercially available fibres, although producing small changes in crystallinity, tends to promote a loss in performance.

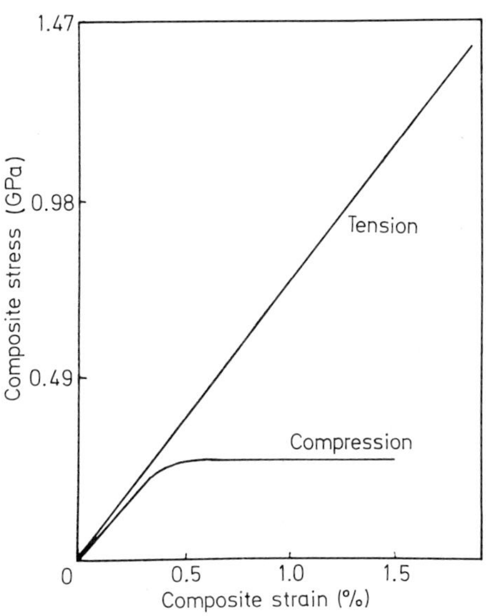

Fig. 10. Stress-strain diagram of a unidirectional epoxy composite with Kevlar 49, tested under tension and compression [80]

4.1.3.3 Optical Behaviour

Measurement of optical anisotropy or birefringence provides structural information about the orientation of a particular molecular system. Aromatic polyamides such as PPT display particularly high refractive indices [84]. For example Kevlar 49 has a refractive index ($n_{\parallel}$) for polarised light vibrating parallel to the fibre axis of about 2.267 compared with a value ($n_{\perp}$) around 1.605 for light vibrating in a perpendicular plane. Thus the overall magnitude of the birefringence, $n_{\parallel} - n_{\perp} = 0.662$, is indicative of high orientation of the molecular chains about the fibre axis.

The aramids also exhibit very unusual properties laterally. Indeed under crossed polars, complete cross sections show an azimuthal intensity distribution which is segmented and takes the form of a Maltese cross. Such behaviour is interpreted in terms of preferred radial orientation (see Sect. 4.1.4.2).

4.1.3.4 Radiative and Chemical Resistance

PPT fibres are sensitive to prolonged exposure to light which tends to have a detrimental effect on the mechanical properties. Ultra violet radiation is particularly effective and discolours the fibres from straw yellow to a dark brown [85].

As might be expected from a highly crystalline material aromatic polyamide fibres such as PPT are generally extremely resistant to chemical attack except by strong acids and bases at high concentrations or elevated temperature. Exposure to superheated steam significantly lowers the strength although boiling water has little effect in the short term. This high corrosion resistance is a highly desirable characteristic especially in engineering applications.

4.1.4 Structure

From the extensive work on the relation between structure and mechanical properties of fibrous polymers it has become abundantly clear that the tensile modulus is largely determined by molecular orientation about the fibre axis and the effective cross-sectional area occupied by single chains which in turn is related to chain linearity. In PPT this is assured by bonding of rigid phenylene rings in the para position and is to be contrasted with the case of Nomex fibres where the phenylene and amide units are linked in the meta position leading to irregular chain conformation and a correspondingly lower modulus.

In PPT the occurence of amide groups at regular intervals along the essentially linear molecules facilitates extensive hydrogen bonding laterally between adjacent chains and leads to efficient chain packing (high crystallinity). The selection of polymers of high enough molecular weight ensures the structural continuity of the bulk system.

4.1.4.1 Crystallography

The high three dimensional order of the molecular packing in PPT fibres is shown by the large number of layer lines (>10) and reflections in the wide angle x-ray and electron diffraction patterns. On the molecular level two crystal modifications have been observed [86]. When the polymer is spun from a highly concentrated anisotropic solution the chains form an essentially monoclinic (pseudo-orthorhombic) [87,88] unit cell ($a = 0.719$ nm, $b = 0.518$ nm, c (fibre axis) $= 1.29$ nm, $\gamma = 90°$) as shown in Fig. 11. There are two molecular chains per cell, one through the centre, the other through a corner, and two monomeric units in the axial repeat, giving a crystallographic density of 1480 Kg m^{-3}. However fibres formed from anisotropic solutions of lower concentration exhibit a different packing equivalent to a lateral displacement (b/2) of chains along alternate 200 planes. Both forms coexist in fibres spun from solutions of intermediate concentration.

Adjacent chains are linked into hydrogen bonded sheets forming the (200) lattice planes which on a higher structural level give rise to an unusual mode of lateral organisation in the fibre. Bond rotation and hence molecular flexibility is inhibited by the presence not only of the aromatic rings, but also by the double bond nature of the amide group arising from resonance effects.

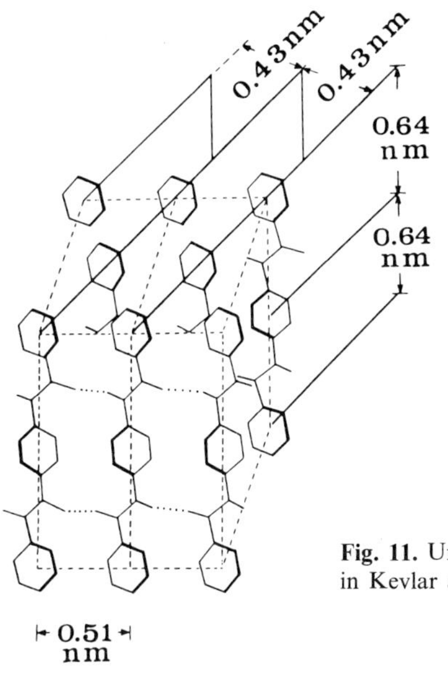

Fig. 11. Unit cell of one crystallographic form of PPT usually found in Kevlar aramid fibres [89]

The apparent crystallite sizes derived from the resolved profiles of the 110 and 200 equatorial x-ray diffraction peaks are of the order 5–6 nm [89] normal to the chain direction. Values of 20–100 nm in the fibre direction have been obtained from analyses of the meridional 006 reflections. Direct observation of molecular structure is unfortunately difficult to achieve in the electron microscope because of sensitivity in the electron beam. However crystallite lattice fringes have been imaged for the first time in an organic polymer [90] in PPT. Micrographs (Fig. 12) show that the individual arrays of equatorial fringes (derived from the 0.433 nm equatorial reflections) are significantly less extensive than the meridional arrays (spacing 0.645 nm). There is little evidence of lattice distortion apart from the occasional appearance of curved layer planes.

Generally it is found that PPT fibres are highly oriented and that those variants with the lowest values of orientation angle exhibit the highest tensile modulus. Indeed the average crystallite orientations derived from azimuthal peak widths at half maximum intensities of the 200 reflections in Kevlar 49 and Kevlar 29 are found to be nine and eleven degrees respectively.

There is no evidence for the small angle two or four point meridional reflections which are traditionally associated with chain folding in polyamide and polyester fibres. Such observations support the existence of an extended chain conformation. However PPT fibres are generally characterised by intense but diffuse small angle scattering on the equator of the x-ray diffraction patterns. This phenomenon is undoubtedly due to the presence of microvoids. Electron microscope and x-ray data indicate that the voids are rod-shaped with their long axes almost parallel to the fibre axis, having typical widths in the range 5–10 nm and length about 25 nm.

Fig. 12. Electron micrograph showing lattice fringes derived from 0.433 nm equatorial reflections (E) and 0.645 nm meridional diffraction reflections (M) [89]

4.1.4.2 Supramolecular Organisation

Although skin-core organisation is not uncommon in conventional textile fibres they are generally considered to be essentially transversely isotropic. However dark field electron microscope images formed using the 200 reflection from sections of PPT fibres cut at 45° to the fibre axis show diffraction intensity confined to two opposite segments of the section rather than distributed randomly throughout. Moreover bright-field micrographs of silver stained transverse sections indicate an unusual radial orientation. Indeed there is overwhelming evidence [91,92] that the aramid fibres posses somewhat imperfectly developed radial orientation which may be regarded as an aggregation of similarly oriented crystallites although exhibiting a restricted distribution of orientations of the hydrogen-bonded planes.

Longitudinal sections of PPT fibres exhibit a distinctive periodic organisation in dark field images derived from meridional or off-meridional reflections. A system of broad bands of spacing approximately 500 nm is associated with the off-meridional diffraction reflections; another system of narrow bands, 30 nm in width spaced at intervals of 250 nm is associated with the meridional 006 reflection. Detailed examination shows that there is a uniform distribution of ordered crystalline material throughout the fibre and that the dark field banding is a manifestation of changes in crystalline orientation and not of crystalline order. Such results provide strong evidence [91] for a regular pleated sheet structure with the alternating components of each sheet arranged at approximately equal but opposite angles to the plane of the section, and that only over a short transitional band are the PPT molecules parallel to the plane of the section. Measurements of the angular extent of the electron diffraction 200 reflection indicate that the angle between adjacent components of the pleat is about 170°. A schematic model of the structure is shown in Fig. 13.

It would appear that although the pleat persists when the fibres are subjected to moderate temperature, heat treatment in air at 500 °C for 10 minutes causes exaggerated pleating (i.e. the angle between adjacent components reduces progressively) prior to subsequent degradation.

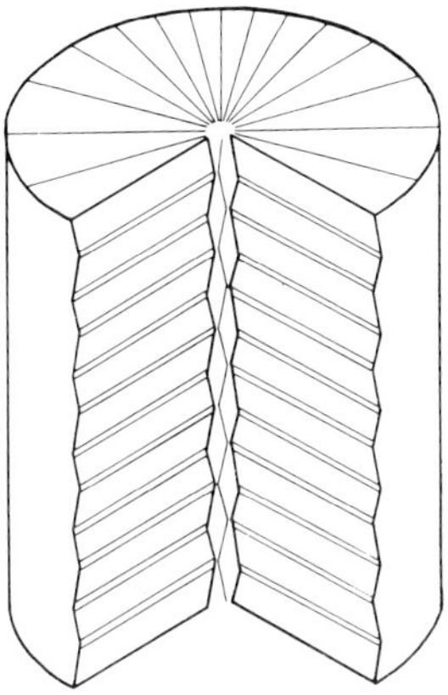

Fig. 13. Schematic model of a PPT fibre showing radially oriented axially pleated sheets [91]

4.2 Poly(*p*-Phenylenebenzobisthiazole)

Poly[(benzo[1,2-d:4,5-d']bisthiazole-2,6-diyl)-1,4-phenylene] (PBT) can be conveniently prepared by the polycondensation of 2,5-diamino-1,4-benzenedithiol dihydrochloride with terephthalic acid in poly(phosphoric acid) (PPA) according to the equation shown in Fig. 14.

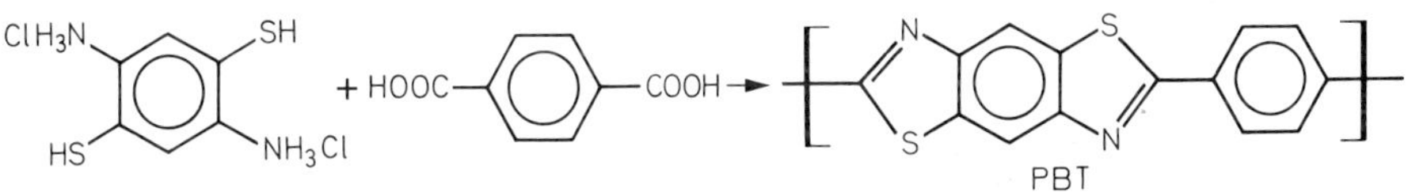

Fig. 14. Synthesis of PBT from intermediates

4.2.1 Synthesis

The monomer 2,5-diamino-1,4-benzenedithiol dihydrochloride may be synthesised, as reported by Wolfe et al. [93], via the intermediates shown schematically in Fig. 15.

According to the authors the best method for completing the last stage of the synthesis consists of transferring compound (XII) under argon as a solid into a large volume of 6M HCl. The resulting solid is collected by filtration and transferred

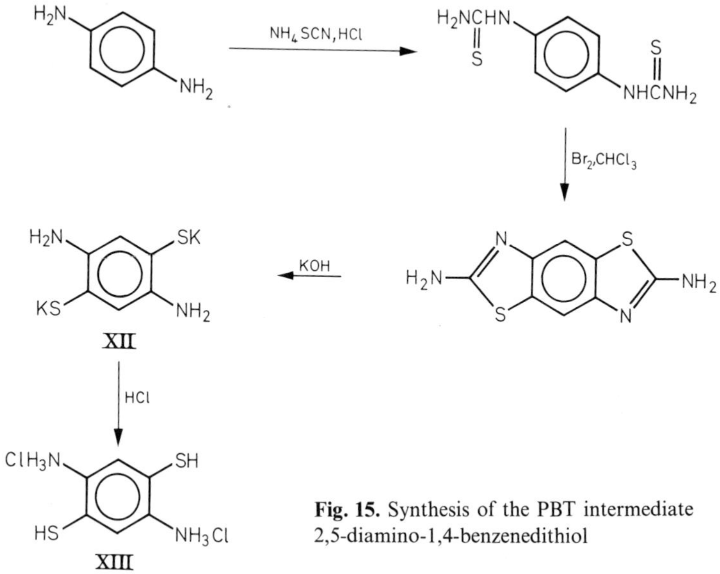

Fig. 15. Synthesis of the PBT intermediate 2,5-diamino-1,4-benzenedithiol

immediately into a large volume of dilute hydrochloric acid. Concentrated HCl is then added slowly to give large hexagonal prisms of the monomer (XIII). In this way oxidation of compound (XII) can be minimised and an appropriate crystal size of monomer (XIII) obtained so that subsequent dehydrochlorination on a large scale can be easily controlled.

Monomer (XIII) (2–15% by weight) is then added to degassed poly(phosphoric acid) (PPA) and stirred for 24 h at room temperature under a stream of argon. The temperature is raised slowly to 100 °C and reduced pressure applied until the mixture becomes clear. In order to obtain the monomer in a reactive form it is essential to remove hydrogen chloride completely. A stoichiometric amount of terephthalic acid is then added followed by additional PPA to adjust the polymer concentration by the desired value. The yellow mixture is then heated under argon to 160 °C within 5 h and reduced pressure applied to remove volatiles.

For polymer concentrations <3% the mixtures remain isotropic during polymerisation and become a rubbery mass that is too viscous to stir in the later stages of heating. In contrast polymer concentrations of 5–10% form yellow green stir-opalescent solutions within 30 minutes of the dissolution of the terephthalic acid. Such solutions remain stirrable throughout the polymerisation despite the higher concentrations — an indication of the liquid-crystalline nature of the medium.

After final heating at 190–200° for 12 h the reaction mixture is cooled and stored.

4.2.2 Fibre Formation

The work reported so far has concerned fibres spun from solutions of either (a) 5–6% polymer in PPA or (b) 10% polymer in a mixture of 97,5% methanesulphonic acid (MSA)/2.5% chlorosulphonic acid. The processing of PBT fibres from PBT/PPA dopes has however two advantages over spinning from PBT/MSA dopes. The spinnability of PBT/PPA dopes is found to be better and the spinning of fibres from these dopes represents fibre production directly from the PBT polymerisation medium. Thus the isolation of the polymer after polymerisation and redissolving in a suitable solvent, which is a necessary step for PPT, is eliminated.

In both cases a dry-jet wet-spinning process is used employing coagulation baths of water or water/methanesulphonic acid.

Fibres formed in this way are subsequently heat treated by drawing them through a tubular oven in a nitrogen atmosphere.

4.2.3 Fibre Properties

Both as-spun and heat treated fibres exhibit high modulus and high strength as can be seen in Fig. 9. Fibres spun from a PBT/PPA dope exhibit moduli up to about 75 GPa with strengths approaching 2.3 GPa. The heat treatment promotes the attainment of a much higher modulus of 250 GPa, a figure well in excess of other organic fibrous polymers such as the commercially available Kevlar 49. As with

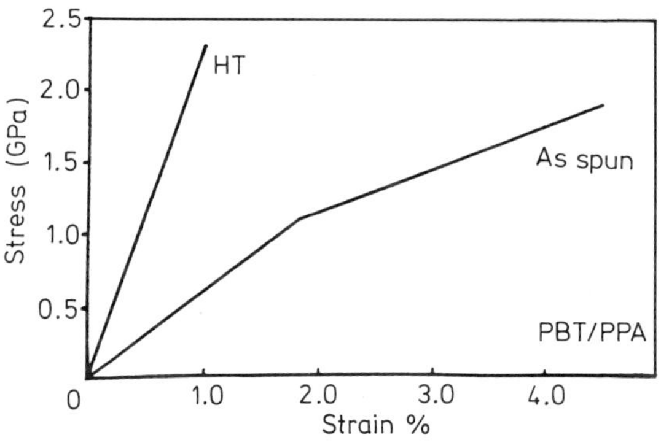

Fig. 16. Stress-strain diagram for as-spun and heat-treated PBT fibres [94]

other high modulus fibres the breaking extension of around 1.5% is very low. The stress-strain behaviour of PBT fibres is illustrated in Fig. 16, where the non-linear character of the as-spun fibre is clearly apparent. Such behaviour has been discussed [94] in terms of residual stresses arising during coagulation which can be relieved by subsequent heat treatment to give essentially linear elastic behaviour. The high modulus of these heat treated specimens results from the high degree of molecular orientation along the fibre axis and a well developed two dimensional order as revealed by diffraction studies.

The limited values in strength have been attributed to the presence of large voids. No doubt considerable improvement can be achieved by more careful attention to processing parameters.

Generally higher values of as-spun fibre modulus are obtained with higher spin/draw ratios and slightly higher values of fibre strength have been found with higher extrusion velocities. Also slower coagulation rates obtained by using baths with high MSA/H_2O

ratios and/or low bath temperatures have tended to produce fibres with higher modulus and strength.

4.2.4 Molecular Structure

The structure of PBT fibres, has not been completely resolved although efforts are continuing using diffraction analysis. For example Adams et al. [95] havc interpreted the x-ray diffraction patterns of oriented PBT fibres in terms of a nematic arrangement of molecules which are treated as periodic cylinders packed in an hexagonal array. The cylinders are oriented parallel to each other but are arbitrarily displaced axially. Unfortunately although the model explains many features of the diffraction pattern it predicts a fibre density which is well below the observed experimental value.

Electron diffraction studies by Roche et al. [96] of annealed fibres have revealed a fibre repeat of 1.24 nm which remains constant even when the processing conditions are varied. This value is in good agreement with the length of 1.247 nm of t-bisthiazole as measured by the C(8) to C(5′)separation by Wellmann et al. [97]. This implies that the PBT chain has the fully extended conformation, since the fibre repeat consists of only one repeat unit. The electron diffraction patterns have also been interpreted in terms of monoclinic and triclinic unit cells but a more definite assignment may be possible with more highly ordered fibres.

4.3 Applications of Lyotropic Polymer Systems

In the case of PPT, the main commercial outlet has been in the form of aramid fibres having the trade names 'Kevlar' and 'Arenka' whereas the development of PBT is still at the experimental stage and evaluation trials, although encouraging, are not very comprehensive.

The use of Kevlar has been confined to specialised applications [98], where high mechanical performance and lightweight properties are essential, because of its present relatively high cost compared with conventional textile materials. These applications can be conveniently divided into two main categories, one where the fibres alone form the final product such as in cables and fabrics and the other where they act as reinforcing elements for the production of composite structures.

Three aramid versions are commercially available each having distinctive properties suitable for different end-uses. Kevlar 29 which has half the modulus of Kevlar 49 with double the elongation is more suited for use in areas where high impact resistance is of primary importance such as body armour and in cables of all types including ropes, parachute shrouds, riggings etc. Kevlar 49, on the other hand, has been specifically designed for plastic reinforcement and is intended more for the aerospace industry where significant reduction in weight can be achieved over metallurgical products without compromising performance. Composites of this type have also been used extensively in a range of sporting goods, particularly tennis racquets, fishing rods and boat hulls.

A third basic type, T950, has been developed especially for the rubber industry and is intended as a bracing in radial tyres. Arenka 900 and Arenka 930 have physical properties similar to those of Kevlar 29 and 49 respectively.

5 Development of Thermotropic Polymers

5.1 Preparative Techniques

Some of the early thermotropic polyesters were made by either interfacial polymerisation or high-temperature solution polymerisation from diphenols and dicarboxylic acid chlorides, but the majority are now made by an ester exchange reaction between acetoxyaryl groups and carboxylic acid groups, with elimination of acetic acid, at a temperature above the crystalline melting point of the polymer produced. Under these conditions the polymer often forms a nematic phase during polymerisation, and exhibits the shear opalescence characteristic of that state. Poly(*p*-oxybenzoyl-*co*-ethylene terephthalate) requires special treatment. Preformed molten poly(ethylene terephthalate) is reacted with *p*-acetoxybenzoic acid and the partially degraded product is repolymerised with elimination of acetic acid. This type of polymer cannot be made from a mixture of reactants including ethylene glycol and *p*-hydroxybenzoic acid because such mixtures produce polymers that contain repeating units derived from *p*-hydroxyethoxybenzoic acid.

Direct reaction of diphenols and dicarboxylic acids in a molten state can be catalysed by adding tin, titanium or antimony compounds such as n-butylstannic acid or tetraoctyl titanate, and values of η_{inh} ranging above 3.0 have been obtained [99].

Melt polycondensation is also the most popular method for other thermotropic condensation polymers, including the polyazomethines where the reaction between aromatic aldehydes or ketones and primary amines with elimination of water leads to azomethine (Schiff's base) formation [48].

5.2 Rheological Behaviour

The rheological behaviour of thermotropic polymers is complex and not yet well understood. It is undoubtedly complicated in some cases by smectic phase formation and by variation in crystallinity arising from differences in thermal history. Such variations in crystallinity may be associated either with the rates of the physical processes of formation or destruction of crystallites, or with chemical redistribution of repeating units to produce non-random sequences. Since both shear history and thermal history affect the measured values of viscosity, and frequently neither is adequately defined, comparison of results between workers and between polymers is at present hazardous.

Some general features can be discerned. The viscosity of a nematic melt falls with increasing duration of shearing to a steady value, and only very slowly returns to the initial value on cessation of shearing. Return to the original value is accelerated by heating the sample until it becomes isotropic then cooling to the original temperature. The steady shear viscosity is shear-rate dependent; the viscosity is higher at low frequency or low shear rate, as if the polymer exhibited yield stress behaviour, and the shear rate at which the behaviour changes is in some cases of the order 0.1 sec^{-1} [100]. Most strikingly, steady shear viscosities are lower for a given polymer in the nematic phase than in the isotropic phase, despite the lower temperatures of measurement in the former case.

The fall in viscosity due to the onset of nematic behaviour has been estimated in one case, by comparing polymers of similar molecular weight but different chemical composition, to be over three orders of magnitude [101].

The effect of thermal history is illustrated by an increase in measured viscosity on holding a sample immediately beforehand at a temperature lower than the temperature of measurement, and a decrease on melting at a temperature above that of measurement [100, 102]. This behaviour affects the orientation and physical properties of extruded samples [35]. Heating a thermotropic polymer for a brief period to a temperature well above the intended extrusion temperature then returning it rapidly to the lower temperature has been used in a patent [103] to reduce the melt viscosity at the extrusion temperature by three orders of magnitude or more. The effect of this treatment may be attributable to destruction of residual two- or three-dimensional order at the higher temperature.

Most high polymers exhibit die swell on extrusion. Thermotropic polymers do not, and indeed values of the ratio of extrudate diameter to capillary diameter as low as 0.9 have been observed [104]. Since die swell is associated with viscoelasticity in the melt, it is interesting to note that even in the isotropic phase little or no elastic behaviour is observed in a thermotropic polymer [105].

Shear history is also of practical importance, since extrusion at a low shear rate can lead to a lumpy homogeneous product [106]. It is useful to convert shear flow, which possesses a velocity gradient transverse to the flow direction, into elongational flow, which possesses a velocity gradient parallel to the flow direction, before extrusion. The resulting solidified extrudate exhibits greater orientation of the polymer molecules and superior mechanical properties. A device for converting shear flow into elongational flow that has been described consists of a grid possessing a large number of converging cone- or trumpet-shaped passages [106].

5.3 Fibres and Films

A feature of the melt-spinning behaviour of nematic thermotropic polymers is the very high orientation achievable at low wind-up speeds. This behaviour contrasts with that of more flexible polymers such as poly(ethylene terephthalate) and nylon 6,6, where wind-up speeds as high as 6000 $m \cdot min^{-1}$ produce degrees of orientation that still fall short of fully drawn yarns. Unlike the more flexible polymers, thermotropic polymers exhibit little or no die swell on emergence from the spinneret hole, because of their very high relaxation time. They therefore retain the orientation introduced by shear in the capillary, and this orientation is reinforced by the process of drawdown. Consequently the spun yarns exhibit tenacities much higher than those of conventional spun yarns, and sometimes similar to those of high tenacity commercial polyester or nylon. At the same time they exhibit moduli substantially higher than and extensions at break substantially lower than these commercial fibres. Extensions at break are usually only of the order 1–2%, so the filaments are rather brittle.

Annealing fibres melt-spun from nematic mesophases can produce very significant changes in some of the physical properties, notably increases in both tenacity and extension to break. Some of the values of tenacity attainable in this way are among the highest recorded. For example, annealing a fibre melt-spun from poly-(chloro-*p*-phenylene terephthalate-*co*-naphthalene-2,6-dicarboxylate) (70/30 molar) raised the tenacity from 5.9 to 27.4 g $dtex^{-1}$ and the extension at break from 1.8% to 4.7% [44]. Annealing filaments melt-spun from the polyazomethine (XIV) raised the tenacity from 6.6 to 34.2 g $dtex^{-1}$ and the extension at break from 1.1% to 4.4% [43]. This annealing process is clearly quite different from that applied to the aramids spun from lyotropic solutions, where the annealing temperatures are in the range 400–550 °C, the durations are of the order 1 sec., the initial and average moduli and the tenacity are raised, and the

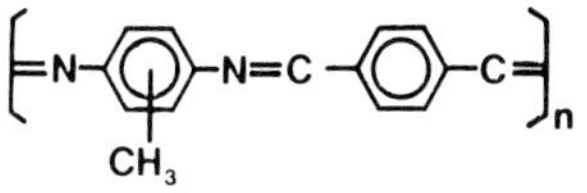

XIV

extensions at break fall. In the case of these thermotropic polymers, the final annealing temperature is close to the original melting or flow temperature, usually in the range 250–350 °C, the duration is typically some hours, the tenacity and extension at break both rise and the moduli are often relatively little affected. In the former case the work of rupture is not increased and often decreased by the thermal treatment; in the latter case the work of rupture is substantially increased by the thermal treatment. As already discussed, in the case of the aramids the annealing produces morphological changes in the fibre. In the fibres produced by melt-spinning, morphological changes do occur on annealing and result, for example, in a reduction in the X-ray orientation angle, but the improvement in physical properties appears to be mainly a consequence of an increase in molecular weight due to further polymerisation in the solid phase at the annealing temperature. This increase may be quite substantial. In one case, that of poly(chloro-*p*-phenylene *co* 2,5-dichloro-*p*-phenylene *co* 2,3-dichloro-*p*-phenylene hexahydroterephthalate) (95/4/1 molar), the logarithmic viscosity number rose from 2.3 to 9.9 dl g^{-1} on annealing for about 4 hr [110]. In many cases solution viscosities cannot be measured after annealing due to lack of solubility.

Figure 17 shows some tensile properties for fibres spun from a series of copolymers

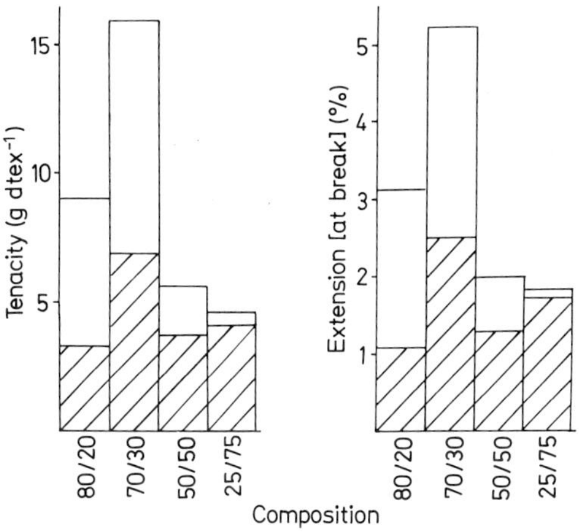

Fig. 17. Tensile properties before and after annealing for copolyesters containing repeating units XV and XVI [44]

Table 3. Tensile data for annealed yarns

Ref.	Polymer structure			Molar ratio*	Tenacity	Initial modulus	Extension at break
	AB Units	AA Units	BB Units	a/b or a/b/c or a/b/c/d/e	g $dtex^{-1}$	g $dtex^{-1}$	%
44)	—	Chloro-*p*-phenylenedioxy (a)	terephthaloyl (b) naphthalene-2,6-dicarbonyl (c)	100/70/30	27.4	474	4.7
107)	*p*-oxybenzoyl (a)	*p*-phenylenedioxy (b)	naphthalene-2,6-dicarbonyl (c) isophthaloyl (c)	75/25/12.5	21.2	394	5.2
41)	*p*-oxybenzoyl (a) 2,6-oxynaphthoyl (b)	—	—	75/25	18	495	5
108)	*p*-oxybenzoyl (a)	anthraquinone-2,6-dioxy (b)	terephthaloyl (c) isophthaloyl (c)	77/23/11.5	17.4	613	3.4
42)	*p*-oxybenzoyl (a)	naphthalene-2,6-dioxy (b)	isophthaloyl (b)	75/25	13.6	518	2.9
109)	*p*-oxybenzoyl (a) 2,6-oxynaphthoyl (b)	*p*-phenylenedioxy (c) *p*-oxyphenyleneimino (d)	terephthaloyl (e)	26.7/40/13.3/20/33.3	20.5	476	4.5
48)	Polyazomethine (XIV)			—	34.2	911	4.4

* Molar ratios have been normalised throughout this review such that the sum of the simplest possible set of *repeating* units is 100

with varying ratios of the same repeating units, (XV) and (XVI) [44]. In this case the best properties, both before and after annealing, are obtained at an intermediate composition, although not at the composition of minimum flow temperature.

Most of the patents that claim thermotropic polymers also claim fibres made from them, although full spinning conditions and physical data for the products are not always provided. Tensile data for the samples of highest tenacity after annealing from some representative patents are listed in Table 3.

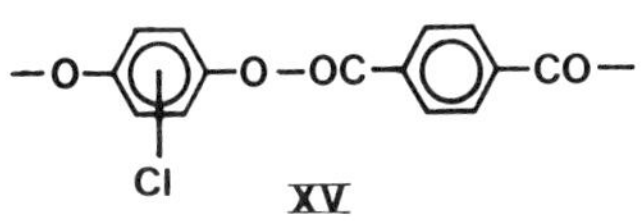

XV

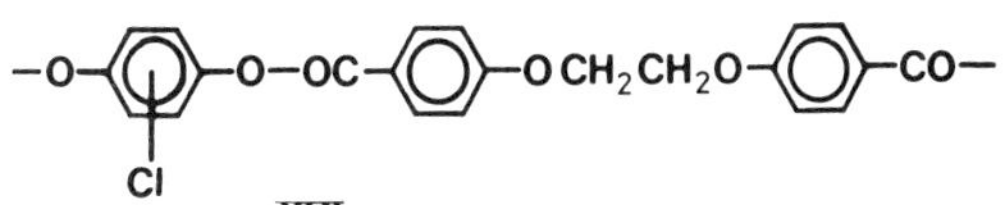

XVI

Commercial hydroxypropyl cellulose has been melt-spun from a liquid-crystalline melt at 180 °C to produce fibres with an orientation factor (f_c) of 0.6 even at zero take-up stress and 0.7 at higher take-up stresses, a tensile strength of 80 MPa, Young's modulus of 3 GPa and extension at break of 2 to 5% [71].

Blending a thermotropic polymer into a fibre-forming polymer that normally forms an isotropic melt produces a remarkable change in spun yarn orientation, particularly when the wind-up speed is high [111]. At a wind-up speed of 4500 m min^{-1} the spun yarn orientation of poly(ethylene terephthalate) was reduced by addition of only 3% of a thermotropic polyester based on *p*-oxybenzoyl and 2,6-oxynaphthoyl units to such an extent that the birefrigence fell from 73.2×10^{-3} to 27.4×10^{-3}, the boiling water shrinkage rose from 11.1% to 31.1% and the extension at break rose from 77% to 142%. The potential commercial value lies in the improvement in productivity arising from combination of a high wind-up speed with a high draw ratio.

In general, fibre-forming thermotropic polymers can also be processed from the melt to produce films and, if the polymer is sufficiently rigid not to chain-fold during annealing, can be annealed under conditions of solid-phase polycondensation to increase the tensile strength. There are relatively few published examples, but poly-(chloro-*p*-phenylene *trans*-hexahydroterephthalate). has been processed [44] in this way to give film with a tensile strength of 0.77 GPa, Young's modulus of 20.7 GPa and extension at break of 3.5%.

The method of modifying the flow pattern in which shear flow is converted into elongational flow immediately before extrusion, already discussed, has been applied to film formation [106]. The tensile strength, Young's modulus and extension at break of the extruded, gas-quenched films were all substantially higher, measured parallel to the machine direction, where an extensional flow pattern had been imposed on the molten polymer. This improvement was maintained after annealing. A film made using this technique from a *p*-oxybenzoyl/2,6-oxynaphthoyl copolymer (75/25 molar) and annealed at 260 °C for 30 h. had a tensile strength of 0.83 GPa, Young's modulus of 28 GPa and extension at break of 3.5%.

possibility, since industrial high-tenacity high-modulus products are accessible. Any such products will compete with the established aramid fibres, relative to which their merits can be summarised as:

+	—
Melt polymerised, no solvent	Lower maximum modulus
Melt-spun, no solvent	Development of best properties involves slow stage. —
Higher maximum tenacity	Lower maximum temperature of use. —
Higher photostability	Dynamic loss process occurs at lower temperature.

Film applications are perhaps less likely, in view of the tendency to fibrillate, although coating technology may make use of the flow properties. In mouldings, there are two likely areas:

i those where anisotropy of physical properties is desired and is built in by shear during moulding followed by rapid quenching, and

ii those where isotropic physical properties are desired with modified toughness arising from the randomly oriented nematic structure. In both cases, but particularly the former, changes in mould design and moulding practice are desirable to take full advantage of the properties. Finally, the unusual electric and magnetic properties suggest possible applications in information storage and retrieval and in undirectional conduction.

6 References

1. Flory, P. J.: Proc. Roy. Soc. *A234*, 60 (1956)
2. Flory, P. J.: Proc. Roy. Soc. *A234*, 73 (1956)
3. Robinson, C.: Trans. Faraday Soc. *52*, 571 (1956)
4. Robinson, C., Ward, J. C., Beevers, R. B.: Disc. Faraday Soc. *25*, 29 (1958)
5. Elliott, A., Ambrose, E. J.: Disc. Faraday Soc. *9*, 246 (1950)
6. Courtaulds Ltd. (Ballard, D. G. H.): B.P. 864,962 (priority 21 April 1958)
7. Courtaulds Ltd. (Ballard, D. G. H., Griffiths, J. D., Watson, J.): U.S.P. 3,121,766 (priority 9 May 1961, U.K.)
8. Hermans, J. J.: Adv. Chem. Phys. *B13*, 707 (1967)
9. Wee, E. L., Miller, W. G.: J. Phys. Chem. *75*, 1446 (1971)
10. DuPont de Nemours & Co., E. I. (DuPont) (S. L. Kwolek): B.P. 1,198,081 (priority 13 June 1966, U.S.A.)
11. DuPont (S. L. Kwolek): B.P. 1,283,064 (priority 12 June 1968, U.S.A.)
12. Imperial Chemical Industries Ltd. (ICI) (Goodman, I., McIntyre, J. E., Stimpson, J. W.): B.P. 989,552 (priority 19 Feb., 1962)
13. ICI (Goodman, I., McIntyre, J. E., Aldred, D. H.): B.P. 993,272 (priority 22 May, 1962)
14. The Carborundum Co.: B.P. 1,303,484 (priority 28 May, 1969, U.S.A.)
15. Roviello, A., Sirigu, A.: J. Polym. Sci., Polym. Letters Ed. *13*, 455 (1975)
16. DuPont (Kleinschuster, J. J., Pletcher, T. C., Schaefgen, J. R.): Belg. P. 828,935 (priority 10 May, 1974, U.S.A.)

17. Jackson, W. J., Kuhfuss, H. F.: J. Polym. Sci., Polymer Chem. Ed. *14*, 2093 (1976)
18. Eastman Kodak Co. (H. F. Kuhfuss, W. J. Jackson): B.P. 1,435,021 (priority 28 Sept. 1972, U.S.A.)
19. Papkov, S. P. et al.: Vysolcomol. Soed., Ser. B, *15*, 357 (1973)
20. Ciferri, A.: Polymer Engng. Sci. *15*, 191 (1975)
21. Blumstein, A., Hsu, E. C.: Liquid crystalline order in polymers with mesogenic side groups, in: Liquid Crystalline Order in Polymers (ed.) Blumstein, A., p. 105, New York, Academic Press 1978
22. Porter, R. S., Johnson, J. F.: The rheology of liquid crystals, in: Rheology, Vol. 4 (ed.) Eirich, F., p. 317, New York, Academic Press 1967
23. Capaccio, G., Ward, I. M.: Nature *243*, 143 (1973)
24. Teijin, K. K.: B.P. 1,501,948 (priority 27 Dec., 1974)
25. Teijin, Co.: Data sheets for HM-50 (1982)
26. Robinson, C.: Tetrahedron *13*, 219 (1961)
27. Uematsu, I., Uematsu, Y.: Adv. Polym. Sci. *59*, 37 (1983)
28. Flory, P. J.: Statistical Mechanics of Chain Molecules, p. 183, New York, Interscience 1969
29. Papkov, S. P. et al.: J. Phys. Chem., Polym. Phys. Edn. *12*, 1753 (1974)
30. Savinov, V. M., Sokolov, S. B., Fedorov, A. A.: Vysokomol. Soed., Ser. B *10*, 111 (1968)
31. Panar, M., Beste, L. F.: Macromolecules *10*, 1401 (1977)
32. Bair, T. I., Morgan, P. W., Killian, F. L.: Macromolecules *10*, 1396 (1977)
33. DuPont (Hartzler, J. D.): B.P. 1,501,948 (priority 23 Sept. 1977, U.S.A.) (U.S.P. 3,642,707)
34. Morgan, P. W.: J. Polym. Sci., Polym. Symp., No. *65*, 1 (1978)
35. Jackson, W. J.: Br. Polym. J. *12*, 154 (1980)
36. Eastman Kodak Co. (Jackson, W. J., Morris, J. C.): G.B.P. 2,002,404 (priority 8 Aug. 1977, U.S.A.) (U.S.P. 4,169,933)
37. Volksen, W. et al.: Polym. Prepr. *20*, 86 (1979)
38. Cottis, S. G.: Mod. Plastics, July 62 (1975)
39. The Carborundum Co. (Cottis, S. G., Economy, J., Wohrer, L. C.): B.P. 1,499,513 (priority 25 Jan. 1975, U.S.A.)
40. Celanese Corp. (Calundann, G. W.): B.P. 1,585,511 (priority 13 May 1976, U.S.A.) (U.S.P. 4,067,852)
41. Celanese Corp. (Calundann, G. W.): G.B.P. 2,006,242 (priority 20 Oct. 1977, U.S.A.) (U.S.P. 4,161,470)
42. Celanese Corp. (Calundann, G. W.): U.S.P. 4,184,996 (priority 15 Feb. 1978)
43. Celanese Corp. (Calundann, G. W., Charbonneau, L. F., East, A. J.): U.S.P. 4,357,917 (priority 6 Apr. 1981)
44. DuPont (Schaefgen, J. R. et al.): B.P. 1,507,207 (priority 10 May 1974, U.S.A.) (cf. Ref. 16)
45. DuPont (Payet, C. R.): U.S.P. 4,159,365 (priority 19 Nov. 1976)
46. DuPont (Harris, J. F.): U.S.P. 4,294,955 (priority 10 Dec. 1979)
47. Eastman Kodak Co. (Jackson, W. J., Gebeau, G. G., Kuhfuss, H. F.): U.S.P. 4,153,779 (priority 26 June 1978)
48. DuPont (Morgan, P. W.): U.S.P. 4,048,148 and 4,122,070 (priority 9 May 1975)
49. Celanese Corp. (Choe, E. W., Calundann, G. W.): G.B.P. 2,050,399 (priority 7 May 1979, U.S.A.)
50. ICI (McIntyre, J. E., Milburn, A. H.): U.S.P. 4,272,625 (priority 24 July 1978, U.K.)
51. McIntyre, J. E., Milburn, A. H.: Br. Polym. J. *13*, 5 (1981)
52. Strzelecki, L., Van Luyen, D.: Europ. Polym. J. *16*, 303 (1980)
53. Roviello, A., Sirigu, A.: Europ. Polym. J. *15*, 61 (1979)
54. Aguilera, C. et al.: Makromol. Chem. *184*, 253 (1983)
55. DuPont (Kleinschuster, J. J.): U.S.P. 3,991,014 (priority 10 May 1974) (cf. Ref. 42)
56. Milburn, A. H.: unpublished results
57. Griffin, A. C., Havens, S. J.: J. Polym. Sci., Polym. Phys. Edn. *19*, 951 (1981)
58. McFarlane, F. E., Nicely, V. A., Davis, T. G.: Liquid crystal polymers. II. Preparation and properties of polyester exhibiting liquid-crystalline melts, in: Contemporary Topics in Polymer Science, Vol. 2 (ed.) Pearse, E. M., Schaefgen, J. R., p. 109, London, Plenum 1977
59. Rhône-Poulenc-Textile (Fayolle, B.): U.S.P. 4,057,597 (priority 27 Aug. 1975, France)
60. Lenz, R. W., Feichtinger, K. A.: Polym. Prepr. *20* (*1*), 114 (1979)

61. Jackson, W. J., Kuhfuss, H. F.: J. Appl. Polym. Sci. *25*, 1685 (1980)
62. Martins, A. F. et al.: Macromolecules *16*, 279 (1983)
63. Van Krevelen, D. W.: Properties of Polymers (2nd Edn.), p. 99, Amsterdam, Elsevier 1976
64. Warner, S. B., Jaffe, M.: J. Crystal Growth *48*, 184 (1980)
65. Blundell, D. J.: Polymer *23*, 359 (1982)
66. Bhadani, S. N., Gray, D. G.: Proc. 28th. Macromolecular Symp., IUPAC, p. 813 (1982)
67. Tseng, S.-L., Laivins, G. V., Gray, D. G.: Macromolecules *15*, 1262 (1982)
68. Tseng, S.-L., Valente, A., Gray, D. G.: Macromolecules *14*, 715 (1981)
69. Aharoni, S. M.: J. Polym. Sci., Polym. Letters Edn. *19*, 495 (1981)
70. Suto, S., White, J. L., Fellers, J. F.: Rheol. Acta *21*, 62 (1982)
71. Shimamura, K., White, J. L., Fellers, J. F.: J. Appl. Polym. Sci. *26*, 2165 (1981)
72. Kesuya, S., et al.: Polym. Bull (Berlin) *7*, 241 (1982)
73. Uematsu, I., Watanabe, J.: Proc. 28th. Macromolecular Symp., IUPAC, p. 821 (1982)
74. Magat, E. E.: Phil. Trans. Roy. Soc. *A294*, 463 (1980)
75. Allen, S. R. et al.: Macromolecules *14*, 1135 (1981)
76. Akzo, N. V.: B.P. 1,547,802 (priority 21 Feb. 1975, Netherlands)
77. DuPont (Blades, H.): B.P. 1,393,011 (priority 28 April 1971, U.S.A.) (U.S.P. 3,767,756)
78. Monsanto Co. (Morgan, H. S.): U.S.P. 3,414,645 (priority 19 June 1964)
79. DuPont (Blades, H.): B.P. 1,391,501 (priority 28 April 1971, U.S.A.) (U.S.P. 3,869,430)
80. DuPont: Information Bulletin No. 6E (1974)
81. Northolt, M. G., Van Aartsen, J. J.: J. Polym. Sci., Polym. Symp. *58*, 283 (1977)
82. Greenwood, J. H., Rose, P. G.: J. Mater. Sci. *9*, 1809 (1974)
83. Dobb, M. G., Johnson, D. J., Saville, B. P.: Polymer *22*, 960 (1981)
84. Hamza, A. A., Sikorski, J.: J. Microscopy *113*, 15 (1978)
85. Carlsson, D. J., Gan, L. H., Wiles, D. M.: J. Polym. Sci., Polym. Chem. Edn. *16*, 2365 (1978)
86. Haraguchi, K., Kajiyama, T., Takayanagi, M.: J. Appl. Polym. Sci. *23*, 915 (1979)
87. Northolt, M. G.: Eur. Polym. J. *10*, 799 (1974)
88. Yabuki, K., Ito, H., Ota, T.: Sen'i Gakkaishi *31*, T524 (1975)
89. Dobb, M. G., Johnson, D. J., Saville, B. P.: J. Polym. Sci., Polym. Symp. *58*, 237 (1977)
90. Dobb, M. G., Johnson, D. J., Saville, B. P.: Polymer *20*, 1284 (1979)
91. Dobb, M. G., Johnson, D. J., Saville, B. P.: J. Polym. Sci., Polym. Phys. Edn. *15*, 2201 (1977)
92. Hagege, R., Jarrin, M., Sotton, M. J.: J. Microscopy *115*, 65 (1979)
93. Wolfe, J. F., Bock, H. L., Arnold, F. E.: Macromolecules *14*, 915 (1981)
94. Allen, S. R., et al.: J. Appl. Polym. Sci. *26*, 291 (1981)
95. Adams, W. W., Azaroff, L. V., Kulshreshtha, A. K.: Kristallogr. *150*, 321 (1979)
96. Roche, E. J., Takahashi, T., Thomas, E. L.: A.C.S. Symp. Ser. No. 141, 303 (1980)
97. Wellman, M. W., et al.: Macromolecules *14*, 935 (1981)
98. Gupta, N.: Textile Inst. and Ind. p. 39, Feb. (1980)
99. DuPont (Elliott, S. P.): U.S.P. 4,093,595 (priority 19 Nov. 1976)
100. Wissbrun, K. F.: Brit. Polym. J. *12*, 163 (1980)
101. Griffin, B. P., Cox, M. K.: Brit. Polym. J. *12*, 147 (1980)
102. Cogswell, F. N.: Brit. Polym. J. *12*, 170 (1980)
103. Celanese Corp. (Ide, Y., Wissbrun, K. F.): U.S.P. 4,093,595 (priority 15 July 1980)
104. Baird, D. G.: The rheology of polymers with liquid-crystalline order, in: Rheology, Vol. 3, Applications (ed.) Astarita, G., Mannicci, G., Nicolais, L., p. 647, New York, Plenum, 1980
105. Wissbrun, K. F., Griffin, A. C.: J. Polym. Sci., Polym. Phys. Ed. *20*, 1835 (1982)
106. Celanese Corp. (Ide, Y.): U.S.P. 4,332,759 (priority 15 July 1980)
107. Celanese Corp. (Calundann, G. W.): B.P. 1,585,512 (priority 13 May 1976, U.S.A.)
108. Celanese Corp.(Calundann, G W., Charboneau, L.F.): U.S.P. 4,224,433 (priority 16 Mar. 1979)
109. Celanese Corp. (Charbonneau, L. F., East, A. J., Calundann, G. W.): U.S.P. 4,357,918 (priority 6 April 1981)
110. DuPont (Luise, R. R. et al.): B.P. 1,508,646 (priority 10 May 1974, U.S.A.)
111. ICI (Brody, H.): Europ. P. Appn. 41,327 (priority 29 July 1980, U.K.)
112. ICI (Cogswell, F. N., Griffin, B. P., Rose, J. B.): Europ. P. Appn. 30,417 (priority 30 Nov. 1979, U.K.)

M. Gordon (Editor)
Received August 3, 1983

Liquid Crystal Side Chain Polymers

H. Finkelmann and G. Rehage
Institut für Physikalische Chemie, Technische Universität Clausthal
Adolf-Römerstr. 2A, 3392 Clausthal-Zellerfeld, FRG

Attaching non amphiphilic or amphiphilic liquid crystalline molecules as side chains to linear, branched or crosslinked polymers yields liquid crystal (l.c.) side chain polymers, which can exhibit the liquid crystalline state analogously to the conventional low molar mass liquid crystals. The l.c.-side chain polymers combine the specific, anisotropic properties of the liquid crystalline state with the specific properties of polymers.

The systematic synthesis of non amphiphilic l.c.-side chain polymers and detailed physico-chemical investigations are discussed. The phase behavior and structure of nematic, cholesteric and smectic polymers are described. Their optical properties and the state of order of cholesteric and nematic polymers are analysed in comparison to conventional low molar mass liquid crystals. The phase transition into the glassy state and optical characterization of the anisotropic glasses having liquid crystalline structures are examined.

The synthesis of amphilic l.c.-side chain polymers and their phase behavior in aqueous solutions verifies that these polymers exhibit hexagonal, cubic and lamellar phases in analogy to the monomeric systems.

Advances in Polymer Science 60/61

Abbreviations for Liquid Crystals Used in the Text

l-l.c. = *l*ow molar mass *l*iquid *c*rystal
m-l.c. = *m*onomeric *l*iquid *c*rystal; e.g. a l-l.c., which is substituted by a reactive, polymerizable substituent
p-l.c. = *p*olymeric *l*iquid *c*rystal

1 Introduction and Classification of Liquid Crystalline (l.c.) Polymers

As can be seen in H. Kelkers [1] excellent review on the history of liquid crystals, investigations on liquid crystalline polymers already exist before F. Reinitzer in 1888 gave the very first description of a low molar mass liquid crystal (l-l.c.). While, however, l-l.c.'s have become an extensive field of research and application during the past decades, these activities on l.c. polymers have come rather late. The research on l.c. polymers during the last years is mainly joined with activities in material science and tries to realize polymers with exceptional properties. These exceptional properties are expected because of the combination of the physical anisotropic behavior of l.c. and the specific properties of macromolecular material.

The definition of the liquid crystalline state of l-l.c.'s and polymer liquid crystals (p-l.c's) does not differ. While in the crystalline state the molecules are three dimensionally long range ordered with respect to their centers of gravity, the l.c. state is characterized by the absence of at least one positional long range order in one dimension. The lack of this positional long range order gives rise to the liquid state, while the remaining positional and/or long range orientational order are the basis for the anisotropic physical properties of the l.c.'s. The orientational order describes the non statistical orientation of the molecules with respect to any molecular axis of the non spherical molecules. While for l-l.c.'s the positional and orientational long range order is related to the single molecule, for polymers the order can be related either to the entire macromolecule or to the monomer unit of the macromolecule, depending on the chemical constitution of the l.c. polymer. This will be discussed

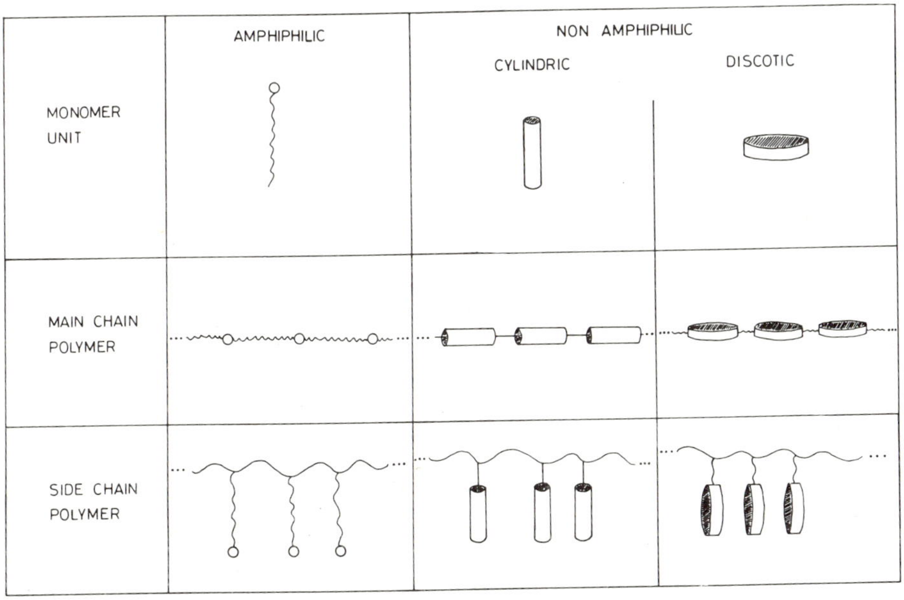

Fig. 1. Classification of liquid crystalline monomers and polymers

in more detail later. At the phase transformation to the isotropic liquid or melt the long range positional and orientational order vanishes.

Systematic investigations on l-l.c.'s have proved that the l.c. state can be directly related to the chemical constitution of the molecules [3-6]. Owing to their chemical constitution, l.c. phases having defined molecular organisations are formed. Following the classification of Gray and Winsor [5], one can distinguish between two types of chemical constitution of l.c.'s (Fig. 1):

1) Non amphiphilic l.c. and
2) Amphiphilic l.c.

Non amphiphilic l.c.'s can be further differentiated into l.c. molecules having a cylindrical molecular shape and those having a disc-like shape. L.c.'s of molecules with cylindrical shape are long range positionally and orientationally ordered in *smectic phases* and only orientationally ordered in *nematic* and *cholesteric phases.* A rich polymorphism is also present in l.c.'s of disc-like molecular shape with regard to the positional and orientational long range order [7].

Amphiphilic molecules associate in organic or aqueous solution to spherical, cylindrical or planar micelles, which form in defined ranges of concentration l.c. structures, e.g. hexagonal and lamellar phases are known [8-11].

The occurrence of the l.c. state of low molar mass substances is, as described, always related to a defined chemical constitution. The idea is obvious to tie up the mesogenic molecules to a macromolecule by a suitable chemical reaction. Assuming the mesogenic structure of the single molecules is uncharged by polymerization and can be found in the monomer unit of the polymer, it can be expected that the macromolecules also exhibit the l.c. state.

For this concept in principle two different possibilities exist to realize a macromolecule with mesogenic reactive monomeric liquid crystals (m-l.c.)
i) the m-l.c.'s are tied up by suitable reactive substituents head to tail. By this way the mesogenic monomer units form the macromolecule (*l.c. main chain polymer*).
ii) the m-l.c.'s are tied up head to head. This can be performed by a polymerizable substituent linked to the mesogenic monomer or by an addition reaction of a reactive monomer to a polymer backbone. In this case the m-l.c.'s are linked as side chains to the macromolecule (*l.c. side chain polymer*).

Following this concept six different types of polymers (Fig. 1) are conceivable. So far the characterization is related to the linkage of the monomer units within the polymer backbone, related to the classification of conventional l-l.c.'s. A further important differentiation has to be considered with respect to the constitution of the polymer main chain. Three different types must be considered:

i) linear,
ii) branched and
iii) crosslinked polymers.

Due to these different primary structures of the main chain, important modifications and a broad variety of systems is realizable. While linear polymers can be essentially characterized by the number of the monomer units, for branched and crosslinked systems e.g. the way of branching and their quantity is of significance for the polymer specific properties. In cases of crosslinked systems the molecular dimension is the macroscopic dimension of the sample.

Linear l.c. main chain polymers, built up by cylindrical monomer units, are of high theoretical and technological interest. While for the corresponding monomers the l.c. state can be theoretically well understood in terms of the classical Maier-Saupe theory [12] by intermolecular dispersion interactions, the theoretical concept for the l.c. main chain polymers is mainly based on packing effects, owing to the form anisotropy of the molecules. This will be explicitly discussed by P. J. Flory in Vol. 59 of this series [13]. In this category of l.c. main chain polymers also polymers can be classified, which have a flexible backbone, but that form rigid, rodlike secondary structures, e.g. helical arrangements of the main chain. A typical example are the poly(benzylglutamates).

L.c.-main chain polymers formed by discotic l.c's are not known up to now. For these systems, however, some interesting aspects can be expected. In case of a rigid linkage of the monomer units within the backbone, the entire macromolecule becomes rigid rod like. A polymorphism in analogy to the cylindrical main chain polymers will result. In case of a flexible linkage between the monomer units a polymorphism following the discotic l-l.c.'s could be expected because only the single monomer units will cause the anisotropic packing.

Linear main chain polymers from amphiphilic monomers have been described for the first time by E. T. Samulski [14]. In the main chain hydrophobic and hydrophilic segments alternate. In case of a parallel arrangement of the backbones lamellar structures in aqueous solutions are possible, whereas no packing of the backbones is conceivable, which might cause rodlike structures. For these systems the continuous change from homopolymers to amphiphilic block copolymers [15] is of high interest in view of an understanding of the phase behavior of these materials. The continuous change to block copolymers can be obtained by a continuous lengthening of the hydrophilic and hydrophobic segments.

While for l.c.-main chain polymers more or less the entire macromolecules build up the anisotropic structure, for the l.c.-side chain polymers the original dimension of the m-l.c. essentially remains, causing the anisotropic arrangement of the mesogenic side chains. This does not necessarily require an anisotropic packing also of the backbone. Owing to the linkage to the backbone rotational and translational motions of the mesogenic moieties are restricted. Consequently for these systems l.c. properties and a polymorphism very similar to the l-l.c.'s are to be expected.

In the first part of this paper we will give a review on our experimental work on non amphiphilic l.c. side chain polymers and will compare their properties with the corresponding l-l.c.'s. In the second part results on amphiphilic side chain polymers will be discussed.

2 Non Amphiphilic L.C. Side Chain Polymers

2.1 Model Considerations and Synthesis

For the investigation of the polymerization in l.c.'s two aspects are of interest. For polymerizable m-l.c.'s, which are in the well defined anisotropic phase, model systems exist to study phenomena that occur in biological systems, e.g. reactions in the anisotropic matrices of membranes. On the other hand, in the liquid crystalline state it can be investigated, how far ordered structures of the initial products influence the kinetics of the reaction process or give rise to the formation of stereospecific polymers [16–20].

For the research activities up to 1978, which are comprehensively reviewed by A. Blumstein [21] and V. P. Shibaev [22], one principal problem becamc obvious. While the polymerizable monomers exist in defined l.c. phases, by polymerization the initial structure of the l.c. phase in all was cases destroyed. Starting with nematic derivatives only polymers resulted, which exhibited smectic phases existing at much higher temperatures, or polymers exhibiting the glassy state at temperatures under the conditions of polymerization. These glassy polymers were either amorphous or showed a l.c. *structure*. This l.c. structure, which is locked in by the polymerization process, does not necessarily imply the existence of a liquid crystalline state of the polymer above T_g. It can be seen as a "memory effect" from the l.c. state of the initial monomer. In most cases these l.c. structures were lost irreversibly by heating the material above the glass transition temperature T_g. Kinetic investigations on the polymerization in l.c. media therefore gave different results. Acceleration as well as retardation of the polymerization, was observed compared with the polymerization in the isotropic state. These effects can be deduced from the heterophasic reaction, where not only the anisotropic orientation of the monomers, but also e.g. diffusion processes to the phase boundaries determined the kinetics.

Two principal aspects can be derived from these investigations:
i) the glass transition temperatures of the polymers investigated lay always at much higher temperatures than for the corresponding polymers without mesogenic side chains.
ii) during the polymerization process the initial l.c. state of the monomers was destroyed, indicating that the initial l.c. packing of the mesogenic side chains is prevented by their linkage to the macromolecule.

Regarding the chemical constitution of these monomers, the polymerizable, functional group was always directly linked to the rigid, voluminous mesogenic moieties. As space filling models indicate, polymers result, having a rigid main chain. This

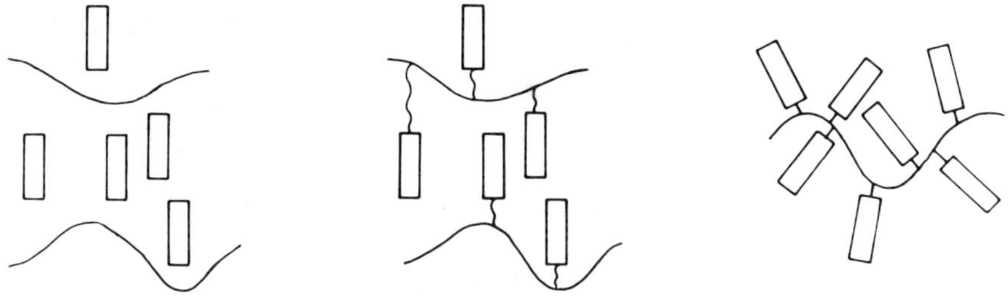

a b c

Fig. 2a–c. Liquid crystalline molecules;
a dissolved in an amorphous polymer; **b** linked to a polymer via a flexible spacer; **c** linked to a polymer without a flexible spacer

explains the high glass transition temperatures. Furthermore it can be derived from the models, that an anisotropic packing of the mesogenic groups is restricted because of steric hindrance. Nematic structures, which presume a statistical distribution of the centers of gravity of the molecules are not realizable under these conditions. Owing to the direct linkage, the backbone rather forces the mesogenic groups to a parallel alignment [22]. Consequently no nematic but only amorphous or smectic ordered polymers result. Therefore it had been speculated by Lorkowski et al. [23] that polymers having a nematic structure could be obtained only, by locking-in the monomeric *structure* by the immobilization of the molecules within a polymer network of high crosslinking density. Nematic polymer *phases* are, of course, not possible in this way.

The existence of a l.c. polymer phase requires that in the polymer melt above T_g the side chains are ordered anisotropically. In order to realize systematically such systems, a simple model consideration can be used [24] (Fig. 2).

Two extreme cases are imaginable. If the rigid, rod-like mesogenic molecules are directly attached to a polymer backbone, as described above, motions of polymer segments and mesogenic groups are directly coupled (Fig. 2c). In the liquid state above T_g the polymer tends to adopt a statistical chain conformation that hinders anisotropic orientation of the mesogenic side chains. A l.c. phase cannot exist. On the other hand, no coupling between main chain and l.c. molecules exists, apart from local anisotropic conformations of the polymer backbone [25] due to the anisotropic matrix, if the polymer is dissolved in a low molar mass nematic l.c. (Fig. 2a). Between these two extreme conditions intermediate states with a more or less strong interaction of main chain and the l.c. molecules can be realized, if the l.c. molecules are linked via "flexible spacers" to the polymer main chain. Under these conditions and depending on the length of the flexible spacer, polymers having a liquid crystalline state should exist, where on the one hand the liquid crystalline order is influenced by the polymer main chain and on the other hand the conformation of the backbone is influenced and changed by the anisotropic orientation of the side chains.

Considerations by V. Shibaev and N. Platé (see this issue) led to a similar conclusion. Investigations on comb like polymers [26], where each monomer unit of the macromolecule carried a non-branched alkyl chain of m methylene groups, have shown that for $m \gtrsim 8$ side chain crystallization takes place independently to the main chain conformation. Consequently, if mesogenic molecules are linked to the side chains, they should occupy a l.c. order without influence of the backbone conformation. Following these considerations, alkyl chain lengths $m > 8$ are necessary for the formation of the l.c. order. As shown later, however, nearly only smectic polymers are possible under these conditions (see Chap. 2.3.3.).

Investigations in the past years have proved that applying the concept of flexible spacer, polymers can be synthesized systematically, which exhibit the l.c. state. Owing to the flexible linkage of the mesogenic molecules to the polymer main chain, very similar relations can be expected with respect to l-l.c., like chemical constitution and phase behavior, or dielectric properties and field effects for the l.c. side chain polymers. This will be in contrast to main chain polymers, where the entire macromolecule, or in case of semiflexible polymers parts of the macromolecules, form the l.c. structure. The introduction of a flexible spacer between backbone and mesogenic group can be performed in a broad variety of chemical reactions. Some arbitrarily chosen examples are listed in Table 1, which also indicates the elements for the synthesis. Detailed synthesis paths can be found in the original literature. An extensive summary of prepared polymers can be found in the article of Shibaev and Platé (this issue).

Table 1. Examples of l.c. side chain polymers containing a "flexible spacer"

MAIN CHAIN	FLEXIBLE SPACER	MESOGENIC GROUP	LIT
CH_3-Si (⋮ –O– ⋮)	$(CH_2)_m$	$O-$◯$-COO-$◯$-R$	29,30
		COO-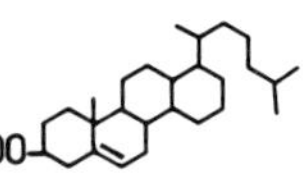	29
		$O-$◯$-$◯$-R$	29
⋮ CH_2 / $(CH_3)H-C-COO$ ⋮	$(CH_2)_m$	$O-$◯$-COO-$◯$-R$	24
		$O-$◯$-$◯$-R$	31,33
		$O-$◯$-CH=N-$◯$-CN$	32

A broad variety of l.c. polymers is conceivable because of the wide range of well known mesogenic molecules, e.g. tabulated in the book of Demus [27], and the different types of polymers. Further variations are possible by copolymers or systems, where each monomer unit carries more than one mesogenic moiety ("en bloc" systems [28]). Furthermore the synthesis of linear, branched and crosslinked systems has to be mentioned. Because of this broad variety a manifold influence on the phase behavior of the systems via the chemical constitution is feasible. In the following chapter we will discuss some basic considerations on the phase behavior of l.c.-side chain polymers.

2.2 General Consideration on the Phase Behavior of L.C.-Side Chain Polymers

From thermodynamic investigations invaluable qualitative and quantitative information is provided with regard to the phase transitions and vicinity of transitions of polymers and conventional low molar mass liquid crystals. Furthermore they give information about the kind of transition and phase stability relations, which are necessary to test theories or to evaluate new theoretical considerations. Owing to their chemical constitution, l.c.-side chain polymers are characterized by their combination of l.c. molecules and conventional macromolecules. Therefore the question is well to the fore, whether the phase behavior of the l.c. side chain polymers resembles that of usual polymers and low molar mass liquid crystals.

In our description of l.c. side chain polymers we will refer to the Ehrenfest scheme of classifying the phase transformations. In general an n-th order phase transformation is determined by a discontinuity at the point of the transition in the n-th derivative of the free enthalpy with respect to temperature T or pressure P.

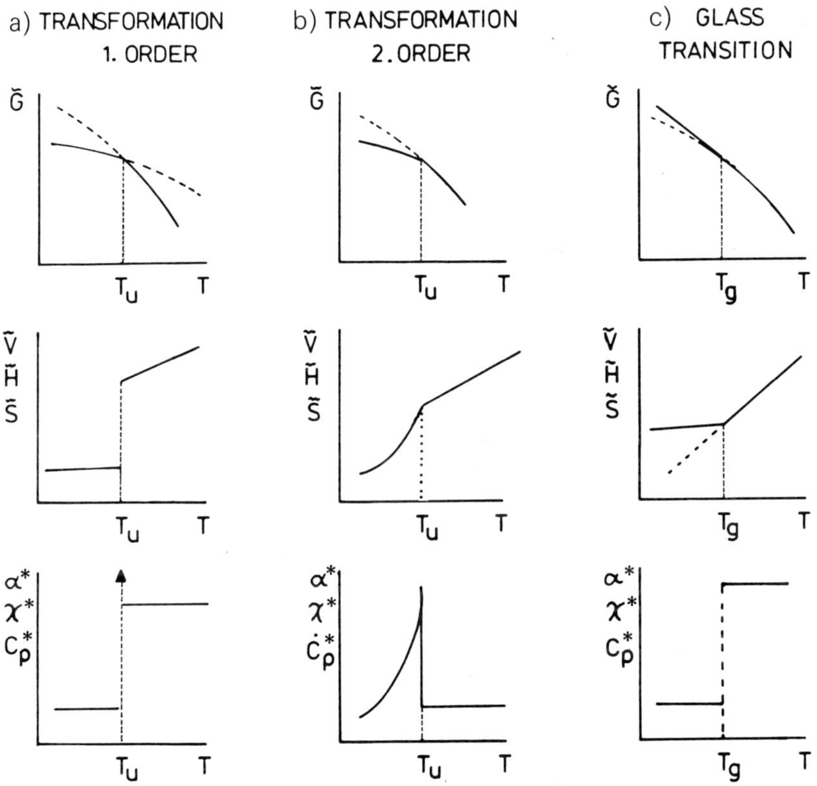

Fig. 3. Scheme of the change of thermodynamic data for transformations of 1st and 2nd order and the glass transition

For the first order transformation the $\tilde{G}(P, T)$ surfaces of the phases, being in an internal equilibrium, intersect over a range of P and T (Fig. 3a). Thus in a first order transformation a finite jump in the first derivative of the specific free energy $\tilde{G}$ the specific entropy $\tilde{S}$ and the specific volume $\tilde{V}$ is observed.

$$\left(\frac{\partial \tilde{G}}{\partial T}\right)_P = -S \qquad \left(\frac{\partial \tilde{G}}{\partial P}\right)_T = \tilde{V}$$

The second derivative of $\tilde{G}$,

$$\frac{\partial^2\tilde{G}}{\partial T^2} = -\frac{c_p^*}{T}; \quad \frac{\partial^2\tilde{G}}{\partial T\,\partial P} = \alpha^*; \quad \frac{\partial^2\tilde{G}}{\partial P^2} = -\chi^*$$

c_p^* = specific heat capacity
χ^* = specific isothermal compressibility
α^* = specific expansion coefficient

becomes infinite at the transformation temperature and pressure. The differential equation of the coexistence curve at the transformation of two phases, the Clausius-Clapeyron equation,

$$\left(\frac{dP}{dT}\right)_{tr} = \frac{\Delta\tilde{S}}{\Delta\tilde{V}} = \frac{\Delta\tilde{H}}{T\Delta\tilde{V}} \tag{1}$$

describes the pressure dependence of the phase transformation temperature of the two phases.

In case of second order transformation (Fig. 3b), where the G(P, T) surfaces are in contact over a range of P and T the entropy changes continuously across the transformation. This transformation must obey the well-known Ehrenfest equations

$$(dP/dT)_{tr} = \Delta\alpha^*/\Delta\chi^* \tag{2a}$$
$$(dP/dT)_{tr} = \Delta c_p^*/T\,\Delta\alpha^* \tag{2b}$$
$$(\Delta\alpha^*/\Delta\chi^*)_{tr} = \Delta c_p^*/T\,\Delta\alpha^* \tag{2c}$$

These equations connect the slope $(dP/dT)_{tr}$ of the transformation line $T_{tr}(P)$ for second order transformations in the PVT-diagram to the steps in the specific volume expansion coefficient, the specific compressibility and the specific heat.

In case of a genuine transformation, the phase considered must be in its internal thermodynamic equilibrium above and below the transformation temperature T_{tr}. This means that the physical properties of a one-component system depend only on two variables and must be independent of the path.
In the following we will contrast the phase behavior of a conventional non crystallizing polymer with that of a conventional, low molar mass liquid crystal. Thereafter we will discuss the experimental results on l.c. side chain polymers.

In Fig. 3c the schematic volume-temperature curve of a non crystallizing polymer is shown. The bend in the V(T) curve at the glass transition indicates, that the extensive thermodynamic functions, like volume V, enthalpy H and entropy S show (in an idealized representation) a break. Consequently the first derivatives of these functions, i.e. the isobaric specific volume expansion coefficient α^*, the isothermal specific compressibility χ^*, and the specific heat at constant pressure c_p^*, have a jump at this point, if the curves are drawn in an idealized form. This observation of breaks for the thermodynamic functions V, H and S in past led to the conclusion that there must be an internal phase transition, which could be a true thermodynamic transformation of the second or higher order. In contrast to this statement, most authors assume to the present, that the glass transition is a kinetically controlled process. Some scientists conclude from theoretical models that the basis of the glass transition is a thermodynamic transformation, which will take place at a temperature T_2 approximately 50 K below T_g. Gibbs and Di Marzio regard this transformation as the underlying cause of the glass transition at normal cooling rates [34–35]. Detailed investigations on the glass transition of amorphous polymers have shown that this phase transition cannot be described by the Ehrenfest equations for a second order transformation. But using the concept of internal ordering parameters in addition to the conventional variables T and P, it has been shown that one internal order parameter in some cases is sufficient to describe the behavior in the glassy state nearly quantitatively [36].

In Fig. 4 the experimental isobaric volume temperature curve of the l-l.c. 4'-hexyloxybenzoic acid 4'-hexyloxyphenylester is shown, which possesses a nematic phase [37,38]. Two phase transformations are indicated by the jumps of the V—T curve: the isotropic to nematic and, at lower temperatures, the nematic to crystalline transformation. As well known, both transformations are of first order and obey the Clausius-Clapeyron equation.

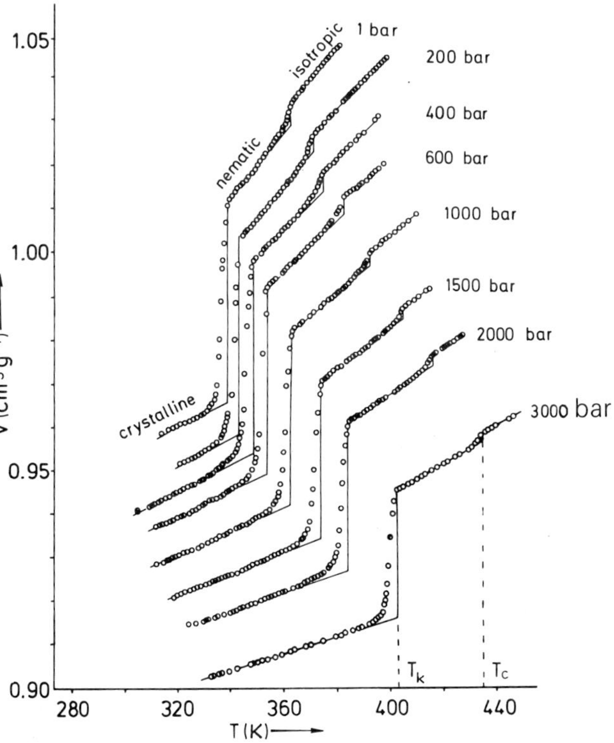

Fig. 4. Specific volume-temperature curves at different pressures of the l-l.c. (isobaric measurements) [37,38]

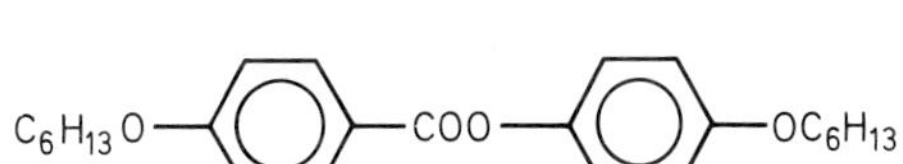

In comparison to a conventional polymer and l.c. mentioned above, we will now discuss the PVT behavior of a l.c. side chain polymer, which has linked mesogenic moieties as side chains, and is very similar to the previous monomer. The experimental results are shown in Fig. 5. It is obvious, that the phase behavior of the l.c. polymer differs from that of a l-l.c. and amorphous polymer. At high temperature we observe a transformation from the isotropic polymer melt into the l.c. phase, indicated by the jump in the V(T) curve. At low temperatures no crystallisation is observed but the bend in the curves signifies a glass transition. Obviously the phase behaviour is determined by the combination of l.c. and polymer properties.

Two questions are of importance for the understanding of the l.c. side chain polymers:

i) Is the phase transformation: isotropic → liquid crystalline of polymers, which have linked the mesogenic molecules as side chains to the main chain, of first order in analogy to the l-l.c. systems and

ii) is the l.c. phase of the polymer a homogeneous phase, which is in its internal, thermodynamical equilibrium?

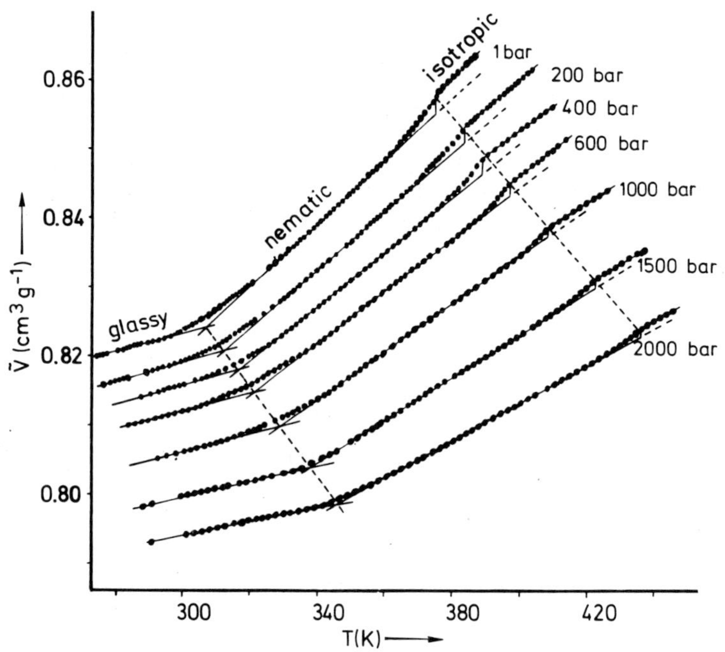

Fig. 5. Specific volume temperature curves at different pressures of the l.c. side chain polymer (isobaric measurements) [37, 38]

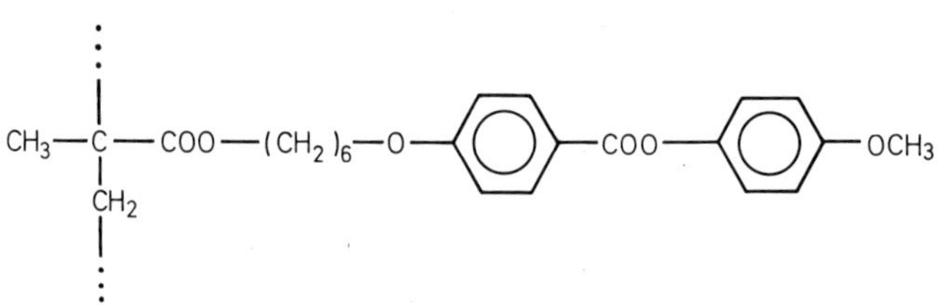

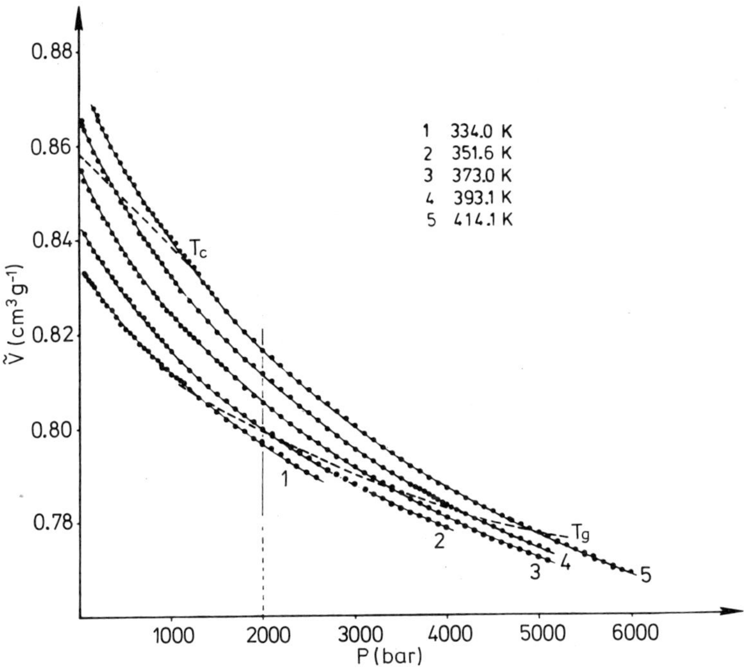

Fig. 6. Specific volume pressure curves for the l.c. polymer shown in Fig. 5. Thin dashed lines: pressure dependence of the phase transformation temperatures l.c. to isotropic, T_c, and the glass transition temperatures, T_g,; full line: specific volume-temperature cut at 2000 bar (isothermal measurements)

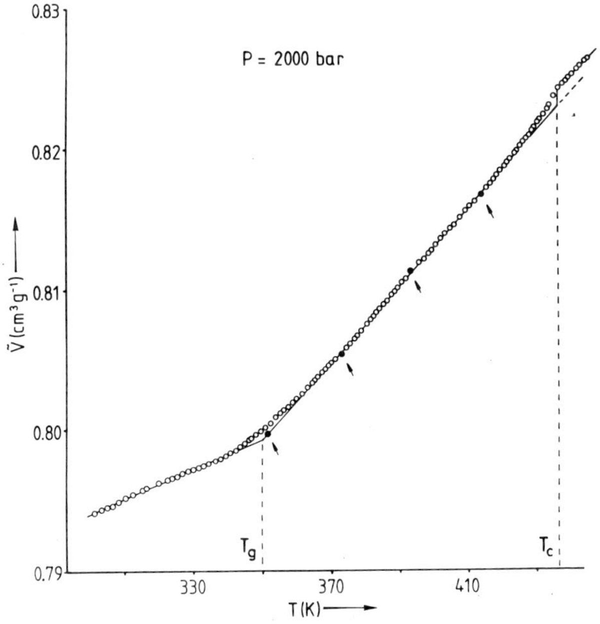

Fig. 7. V(T)-diagram of the l.c. polymer from Fig. 5; isobaric measurements (○); isothermal measurements (●), obtained from the cut at 2000 bar in Fig. 6

To check the phase transformation isotropic → nematic, the validity of the Clausius Clapeyron equation is examined. It has been shown [38], that within the experimental error the results fulfill Eq. 1 in analogy to the low molar mass l.c. The phase transformation isotropic to l.c. is therefore of first order with two coexisting phases at the transformation point. Optical measurements on the polymers confirm these thermodynamical measurements (refer to 2.3.1.3).

As there exists a phase equilibrium both phases must have reached in the internal thermodynamic equilibrium with respect to the arrangement and distribution of the molecules the measuring time. Therefore, no time effects or path dependencies of the thermodynamic properties in the liquid crystalline phase should be expected. To check this point for the l.c. polymer, a cut through the measured V(P) curves at 2000 bar has been made (Fig. 6) and the volume values are inserted at different temperatures in Fig. 7, which represents the measured isobaric volume-temperature curve at 2000 bar [38]. It can be seen from Fig. 7 that all specific volumes obtained by the cut through the isotherms in Fig. 6 lie on the directly measured isobar. *No* path dependence can be detected in the l.c. phase. From these observations we can conclude that the volume as well as other properties of the polymers depend only on temperature and pressure. The liquid crystalline phase of the polymer is a homogeneous phase, which is in its internal thermodynamic equilibrium within the normal measuring time.

At low temperatures the bend in the V(T) curves indicates a glass transition. It is of interest, whether this transition can be described in analogy to amorphous polymers on the basis of the concept of using one internal order parameter and whether the liquid crystalline structure is maintained in the glassy state. This will be discussed in Chap. 2.3.4.

The synthesis of the l.c. polymers has proved that the polymers exhibit a polymorphism in analogy to conventional l.c. This means that phase transformations within the l.c. state are observed. Furthermore it has been shown, that at low tem-

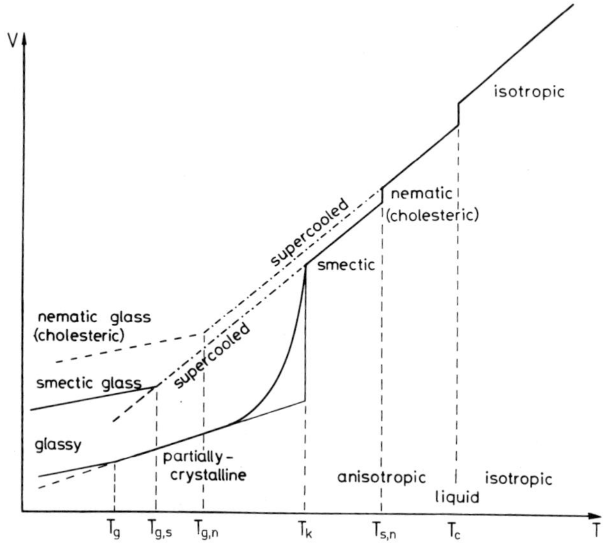

Fig. 8. Schematic volume-temperature diagram of a l.c. polymer

peratures not only glassy polymers exist but also crystalline or partially crystalline polymers which are described in the literature [39,40]. In conclusion, the phase behavior of l.c. side chain polymers is schematically summarized in the V(T) diagram shown in Fig. 8, which exhibits besides a nematic or cholesteric, a smectic liquid crystalline phase [41]. The transformation from isotropic to liquid crystalline is indicated by a volume step which characterises the first order transformation. The transformation to smectic normally is also of first order. Cooling the polymer, crystallization can occur, or, if a crystallization is suppressed, supercooled l.c. phases are formed. These liquid crystalline phases freeze in at lower temperatures and form a glass below the glass transition temperature T_g, indicated by the bend in the V(T) curve.

Effect of the Polymer Main Chain on the Phase Behavior

If we compare the liquid crystalline phase behavior of a polymerizable mesogenic monomer with that of the corresponding polymer, in all cases observed hitherto, a principal rule can be established: Owing to the polymerization and linkage of the mesogenic moieties to the polymer backbone, the liquid crystalline state is always shifted towards higher temperatures [24]. A more quantitative insight into this behavior is given by the investigations of the phase behavior as function of the degree of

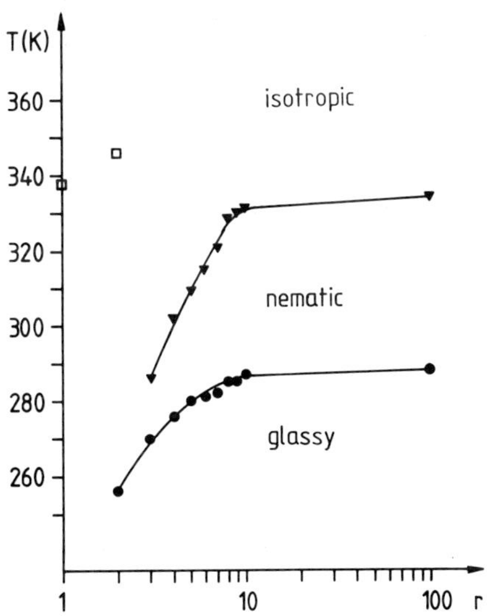

Fig. 9. Phase transition temperatures T_c, T_g and T_k of monodisperse oligomers as function of the degree of polymerization r (□ phase transformation crystalline to isotropic)

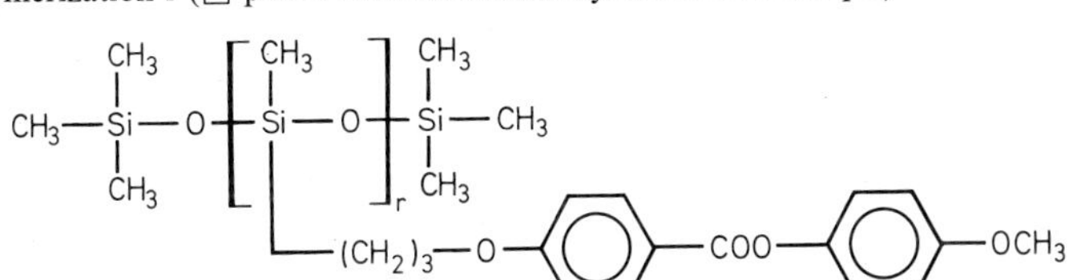

polymerization r of the polymer backbone [42]. Quite recent results obtained for a nematic system are shown in Fig. 9. The monomer and dimer are not liquid crystalline but crystalline. Whereas for the monomer a rapid crystallization is observed, the dimer can be supercooled and a transition into the glassy state is found at 255 K. An additional monomer unit produces for the first time a nematic phase in a temperature interval of 15 K. With each additional monomer unit the phase transformation nematic to isotropic increases steeply causing a strong broadening of the l.c. state. For $r > 10$ nearly no change in the phase behavior is any longer observed. This behavior could also be established for polymers having a different chemical constitution [43].

Two aspects of the investigations have to be emphasized:

i) a strong change in the phase behavior as function of r is only observed for oligomers and

ii) for $r > 10$ the liquid crystalline phase behavior is hardly any more affected by r.

This change of the phase behavior as function of the degree of polymerization can be understood by the restriction of translational and rotational motions of the mesogenic molecules due to their linkage to the backbone. This shift should vanish, when a chain length is reached, for which no correlation exists between the motions of the initial monomer and the least added monomer. Macroscopically, the increase of the density, going from monomer to polymer, indicates the restriction of motions.

2.3 Linear L.C. Side Chain Polymers

2.3.1 Nematic Polymers

Nematic phases are characterized by an unordered statistical distribution of the centers of gravity of molecules and the long range orientational order of the anisotropically shaped molecules. This orientational order can be described by the Hermans' orientation function [44], introduced for l.c.'s as order parameter S by Maier and Saupe [12],

$$S = \frac{3}{2}\left(\overline{\cos^2\theta} - \frac{1}{3}\right). \tag{3}$$

The angle denotes the mean deviation of a molecular axis with respect to the director. The director is the symmetry axis of the orientational distribution function of the molecular axes. For the l.c. side chain polymers the nematic order therefore can not be related to the entire macromolecule but to the mesogenic side chains. With this description, the classical theories of nematic phases [12] can be applied to the side chain polymers, neglecting the backbone and only considering the mesogenic side chains. Because of their linkage to the polymer, however, characteristic interactions of side- and main chains are to be expected, which should be reflected in the physical properties, e.g. phase behavior, optical properties and state of order. The nematic state of the mesogenic side groups on the other hand should be, apart from local deformations, consistent with a statistical chain conformation. Therefore the influence of the backbone on l.c. properties is to be expected to be less pronounced than in the highly ordered smectic state.

2.3.1.1 Chemical Constitution of Nematic Polymers

The structural conditions on molecules, to obtain nematic phases, are extensively reviewed by G. Gray [3–6]. It has been observed that within a homologous series of l-l.c.'s normally nematic phases are obtained, when the rigid mesogenic cores of the molecules are substituted only by short substituents. If these substituents are lengthened, smectic phases become favoured (refer to Chap. 2.3.3). Systematic investigations over the past years have proved that these principles of l-l.c.'s can be transferred to the l.c. side chain polymers. Within a homologous series nematic polymers are only observed for systems, where the mesogenic group is substituted by a *short* flexible spacer and a short substituent at the end of the mesogenic moiety. In Table 2 some typical arbitrarily chosen examples a summarized, where the mesogenic groups are similar to l-l.c. [27]. As discussed in Chap. 2.2 the polymer-fixation of the l-l.c.
i) always produces a shift of the phase transformation l.c. to isotopic towards higher temperatures and ii) often favours the smectic state. For the planning of synthesis of nematic polymers these aspects have to be considered. It turned out to be most practicable, to polymerize monotropic nematic monomers ($T_c < T_K$, T_K = phase transformation crystalline to isotropic), which become enantiotropic nematic by polymerization [45]. If the monomer is already nematic, in most cases smectic polymers result. Only very few examples are known, where the monomer as well as the polymer is nematic.

So far we have not considered the influence of the constitution of the polymer main chain on the formation of the nematic phase. If the same mesogenic group is linked to different backbones, the nematic phase can be preserved, as shown for one example in Table 3. Owing to the different flexibilities of the backbones, the nematic state is shifted with respect to the temperature. With falling flexibility of the main chain, as indicated by the increasing glass transition temperature, the phase transformation temperatures nematic to isotropic are shifted towards higher temperatures. This clearly indicates that the restriction of motions, due to the polymer-fixation, directly reflects on the phase transformation temperature. If this restriction

Table 2. Examples for some nematic side chain polymers
(g = glassy; s = smectic; n = nematic; i = isotropic)

		PHASE TRANSITIONS (K)	LIT
1	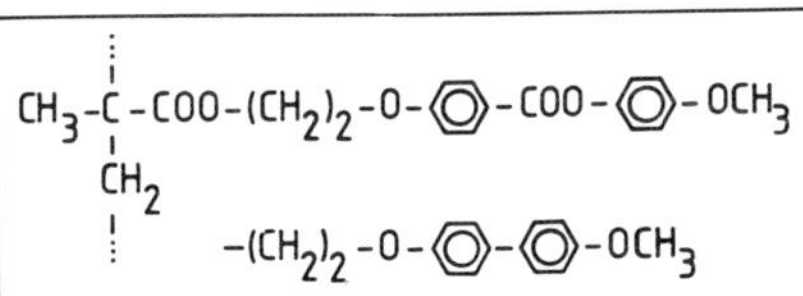	g 369 n 394 i	45
2		g 393 n 452 i	31
3	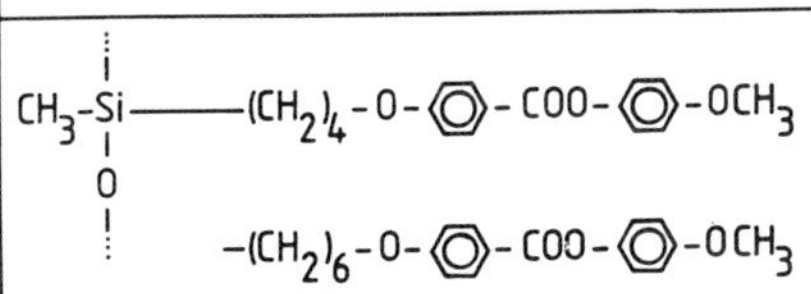	g 288 n 376 i	29
4		g 278 s 319 n 381 i	29

Table 3. Phase behavior of polymers having the same mesogenic group linked to the different main chains (ΔT = extent of the nematic phase)

POLYMER	TRANSITIONS (K)	ΔT (K)
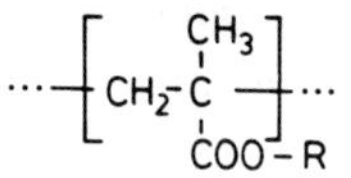	g 369 n 394 i	25
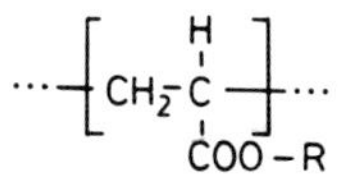	g 320 n 350 i	30
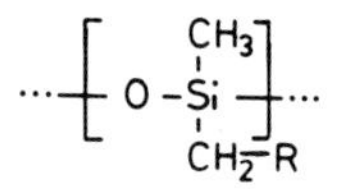	g 288 n 334 i	46
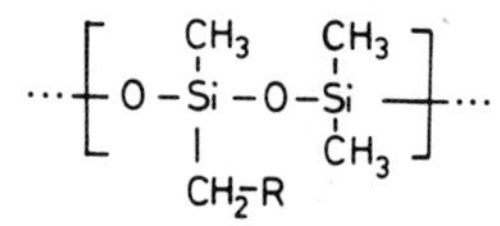	g 276 n 294 i	18

R = 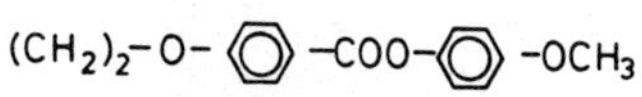

of motions is released, the phase transformation temperatures fall. This is realized for the copolymer in Table 3, where on average each second monomer unit carries a mesogenic moiety, causing a lower T_g and consequently T_{cl}, compared with the homopolymer.

2.3.1.2 X-Ray-Investigations

X-ray investigations on nematic systems show scattering diagrams that are very similar to isotropic liquids with a broad halo in the wide angle area, provided that no macroscopically ordered samples are present. For nematic polymers scattering curves similar to monomeric nematics are obtained [46,47]. If the polymer is macroscopically ordered, e.g. by spinning a fiber, distinct equatorial reflexes are observed, which are perpendicular to the fiber axis [48]. This indicates, that the mesogenic groups are ordered parallel to the fiber axis. Careful investigations of small angle X-ray scattering have shown, that the scattering intensity at small angles is independent of the absolute value of the scattering vector s ($s = |\vec{s}| = 4\pi \sin \theta/\lambda$; θ = scattering angle, λ = wavelength). This indicates that no supermolecular structure for normal nematic polymer phases exists, which might be due to the linkage of the mesogenic moieties to the polymer main chain [49]. A model, proposed by Cser [50] for nematic polymers, which describes the nematic polymers as consisting of nematically ordered aperiodical helices, cannot be supported by the X-ray analysis. (see also 2.3.1.6).

2.3.1.3 Optical Properties

Monomeric l.c.'s show in the polarizing microscope under crossed polarizers characteristic textures, owing to their optical anisotropy [51]. Examining a nematic phase, which is sandwiched between untreated glass plates, typical interferences are observed, because of the variations of the optical axis with respect to the incident of light. The nematic polymers exhibit a similar bevahior. In Fig. 10a a typical picture of the texture of a polymer is shown. While for l-l.c.'s the texture can be observed immediately after preparation because of their low viscosity, in most cases the polymers samples

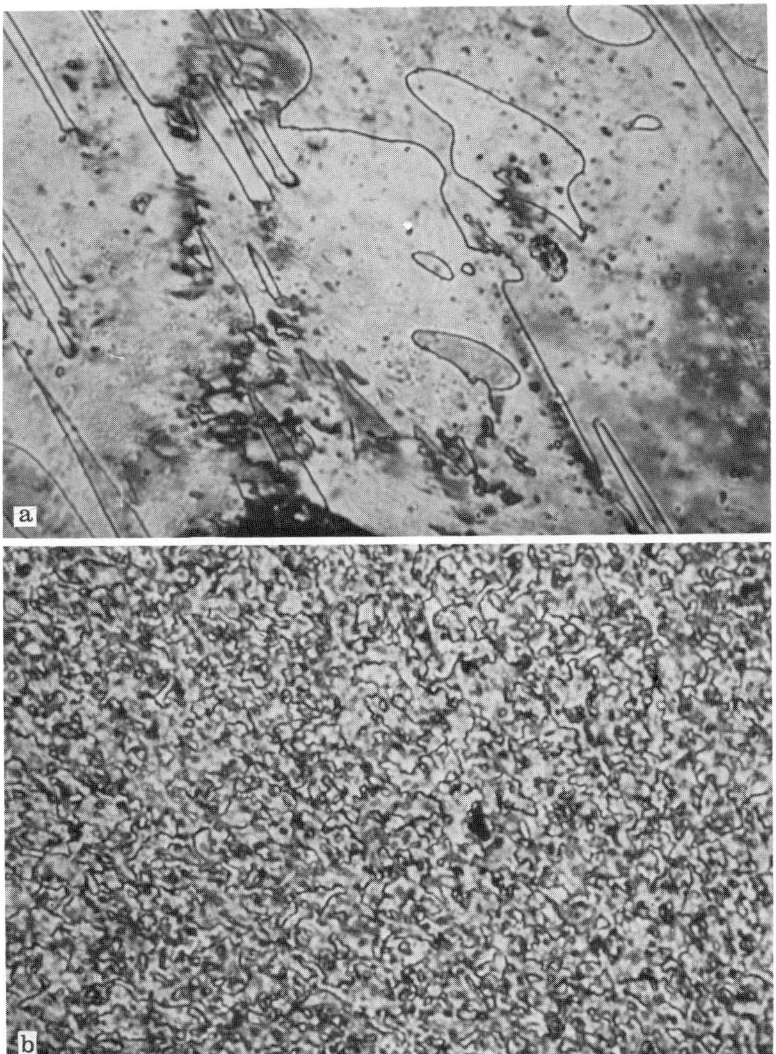

Fig. 10a and b. Textures of a nematic polymer (crossed polarizers, magnification 40); **a** after annealing of the sample; **b** after fresh preparation

have to be annealed. Because of their high viscosity, it takes more time for the formation of uniform textures and defects only slowly vanish. A picture of a freshly prepared polymer is shown in Fig. 10b. Especially the formation of uniform textures by using the well known techniques of treated surfaces [51], normally takes hours or days of annealing time. However, homogeneous and homeotropic orientations of the polymers can be realized, using the conventional techniques described for l-l.c.'s, where the optical axis is parallel or perpendicular to the boundaries of the optical cell respectively. With these uniform textures the uniaxial positive character of the nematic polymers can be easily proved, using a conoscopic path of rays in the microscope. Similar to l-l.c.'s the ordinary refraction index n_0 is smaller than the extraordinary refraction index n_e, causing a positive birefringence $\Delta n (\Delta n = n_e - n_0)$.

Comparing the formation of a uniform texture with polymers, which only differ in the length of their flexible spacer, it is noticeable that the polymer having the longer spacer always needs less time for the uniform texture. From these observations it can be qualitatively concluded that in case of the short flexible spacer the backbone disturbs the macroscopic formation of the texture. Actually, in case of short flexible spacers some anomalies of the textures have been observed and were described for the first time by Kelker [52]. Although these polymers are nematic following X-ray and thermodynamic investigations, no typical textures can be observed [48]. In Fig. 11 a texture of such polymer is shown. No inclination lines can be seen. Conoscopic

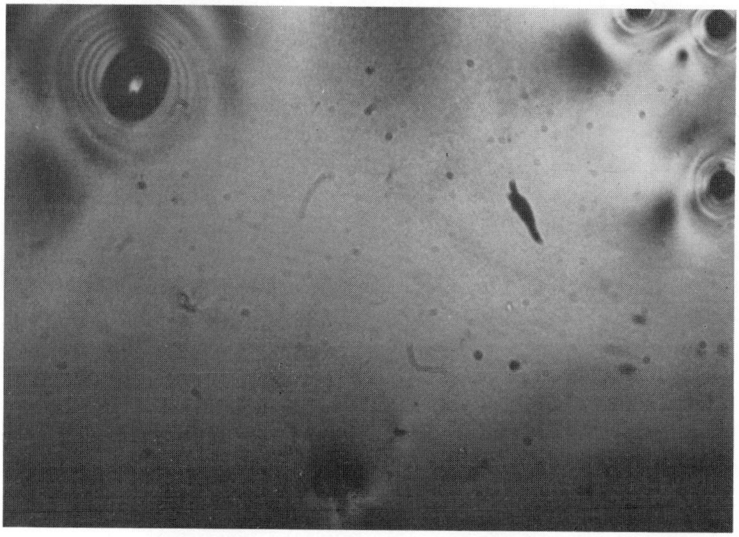

Fig. 11. Anomalous texture of the nematic polymer

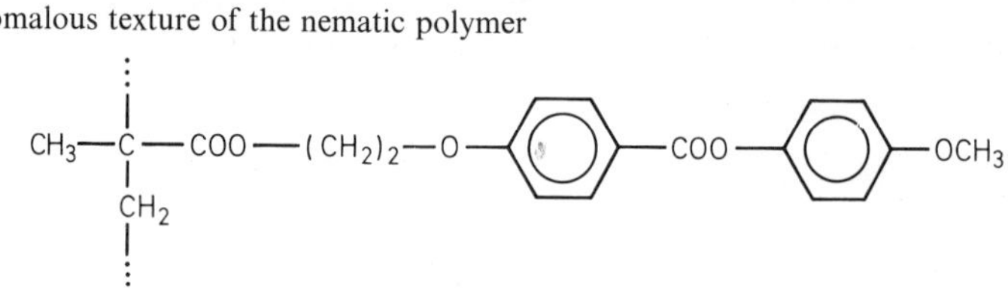

observation even indicates a negative birefringence ($n_e < n_0$), which is confirmed by birefringence measurements. In Fig. 12 the birefringence of two polymers is compared. Curve 1 shows the birefringence of a nematic polymer with a "normal" texture, curve 2 that of a polymer having the "anomalous" texture, shown in Fig. 11. While the absolute value of Δn only differs slightly, the sign of Δn is opposite. Before this result is discussed in more detail, some general remarks on the birefringence of nematic phases will be given.

The birefringence Δn of a nematic phase depends on the anisotropic polarizabilities $\alpha_{\parallel}$, $\alpha_{\perp}$ and the degree of order S (Eq. (3), p. 114). The polarizabilities α_e and α_0 of a nematic phase parallel and perpendicular to the director respectively obey the following relations [53)]

$$\alpha_e = \bar{\alpha} + \frac{2}{3} S \, \Delta\alpha \tag{4a}$$

$$\alpha_0 = \bar{\alpha} - \frac{1}{3} S \, \Delta\alpha \tag{4b}$$

where S is defined in Eq. (3), and

$$\bar{\alpha} = \frac{1}{3} (\alpha_e + 2\alpha_0)$$

$$\Delta\alpha = \alpha_{\parallel} - \alpha_{\perp}$$

$\alpha_{\parallel}$ and $\alpha_{\perp}$ are the molecular polarizabilities parallel and perpendicular to the long molecular axis of the mesogenic molecules. The relation between mean polarizabilities of the nematic phase and the mean refraction indices n_i can be described with a modified Lorentz-Lorenz Equation [54, 55)]

$$\frac{4}{3} \pi N_A \alpha_i = \frac{M}{\varrho} \frac{(n_i^2 - 1)}{\bar{n}^2 + 2} \tag{5}$$

α_i = polarizability tensor

where N_A = Avogadro's number, ϱ = density, M = molar mass and $\bar{n}^2 = \frac{1}{3} (n_e^2 + 2n_0^2)$.

Presuming Eq. (5) is still valid for the polymers, and the molecular polarizabilities are related to the mesogenic side chains, the positive birefringence of curve 1 in Fig. 12 of the nematic polymer is understandable with regard to sign as well as to magnitude, compared with m-l.c.'s. Not consistent, however, is the result for the polymer having the flexible spacer and the "anomalous" texture. If a structure is supposed similar to cholesterics because Δn is negative, where the long molecular axes are perpendicular to the optical axis, the magnitude of the birefringence has to be smaller. Because

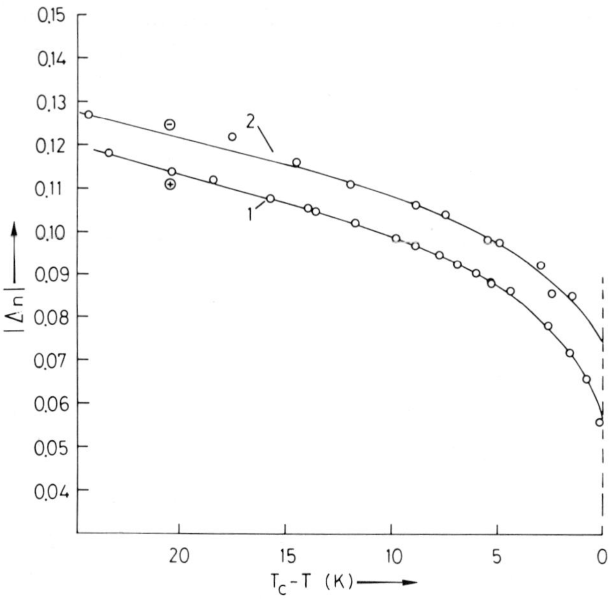

Fig. 12. Birefringence of nematic polymers:
1) "normal" nematic polymer ($\Delta n > 0$)

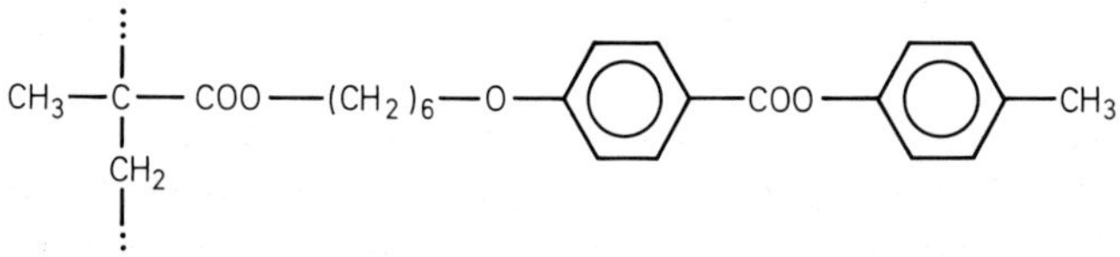

2) "anomalous" nematic polymer having a short flexible spacer (structure refer to Fig. 11) ($\Delta n < 0$)

both polymers differ in their chemical constitution only very slightly, it has to be concluded that Eq. (5) is not applicable for this "anomalous" system. An explanation of this effect might be given by a strong disturbing effect of the polymer backbone on the formation of a macroscopically ordered structure, owing to the short flexible spacer. Under these conditions only strongly disturbed nematic phases are conceivable, having a short correlation length of orientationally ordered domains. Under this condition the assumption of a Lorentz field of Eq. 5, and consequently Eq. 5 itself is no longer valid. Macroscopically measured birefringence and molecular polarizabilities can no longer be correlated. On the other hand, if the flexibility of the "anomalous" polymer is enlarged by softening agents (e.g. by l-l.c.'s) normal nematic textures can be observed. These experiments prove the nematic state of the polymer (see 2.3.1.6).

From the measurements of the birefringence in Fig. 12a further important aspect has to be mentioned. With increasing temperature Δn does not continuously tend to zero at the phase transformation temperature T_c but vanishes discontinuously. At the phase transformation the nematic phase, having a finite birefringence, coexists with the isotropic melt, having only one refraction index. These optical measurements confirm the thermodynamic investigations described in Chap. 2.2 and prove the first order transformation nematic to isotropic for the side chain polymers.

2.3.1.4 State of Order

As already mentioned, in the simple case of cylindrical symmetry of the mesogenic molecules, the long range orientational order of the nematic polymers can be described by Eq. (3):

$$S = \frac{3}{2}\left(\overline{\cos^2\theta} - \frac{1}{3}\right)$$

While in case of main chain polymers S relates to the backbone or rigid segments of the backbone, for side chain polymers only the rigid mesogenic moieties of the side chains will be covered neglecting any possible anisotropic orientations of the backbone.

While for m-l.c.'s the state of order is only determined by the anisotropic interactions of neighbouring molecules, for the polymers additionally a disturbing effect of the backbone via the flexible spacer on the anisotropic order of the mesogenic side chains is to be expected and vice versa. Therefore it is of interest to investigate whether
i) the absolute value and the temperature dependence of S of a polymer differ from that obtained for a chemically corresponding monomer and
ii) the length of the flexible spacer, which couples main and side chain, influences the state of order.

At first we will discuss the absolute value of S and S(T) for a l-l.c. and a corresponding p-l.c. A method to determine the order parameter are measurements of linear dichroism of dichroitic dye probes, having a mesogenic molecular structure, which are dissolved in the l.c. host phase. The dye probes adopt the anisotropic orientation of the host phase and allow to draw conclusions about the state of order of the host. In principle, by this method only the order parameter of the guest molecules can be determined. Presuming, however, the molecular structure of the dye is very similar to that of the host phase, no severe disturbing effects of the surrounding nematic host phase is to be expected. We will therefore compare below results of linear dichroism measurements with direct measurements of S by birefringence. The absorption of the dye probe in the host phase depends on the direction of the incident linearly polarized light and on the order parameter: light polarized parallel to a short (long) axis of the dichroitic dye causes a minimum (a maximum) of the absorption, assuming the direction of the transition moment is polarized parallel to the long axis of the dye. Hence it is possible to determine S of the dye molecule by the measurements of the anisotropy of absorption. The evaluation of the measurements of the linear dichroism bases on the relation between the anisotropic absorption and the order parameter and has been derived for l-l.c.'s by Saupe and Maier [53] and is developed more rigorously by Daniel et al. [56]. The order parameter can be determined by

$$S = \frac{\varepsilon_{\parallel} - \varepsilon_{\perp}}{(\varepsilon_{\parallel} + 2\varepsilon_{\perp}) \cdot \left(1 - \frac{3}{2}\sin^2\alpha\right)} \tag{6}$$

where $\varepsilon_{||}$ and $\varepsilon_{\perp}$ are the absorption coefficients along and perpendicular to the long molecular axis respectively. The angle α denotes the angle between the long molecular axis and the transition moment. It is presumed that the director of the nematic phase is perfectly parallel to the surface of the cell, and the incident light perpendicular to the surface. If we suppose that four our dye molecules the long molecular axis is parallel to the polarization direction of the considered transition moment, Eq. (6) reduces to

$$S = \frac{\varepsilon_{||} - \varepsilon_{\perp}}{\varepsilon_{||} - 2\varepsilon_{\perp}} \tag{6a}$$

Table 4. Mixtures **(a)** and copolymers **(b)** for the determination of the orientational order parameter [57)]

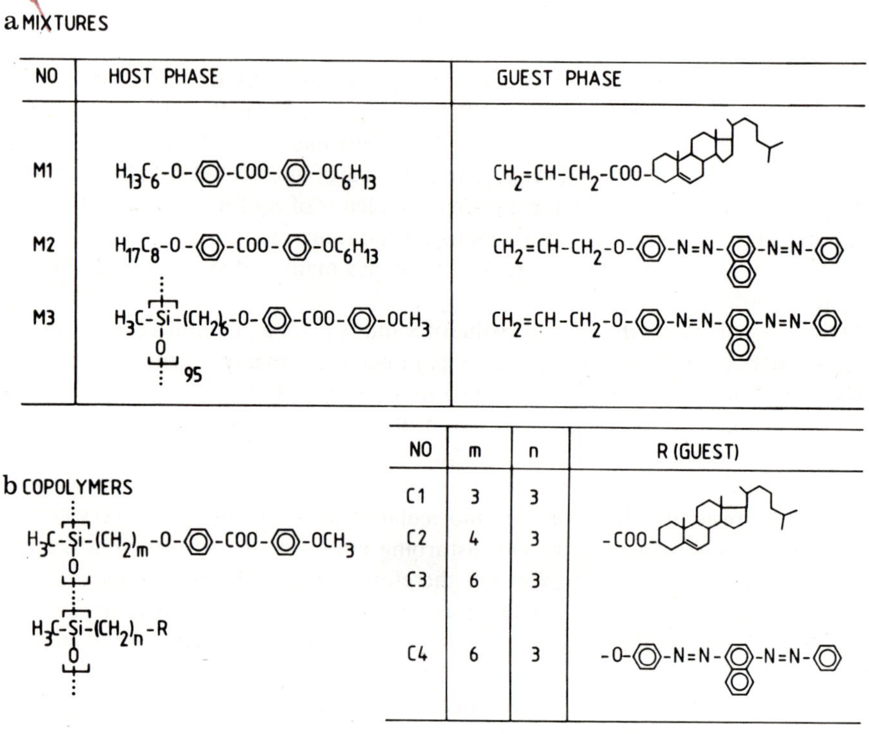

In the following we will compare the order parameter of the dye probe (see Table 4) which is an azoderivative, dissolved in the l-l.c. (M2 in Table 4) and dissolved in the side chain polymer (M3), which has a very similar chemical constitution to the l-l.c. [57)]. The linear dichroism measurements for the polymer/dye system, M3, is shown in Fig. 13, measured at $\lambda = 430$ nm, which is the absorption of the dye having the longest wavelength and which is assumed to be parallel to the long molecular axis. In the nematic phase the absorption $E_{||}$ parallel to the long molecular axis decreases with increasing temperature, whereas the absorption perpendicular, $E_{\perp}$,

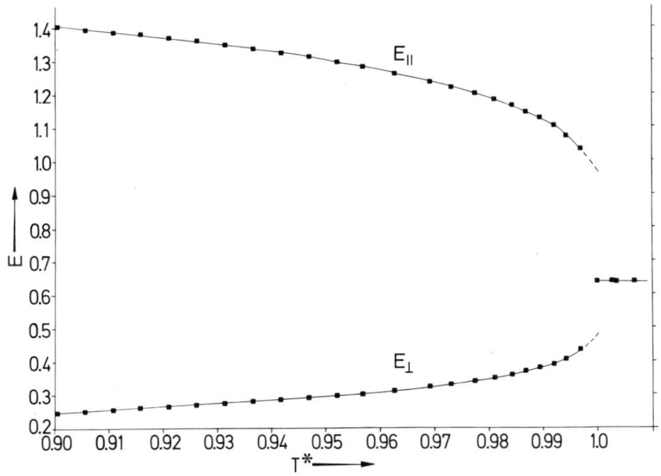

Fig. 13. Temperature dependence of the absorption E of the polymer/dye mixture M3 (refer to Table 4) (λ = 430 nm, c = 0.5 mol% dye, related to the monomer unit of the polymer, d = 10 μm, $E_{\parallel}$ and $E_{\perp}$ are the absorption parallel and perpendicular to the director of the nematic phase, $T^* = \frac{T_m}{T_c}$; T* = reduced temperature, T_m = measuring temperature, T_c = clearing temperature

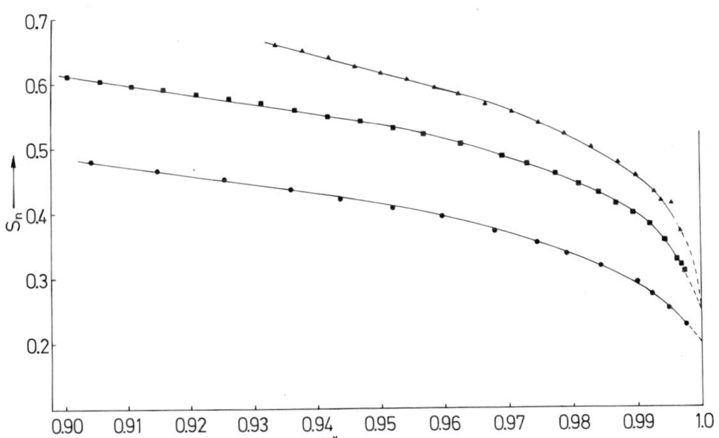

Fig. 14. Temperature dependence of the order parameter S_n of dye probes; (▲) dissolved in a monomeric l.c. (M2); (■) dissolved in a polymeric l.c. (M3); (●) linked to a polymer yielding a copolymer (C4)

increases, indicating falling anisotropic orientation of the dye. Above T_c ($T^* > 1$; $T^* = T_m/T_c$; T_m = measuring temperature) in the melt, the absorption is isotropic.

In Fig. 14 the calculated order parameters according to Eq. 6 are plotted vs. temperature for the three systems M2, M3, C4 (refer to Table 4). For the l-l.c. the well known curve is observed. At the reduced temperature of T* = 0.95 we find S = 0.62. Comparing this order parameter of the dye with the order parameter of the host phase, determined by using another method the same result is observed within the experimental error [58]. This indicates, that for this system the absolute value of S of the dye directly reflects the order parameter of the host phase. It has to be noted, however, that this agreement of the order parameter of host phase and guest molecule normally is not observed if the chemical constitution of the dye differs too much from that of the host phase.

When the dye molecules are dissolved in the p-l.c. M3, where the chemical structure of the mesogenic side chains is very similar to the structure of the l-l.c., another result for S is obtained (Fig. 14). It is obvious, that the temperature dependence of S corresponds to the temperature dependence of the monomer system M2, whereas the absolute value of S is lower. If we determine S at the reduced temperature $T^* = 0.95$, we find $S = 0.54$. This value is reduced by about 10% compared to the l-l.c.

In the third curve shown in Fig. 14 the dye molecule is also linked via a short flexible spacer to the main chain (C4). In this case, the order of the dye is much smaller than the order of the host phase. Because of the linkage to the backbone, the dye can no longer adopt the same anisotropic orientation as the surrounding host phase, which can be understood by the direct linkage of the dye to the backbone and the resulting reduced mobility.

From these measurements of linear dichroism, it can be concluded, that the linkage of l-l.c.'s to a polymer backbone generally reduces the nematic order. This effect has been found so far for poly(acrylates), poly(methacrylates) and poly-(siloxanes) and is established by NMR [59,60], ESR [61] and birefringence measurements.

The second question is, whether the length of the flexible spacer, which couples main chain and side chain, influences the state of order. This will be discussed by birefingence measurements for an example, where the polymer backbone as well as the mesogenic side chains remain constant and only the length of the flexible spacer is varied. The system investigated is also shown in Table 4. The mixture M1 and the copolymers C1, C2 and C3, which differ in the spacer length m = 3, 4 and 6, will be compared. In order to obtain reproducible, uniform textures of the systems, which is a pre-condition for the birefringence measurements, the chiral component of the mixture or comonomer of the copolymers, respectively, converts the nematic host phase (benzoic acid phenylester) into a chiral nematic (cholesteric) phase (see also Chap. 2.3.2). The cholesteric systems form uniform textures, without applying any surface effects. From measurements of the birefringence of these systems, the birefringence of the corresponding untwisted nematic phase can easily be calculated [62]. Then the determination of S is straightforward. The combination of Eqs. (4) and (5) (page 119) yields [63,64] the Lorentz-Lorenz Equation and the corresponding Equation for the anisotropic case:

$$\frac{4}{3}\pi N_A \bar{\alpha} = \frac{M}{\varrho}\,\frac{\bar{n}^2 - 1}{\bar{n}^2 + 2} \tag{7a}$$

and

$$\frac{4}{3}\pi N_A S \Delta\alpha = \frac{M}{\varrho}\,\frac{n_e^2 - n_0^2}{\bar{n}^2 + 2} \tag{7b}$$

where S is directly proportional to the mean refractive indices n_e and n_0. In order to get absolute values of S, an extrapolation method can be used. It is based on the experimental observation that S varies linearly with $-\log(1 - T^*)$, beginning some degrees below T_c. In a log-log-plot the ratio of Eqs. (7b) and (7a).

$$\frac{S\Delta\alpha}{\bar{\alpha}} = \frac{n_e^2 - n_0^2}{\bar{n}^2 - 1} \tag{8}$$

is plotted versus $-\log(1 - T^*)$. A straight line results and by extrapolation to $T \to 0$ K a scaling factor A is obtained [63]. Then S can be calculated by

$$S = \left(\frac{n_e^2 - n_0^2}{\bar{n}^2 - 1}\right)^A \tag{9}$$

With this extrapolation method S can be determined within an absolute error of about ±10%.

Using this method in Fig. 15, the order parameters vs. temperature of the polymers are compared with the corresponding monomer mixture M1. The monomer M1 exhibits a higher value of S than the polymer by about 10% and confirms the previous linear dichroism measurements. If we compare the magnitude and temperature dependence of S for the polymers C1 to C3, which differ in the length of the flexible spacer, *no* difference in S can be found within the experimental error.

From these experiments we can conclude that the length of the flexible spacer has *no* measurable influence on the orientational long range order of the mesogenic side chains. If on the contrary l-l.c.'s are attached to the polymer backbone, the state of order is reduced.

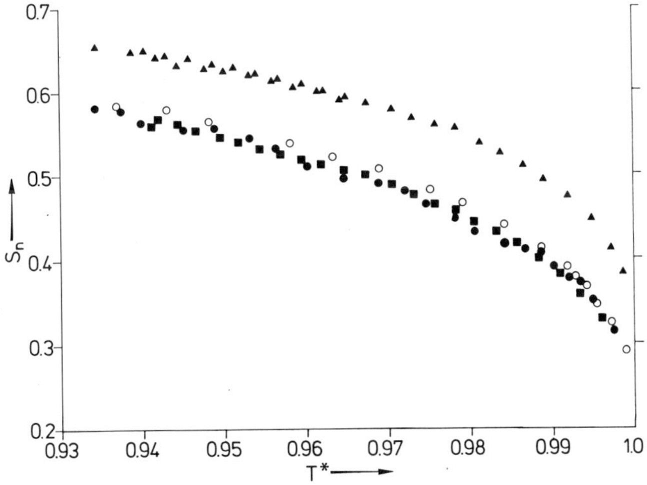

Fig. 15. Temperature dependence of the order parameter for (▲) monomeric l.c. (M1); polymeric l.c.: (●) Cl, spacer length m = 3; (○) C2, spacer length m = 4; (■) C3, spacer length m = 6

A very interesting result has been observed very recently by Boefel et al. [60]. By NMR-technique not only the order of the mesogenic side groups has been determined, but also the order of deuterated C atoms within the spacer chain. The methylene group neighbouring the mesogenic group gives the same order parameter related to the director as the mesogenic groups themselves. For the methylene group, six $-CH_2$-groups apart (C6-position), the order parameter is reduced by about 50%. On the other hand, for a homologous l-l.c. in the C6-position S is only reduced by 25% related to the mesogenic group. These results clearly demonstrate the spacer model of decoupling main chain and side chain. However, they also demonstrate that the decoupling is not complete, which in principle is indicated by the reduced order parameter of the polymers.

2.3.1.5 Field Effects

The ease of orientation of l-l.c.'s in the electric and magnetic field and their response in optical properties are widely investigated in view of theoretical aspects and technological application. This is reflected in numerous reviews and articles [65]. Especially the technological application of l.c.'s for display devices in optoelectronics pushed forward the development of l.c.'s. By measuring electric and magnetic field effects powerful methods exist, to characterize the elastic and viscous behavior of l.c.'s.

With the l.c. side chain polymers a new class of l.c. material was realized and this poses the question, whether this new material also exhibits field effects in analogy to conventional l.c.'s and whether the covalent bonding of the mesogenic groups to the polymer backbone changes the material parameters. In this chapter we therefore will compare the behavior of l-l.c.'s and chemically very similar polymers in the electric and magnetic field.

Orientation Effects by Surfaces

A precondition for the investigation of deformation of a l.c. in an external field is a uniform alignment of the l.c. with respect to the measuring cell, in order to get quantitative informations. Normally the l.c. is aligned by surface effects in the measuring cell, which usually consists of two glass plates separated by a distance of about 10 μm. We will consider three principal modes of alignment of the l.c. (Fig. 16):

i) homogeneous alignment, where the long molecular axes of the l.c. molecules are parallel to the cell surfaces (Fig. 16a)

ii) homeotropic alignment where the axes are perpendicular to the cell surfaces (Fig. 16b) and

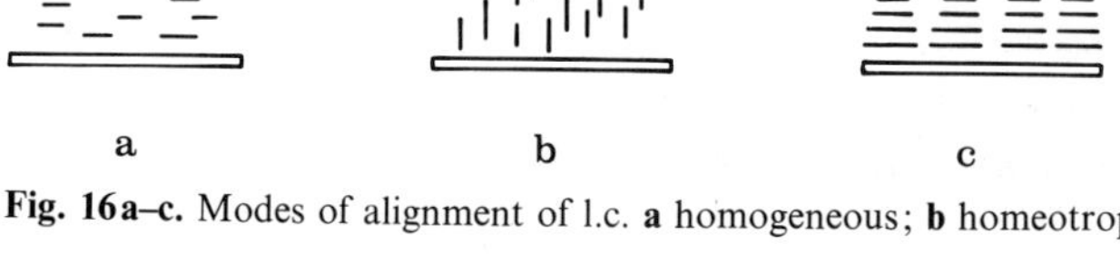

Fig. 16a–c. Modes of alignment of l.c. **a** homogeneous; **b** homeotropic; **c** twisted nematic

iii) the twisted nematic alignment where the director of the homogeneous alignment is twisted by 90° from the lower to the upper glass plate (Fig. 16c).

By using the standard techniques known for l-l.c., these alignments could also be achieved for the polymers [66]. However, because of the high viscosity of the polymers, the orientation requires annealing times up to several hours at temperatures some degrees below T_c depending on the chemical constitution of the material (refer to 2.3.1.3). In principle, no orientation can be observed, if the preparation of the material is performed in the l.c. state in the vicinity of a glass transition.

Electric Field Effects

Already in 1979 qualitative experiments proved that l.c. side chain polymers behave very similarly to l-l.c.'s [67]. To get a more detailed insight into the behavior of the polymeric l.c.'s in the electric field, we will discuss their field effects in a twisted nematic configuration (Fig. 16c) [66]. This configuration is widely applied in display technology (TNLCD = twisted nematic l.c. display). For conventional l.c.'s as well as for the polymers the initial orientation is deformed by an electric field to a homeotropic alignment (Fig. 16b), if the applied voltage is larger than the threshold V_{th} and presuming that the dielectric anisotropy $\Delta\varepsilon > 0$ ($\Delta\varepsilon = \varepsilon_{\|} - \varepsilon_{\perp}$; $\varepsilon_{\|}$ and $\varepsilon_{\perp}$ = dielectric constants parallel and perpendicular to the director).

It has been shown [65,68] that the threshold voltage is a function of the dielectric anisotropy $\Delta\varepsilon$ and the elastic constants of splay (k_{11}), twist (k_{22}) and bend (k_{33}) deformation of the nematic phase (Fig. 17):

$$V_{th} = \pi \left(\frac{k_{11} + (k_{33} - 2k_{22})/4}{\varepsilon_0 \, \Delta\varepsilon} \right)^{0.5} \tag{10}$$

ε_0 = permittivity of vacuum

presuming that no pretilt exists (i.e. a complete parallel alignment of the mesogenic molecules with respect to the cell surfaces).

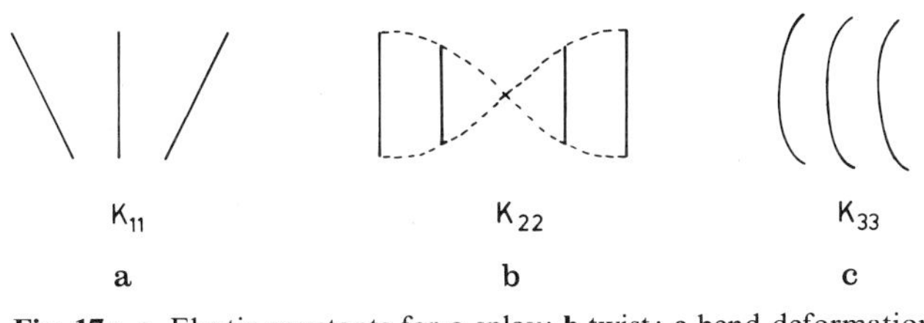

Fig. 17 a–c. Elastic constants for **a** splay; **b** twist; **c** bend deformations of a nematic phase. The full lines represent the director

According to Eq. (10) the threshold voltage V_{th} is determined by $\Delta\varepsilon$ and the elastic constants. If we compare the behavior of a monomeric l.c. and the corresponding polymeric l.c., no principal change in $\Delta\varepsilon$ has to be assumed. Therefore the threshold voltage should directly reflect the change in the elastic constants, when the mesogenic

Table 5. Low molar mass mixtures **(a)** and copolymers **(b)** for the investigations of field effects [66]

a

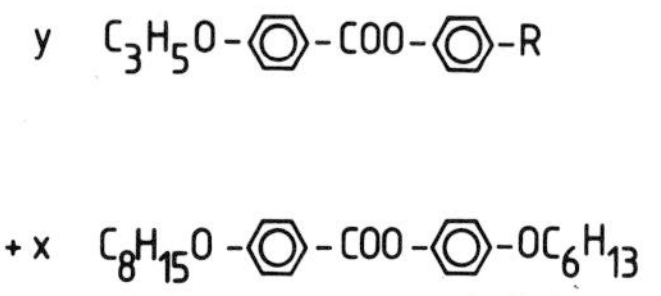

b

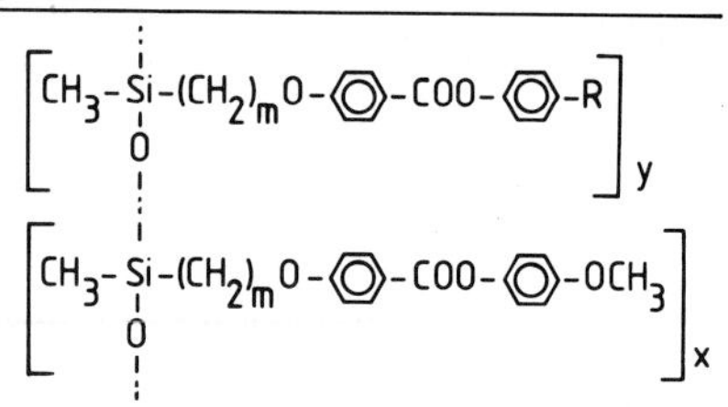

MIXTURES	x/y	R
M1	0.9	Cl
M2	0.85	Cl
M3	0.8	Cl

COPOLYMERS	x/y*	R	m
C1	0.9	Cl	6
C2	0.85	Cl	6
C3	0.8	Cl	6
C4	0.9	Cl	5
C5	0.9	Cl	4
C6	0.9	Cl	3

* (x + y = 95)

molecules are linked to the backbone. A suitable monomer/polymer system, which fulfills these conditions for measuring the influence of the polymer backbone is shown in Table 5. The monomers are mixtures of benzoic acid phenylesters with a polar Cl-derivative, in order to get positive $\Delta\varepsilon$ values. In a first approximation from M1 to M3, $\Delta\varepsilon$ should increase nearly linearly with concentration of the —Cl derivative. In the copolymer C1 to C3 the same mesogenic moieties are linked to the siloxane backbone and with respect to $\Delta\varepsilon$ nearly the same absolute values can be assumed as for the monomers. For these polymers it has to be emphasized that the mesogenic moiety is linked via a long flexible spacer of six methylene groups to the siloxane backbone.

In Fig. 18 the threshold voltage V_{th} is compared for the monomers M1 to M3 and the polymers C1 to C3 as function of the reduced temperature. Two principle aspects can be obtained from these measurements:

i) With increasing $\Delta\varepsilon$, owing to the increasing concentration of the polar guest molecules, for the l-l.c.'s as well as for the polymers V_{th} falls, referring to a defined reduced temperature. This is in accordance with Eq. (10).

ii) the absolute value of V_{th} of the l-l.c.'s and the polymers having approximately the same $\Delta\varepsilon$ is of comparable magnitude. This also indicates that the elastic constants of monomer and polymer are very similar. The polymers even exhibit lower V_{th}.

The strong increase of V_{th} at lower temperatures indicates the formation of a smectic phase at low temperatures for both systems. This effect is well known for conventional l.c.'s. If we compare the ratios of the different threshold voltages for l-l.c.'s and polymers according to Eq. (10), the same ratio of V_{th} is found by varying $\Delta\varepsilon$ (Table 6). This indicates that Eq. (10) is valid for the l-l.c.'s as well as for the polymers.

From electric field effects another important information can be obtained concerning the efficiency of the flexible spacer. The chemical construction of l.c. side

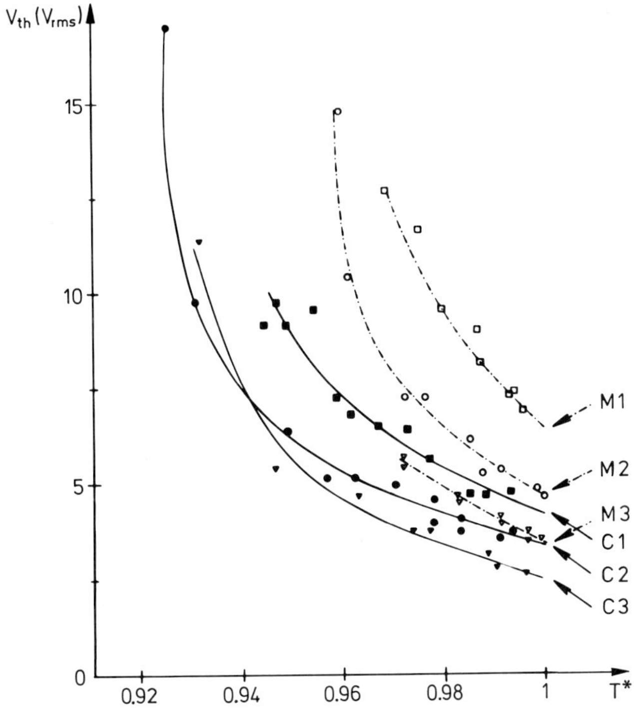

Fig. 18. Threshold voltages V_{th} vs. temperature for low (— — —) and high (———) molar mass systems M1, M2, M3 and C1, C2, C3 respectively. (For chemical structure of M1 to C3 refer to Table 5; V_{rms} = average voltage of alternating current)

chain polymers is based on the linkage of the mesogenic side groups to the polymer main chain via the flexible spacer. According to the model consideration mentioned in Chap. 2.1, a direct linkage of the rigid, rod like side groups to the backbone results in a strong hindrance with the consequence that in most cases no l.c. phase occurs. If the spacer is sufficiently long and therefore rather flexible, one can except that the motions of side and main chain are more or less decoupled. This should be directly reflected by the elastic constants. The properties of the polymers with the long flexible spacers should be very similar to the properties of m-l.c's. This has already been proved by the mesurements mentioned in Fig. 18. On the contrary, polymers with short flexible spacers should be strongly influenced by the polymer main chain. Actually, this can be corroborated by the experiments. In Fig. 19 V_{th} is compared for polymers which *only* differ in the length of the flexible spacer. For a spacer length of m = 6 (copolymer C1 of Table 5), V_{th} = 4 V_{rms}[1] which increases to V_{th} = 7,3 V_{rms} (copolymer C4 of Table 5) for m = 5 at the same reduced temperature of 0.995. For m = 4 (copolymer C5 of Table 5) V_{th} > 50 V_{rms} whereas for the polymer having m = 3 (copolymer C6 of Table 5) no field effects could be observed under the experimental conditions.

1 V_{rms} = average value of voltage for AC

where d is the sample thickness, $\Delta\chi$ the diamagnetic anisotropy and k_{11} the splay elastic constant (see Fig. 17). In Table 7 the results for the polymer having the long flexible spacer of m = 6 are compared with l-l.c.'s, which have a chemical constitution similar to the mesogenic group of the polymer. Here it has to be noted that the elastic constant of the polymer is of nearly the same magnitude as for the corresponding l-l.c.'s at the reduced temperature of 0.95. From the dynamics of the transition the twist viscosity coefficient γ_1 could be determined which, in the weak distortion approximation, is related to the relaxation time τ of the Fredericks transition [114]. From Table 7 it is obvious that the relaxation time for the Fredericks transition τ is larger by orders of magnitude for the polymer, due to the high magnitude of the twist viscosity coefficient γ_1. As already mentioned concerning the electric field effects, these measurements clearly indicate that owing to the high viscosity coefficient of the polymer, large response times result.

2.3.1.6 Mixtures of Nematic Polymers

Extensive studies have been made on binary mixtures of low molar mass l.c. phases [72–74]. The results of these measurements always obey the rule of Arnold and Sackmann, according to which l.c. phases are miscible over the whole concentration range if they have the same phase structure [72]. Miscibility experiments are therefore an important method to classify the l.c. phase of an unknown l.c. material. It is obvious to examine, whether the miscibility rule of Arnold and Sackmann is also applicable for mixtures of l.c. polymers with well characterized l-l.c.'s in order to determine the phase structures of the polymers.

It is straightforward to examine the miscibility rule with simple systems, i.e. polymers having a well defined nematic phase and corresponding 1–l.c.'s. Without going into detail of the phase diagrams, with respect to the problems concerning the crystallisation or glass transition of monomer polymer mixtures, we will focus solely on the question of miscibility in the nematic state. A selected diagram for a nematic polymer and a nematic l.c. is shown in Fig. 20 [75]. The siloxane polymer has linked as side chain a mesogenic benzoic acid phenylester of similar chemical constitution as the 1–l.c. The phase diagram indicates the miscibility of the liquid crystalline components in the isotropic and the nematic state over the whole concentration range. Miscibility experiments with systems having another chemical constitution, which also exhibit the nematic state, confirm these results [76]. This indicates that the rule of miscibility for l.c. phases of binary low molar mass mixtures is also applicable for polymer/1–l.c. mixtures.

In 2.3.1.3 it has been mentioned that polymers having a short flexible spacer sometimes show unusual optical properties, although a nematic phase structure has to be assumed. This follows from X-ray measurements and thermodynamic investigations. A typical example of these polymers is the poly(methacrylate)

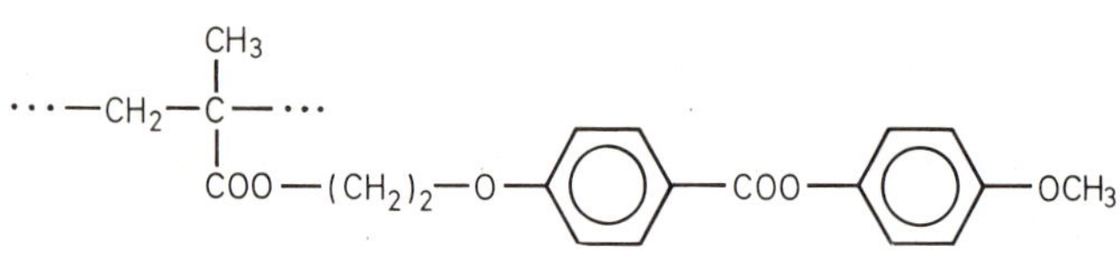

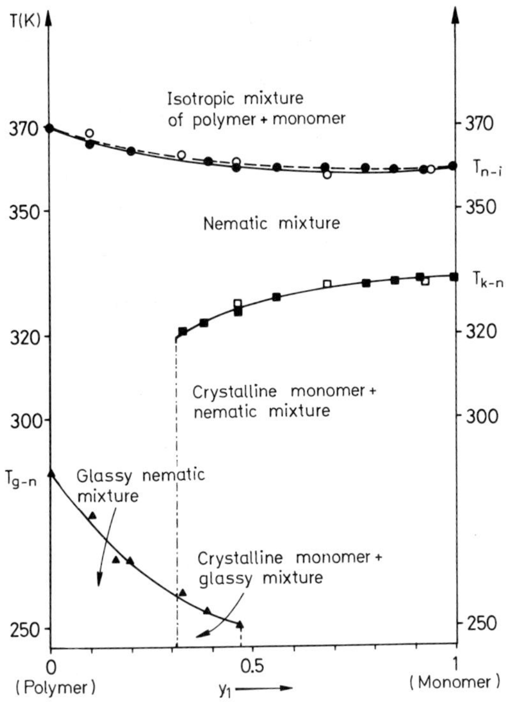

Fig. 20. Phase diagram of mixture of the polymer

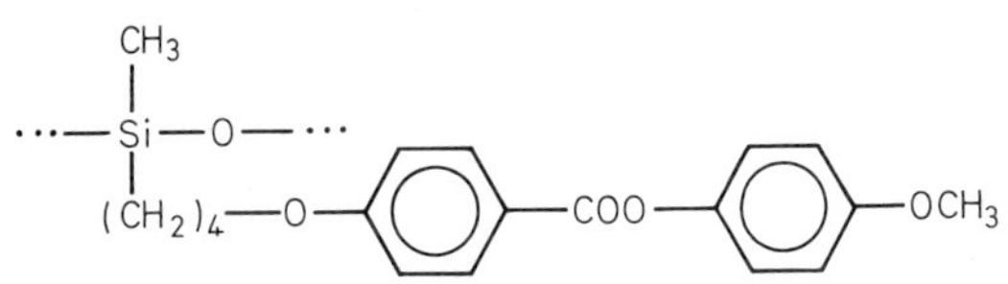

and the monomer

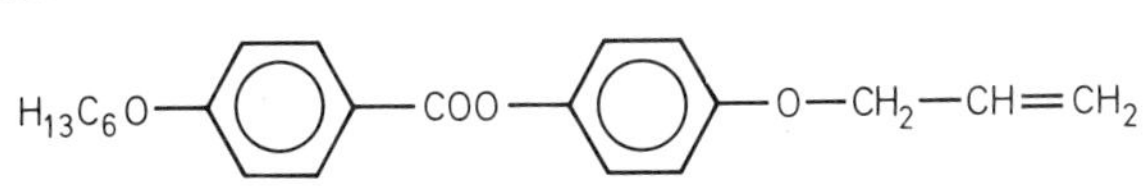

y_1 = weight fraction; full symbols: DSC measurements; open symbols: polarizing microscopic measurements

Owing to the anomalous macroscopic properties of this polymer, Cser et al. [50] postulated a nematic structure built up by aperiodical helices of the entire macromolecule where the mesogenic groups are perpendicular with respect to their long axes to the backbone. Thus the entire molecule forms a cylinder. The cylinders are postulated by the authors to be nematically arranged. Polymers having this symmetry should not be miscible with nematic l-l.c.'s. Actually, miscibility experiments by Cser indicated no miscibility of some l-l.c.'s with this polymer in the whole temperature region. Recently, however, it could be proved [49], that the polymer is miscible with carefully selected nematic l-l.c.'s over the whole concentration range, giving a phase diagram, which is very similar to the diagram shown in Fig. 20. Following the rule of miscibility of l.c.'s, the polymer exhibits a normal nematic phase, which is consistent with the nematic poly(siloxanes) mentioned above. Actually, the poly-(methacrylate) exhibits textures well known for nematics by adding some weight percent of the nematic l-l.c.

On the other hand from these investigations it has to be emphasized that although the polymer and the l-l.c. are nematic, a miscibility of both components *need not* to be observed over the whole concentration range. Miscibility gaps, well known for conventional polymer/polymer or polymer/solvent systems, can also occur in the l.c. mixtures. Recently, a nematic l-l.c./polymer system has been described, where no homogeneous nematic phase occurs in a broad concentration range [76,77].

From the phase diagram shown in Fig. 20 additionally two aspects have to be pointed out. At high polymer concentration the nematic region is substantially enlarged towards lower temperatures, owing to the *plasticising effect* of the l-l.c., which shifts the glass transition towards lower temperatures. Secondly, by adding polymers to the nematic l-l.c., the phase transformation nematic to crystalline can be suppressed. No crystallization occurs, but instead of this the transition into a polymer glass. These effects are of importance for the purpose of variation of the phase behavior of l.c. polymers.

2.3.2 Cholesteric Polymers

The cholesteric mesophase is a helically disturbed nematic phase. As in the nematic phase, the centers of gravity of the mesogenic molecules are statistically disordered, whereas the long molecular axes possess an orientational long range order with respect to the director. The director, however, is not constant in space, but continuously

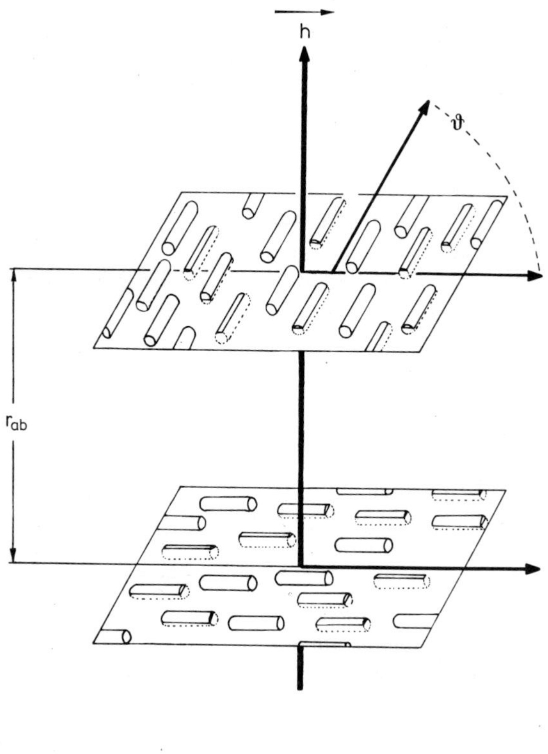

Fig. 21. Schematic representation of the cholesteric phase: ϑ = twist angle, $r_{a,b}$ = distance of molecules perpendicular to the director, $\boldsymbol{h}$ = pitch axis

twisted about an axis perpendicular to the director (see Fig. 21), resulting in a helical structure. The distance, where the director is twisted by $\theta = 2\pi$ is called the pitch p of the cholesteric phase, having a positive (negative) sign for a right handed (left handed) helix. The spontaneous twist of cholesterics is originated by the presence of chiral molecules, and was first theoretically explained by Goossens [78,79]. Expanding the Maier Saupe theory not only to the dipole-dipole but also the dipole-quadrupole interactions of chiral molecules, Goossens could demonstrate that the latter term gives rise to the helical twist.

The realization of nematic side chain polymers implies the possibility of the existence of cholesteric side chain polymers, presuming the mesogenic molecules, which are linked to the backbone, are chiral. For these polymers it is of interest, whether the polymer fixation influences the helical twist and therefore the optical properties of the cholesteric phase. This will be discussed in 2.3.2.2.

2.3.2.1 Chemical Constitution

The best known low molar mass substances forming cholesteric phases are esters of cholesterol, which are responsible for the name "cholesteric" phase. Therefore it was straightforward to attach these derivatives to a polymer in order to obtain cholesteric polymers. These experiments, however, failed because in all cases not cholesteric but isotropic, smectic or crystalline polymers resulted [22] (refer also to Shibaev and Platé, this issue).

According to the change in the phase behavior of mesogenic molecules converting them to polymers, the following principle was established (see Chap. 2.2): The polymerization of mesogenic molecules always stabilizes the l.c. state. This implies: if a cholesteric monomer is polymerized, the polymer generally exhibits the higher ordered smectic phase. For chiral mesogenic monomers, which can form a l.c. phase due to their structure, but which exhibit an isotropic or metastable l.c. phase with regard to the crystalline state, cholesteric polymers should be obtained. Following these principles, recently the first cholesteric *homopolymers* could be realized [80], which are listed in Table 8. These polymers have benzoic acid phenylesters as side groups, which are esterified with the chiral S(—)-2 methyl-1-butanol. The chiral mesogenic moieties are linked via flexible spacers to the poly(siloxane) main chain.

Another possibility to obtain cholesteric phases is well established for l-l.c.'s. Nematic phases can be converted into cholesteric phases by the addition of chiral molecules, which must not necessarily have a mesogenic chemical constitution (induced cholesteric phases). With increasing amount of the chiral derivative an increasing helical twist is induced. This principle can also be applied to obtain cholesteric polymers [81-83] in form of

i) copolymers of nematogenic and chiral comonomers

ii) mixtures of nematic polymers with chiral polymers or chiral monomers.

Some typical examples of copolymers are summarized in Table 8. The phase transition temperatures of the copolymers having different amounts of monomer units can be found in the literature cited.

2.3.2.2 Optical Properties

According to the helical structure, the cholesteric phase (n*) is optically uniaxial negative, where the ordinary refractive index $n_{o,n*}$ is larger than the extraordinary

Table 8. Chemical constitution of some homo- and copolymers having a cholesteric phase (C* denotes an asymmetric carbon atom)

NO		LIT
	HOMOPOLYMERS	
1	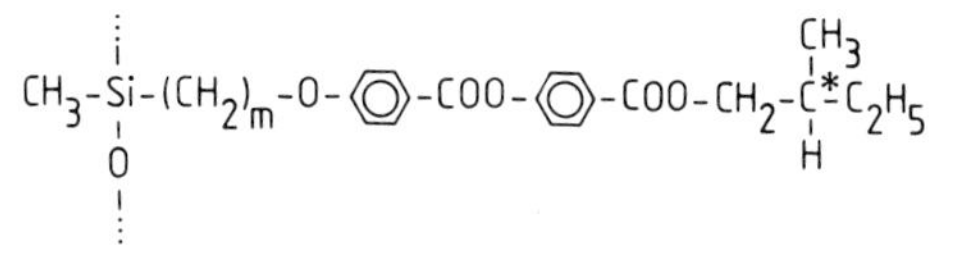	80
	COPOLYMERS	
2	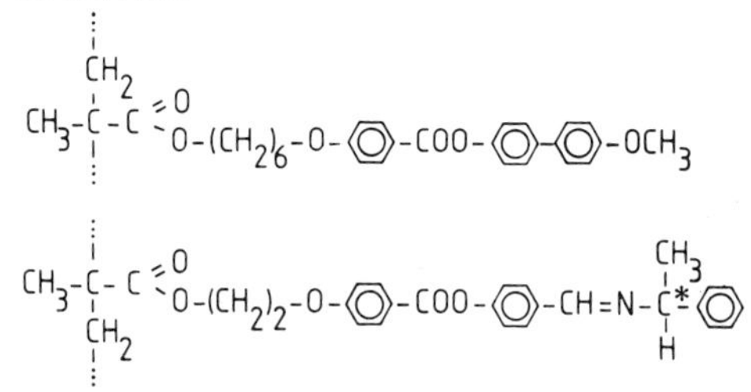	81
3	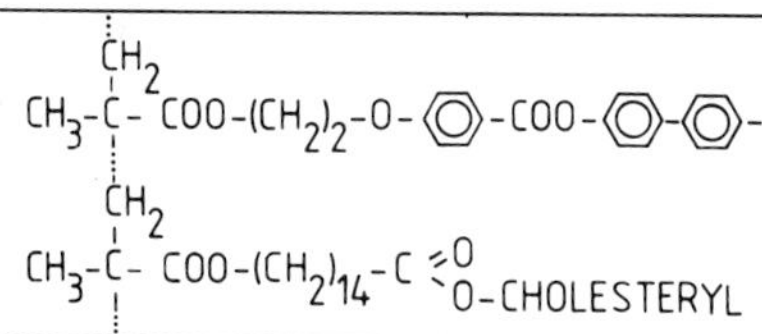	83
4	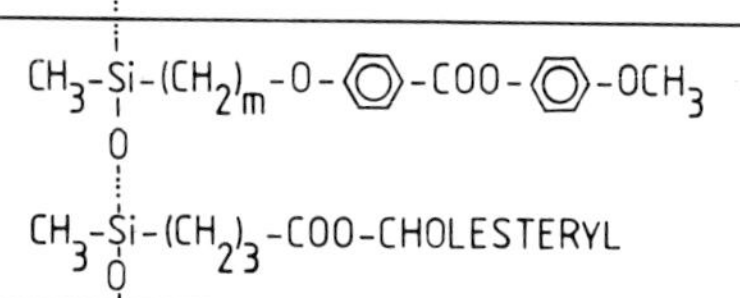	82
5	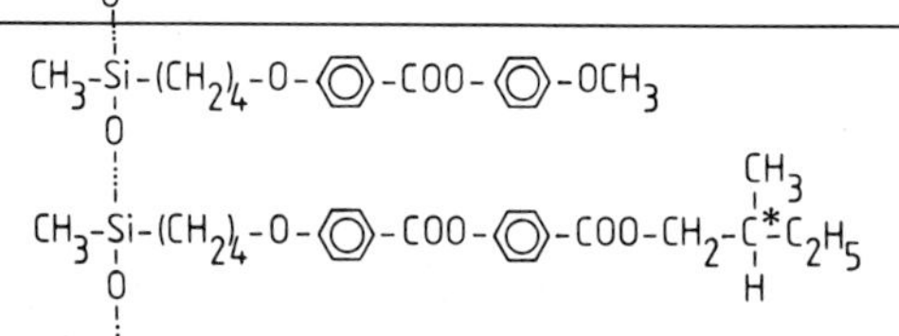	

refractive index $n_{e,n*}$. Therefore the birefringence $\Delta n = n_{e,n*} - n_{o,n*}$ is negative. As an example in Fig. 22 the refractive indices of a copolymer are shown as function of the reduced temperature. For $T^* > 0.91$ the polymer is cholesteric. In analogy to the nematic systems, the first order phase transformation cholesteric to isotropic at $T^* = 1$ is indicated by the discontinous disappearance of the birefringence. Here the cholesteric phase coexists with the isotropic phase. With respect to the refractive index of the isotropic phase n_{is} and according to Eqs. (13) and (5) (p. 119), $n_{e,n*} \approx n_{is} - \frac{2}{3}\Delta n$ and $n_{o,n*} \approx n_{is} + \frac{1}{3}\Delta n$ is fulfilled, provided that the mean refractive index $\bar{n}_{n*}$ of the cholesteric phase is approximately equal to n_{is} (neglecting

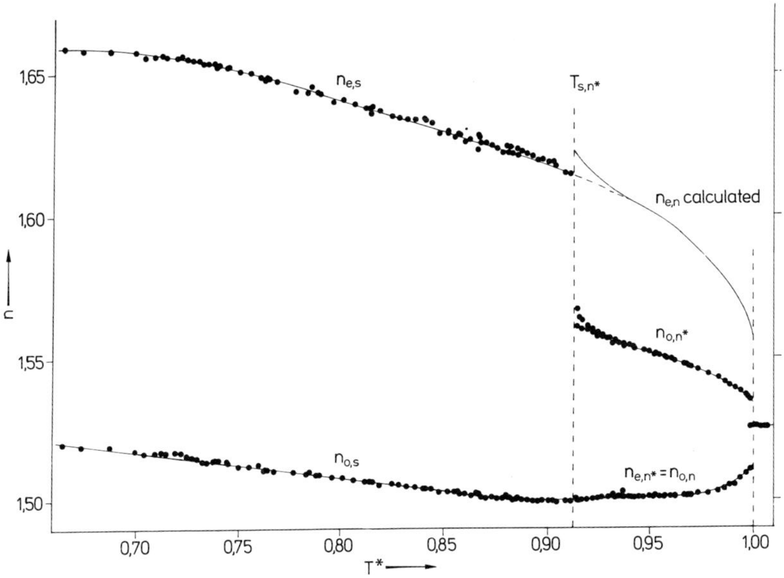

Fig. 22. Temperature dependence of the refractive indices n of the polysiloxane No. 4 of Table 8 (m = 6) in the smectic ($T^* < 0.91$) and in the cholesteric state ($T^* > 0.91$) ($T^* = T_m/T_c$; $n_{e,s}$, $n_{o,s}$, $n_{e,n*}$, $n_{o,n*}$ and $n_{e,n}$, $n_{o,n}$ are the extraordinary and ordinary refractive indices of the smectic, cholesteric and nematic phase)

the small volume jump at the phase transformation cholesteric to isotropic). As the cholesteric phase is a twisted nematic phase, the refractive indices of the corresponding untwisted nematic phase can be calculated [62] from

$$n_{o,n} = n_{e,n*} \tag{13a}$$

$$n_{e,n} = (2n_{o,n*} - n^2_{e,n*})^{1/2} \tag{13b}$$

In Fig. 22 the calculated refractive indices $n_{e,n}$, $n_{o,n}$ of the corresponding untwisted nematic phase of the cholesteric copolymer are also indicated. These calculated refractive indices are the basis for the calculation of the order parameter S (refer to 2.3.1.4). At $T^* < 0.91$ the copolymer becomes smectic, which will be discussed in Chap. 2.3.3. Owing to the helical structure, the cholesteric phase exhibits an unusual optical rotation and a circular dichroism because of a selective reflection of circularly polarized light. This is schematically shown in Fig. 23. If a homogeneous orientated cholesteric phase (Grandjean-texture) is radiated with light parallel to the optical axis, a selective reflection of light having the wavelength λ_R is observed. The reflected light (50% of the initial intensity) is right-handed circularly polarized (rcp) if the cholesteric phase forms a d-helix ($p > 0$) and the passed light (50%) is lcp. The reverse holds for a cholesteric phase with l-helix. The wavelength of reflection is related to the pitch p of the cholesteric phase by [84]

$$\lambda_R = \bar{n}p \tag{14}$$

where $\bar{n}$ is the average refractive index of the cholesteric phase $\bar{n} = (n_{e,n} + n_{o,n})/2$ ($n_{e,n}$, $n_{o,n}$ are the refractive indices of the corresponding untwisted nematic phase [84]).

The selective reflection of circularly polarized light on radiation of normal light is also exhibited by the cholesteric polymers. Like the l-l.c. systems, the Grandjean-texture is formed spontaneously, if the polymer is sandwiched between glass plates as shown in Fig. 23. The measurements of λ_R indicate no difference to low molar mass systems. In Fig. 24 $\lambda_R^{-1}(T)$ is shown for the induced cholesteric polymers, whose birefringence was discussed above (refer to Table 8, copolymers No. 4, m = 6). The different curves refer to different mole fractions of the chiral comonomer. With

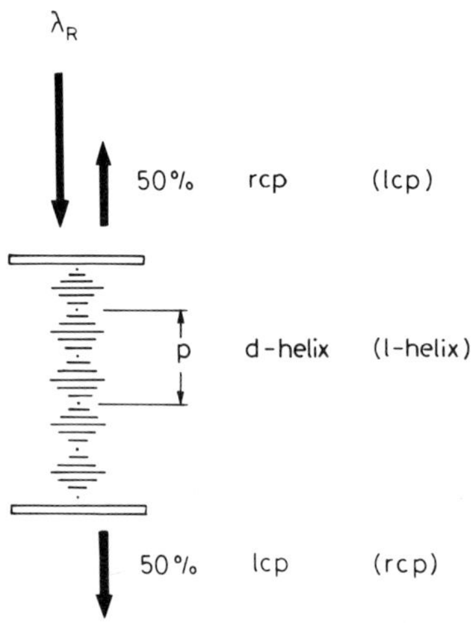

Fig. 23. Circular dichroism of a cholesteric phase having a d-helix (l-helix) owing to the selective reflection of circularly polarized light of the wavelength λ_R. For a cholesteric phase having a d-helix, 50% of the initial light is reflected, being right handed circularly polarized (rcp). 50% passes the sample, being left handed circularly polarized (lcp). For a cholesteric phase having a l-helix, lcp light is reflected and rcp light passes the sample

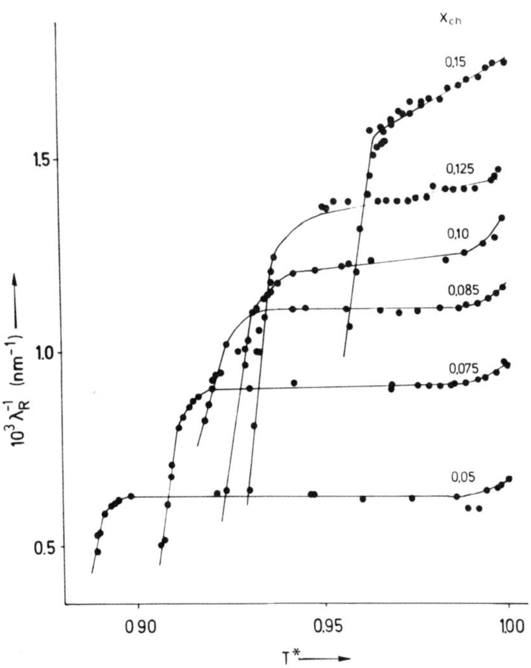

Fig. 24. Inverse wavelength of reflected circularly polarized light λ_R^{-1} vs. the reduced temperature T* for copolymers No. 4 of Table 8 at different mole fractions x_{ch} of the chiral comonomer

increasing amount of the chiral component the cholesteric twist increases, which is reflected in an increase of $^1/\lambda_R$ (falling λ_R). Looking at one curve, at high temperatures only a slight change of λ_R is observed with temperature, whereas at lower temperatures the curve diverges to $\lambda_R \to \infty$. ($1/\lambda_R \to 0$) indicating an untwisting of the helical pitch. This pretransformational effect is well known for conventional cholesterics and is observed near the phase transformation cholesteric to smectic. It is applied in thermotopography for temperature indications. While for l-l.c.'s a change in the temperature

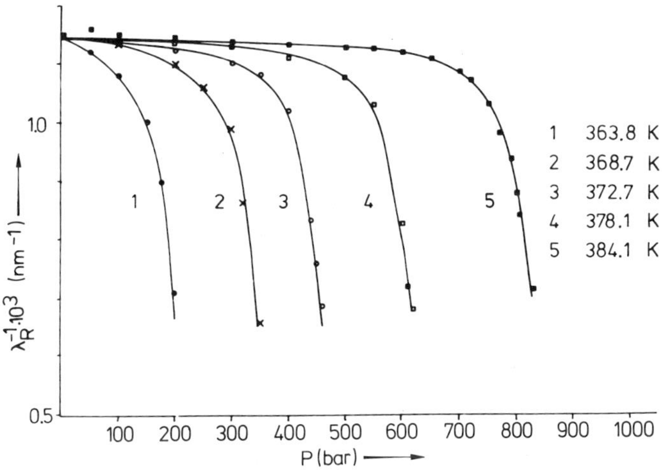

Fig. 25. Inverse wavelength of reflection λ_R^{-1} vs. pressure at different temperatures for copolymer No. 4 of Table 8 (x_{ch} = 0.17)

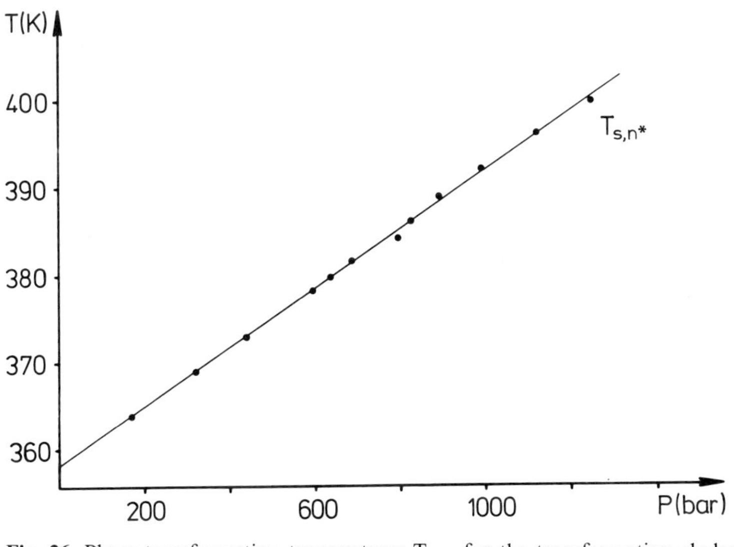

Fig. 26. Phase transformation temperatures T_{s,n^*} for the transformation cholesteric to smectic as function of the pressure

near $T_{s, n*}$ is immediately indicated by the change in λ_R, for the polymer λ_R very slowly adapts to the corresponding temperature. This indicates, that the decrease of λ_R with falling temperature is not exclusively due to an untwisting of the helical pitch. An untwisting of the helical pitch would not require a rearrangement of the polymer backbone, because of the very small changes of the twist angles of the mesogenic side chains. The strong time dependence rather indicates that near the phase transformation cholesteric to smectic the divergence of λ_R^{-1} with falling temperature also involves a rearrangement of the polymer backbone. This can be understood by a formation of smectic clusters, which enlarge the pitch and which, on the other hand, change the conformation of the backbone due to symmetry effects [82, 85] (see Chap. 2.3.3).

The divergence of λ_R in the pretransitional region, by lowering the temperature towards the phase transformation temperature $T_{s, n*}$, can also be observed at constant T by varying the pressure. This effect, well known from low molar mass cholesterics, is also observed for the cholesteric polymers. In Fig. 25 λ_R^{-1} of the copolymer *No 4*, Table 8, is plotted versus pressure for different temperatures. With increasing temperature, the divergence of λ_R is shifted towards higher pressure, due to the shift of the phase transformation. Extrapolating these curves to $1/\lambda_R \to 0$, the phase transformation cholesteric to smectic can be determined. With this optical method, the pressure dependence of the phase transformation cholesteric to smectic can be easily obtained and is shown in Fig. 26 giving a straight line of $T_{s, n*}$ versus pressure.

2.3.2.3 State of Order of Cholesteric Polymers

As the cholesteric phase is a twisted nematic phase, the orientational long range order of the mesogenic molecules, characterized by Eq. (3):

$$S = \frac{3}{2}\left(\cos^2\theta - \frac{1}{3}\right)$$

is equivalent to the corresponding untwisted nematic phase. The change of S comparing the monomer and polymer phase, and the influence of the flexible spacer

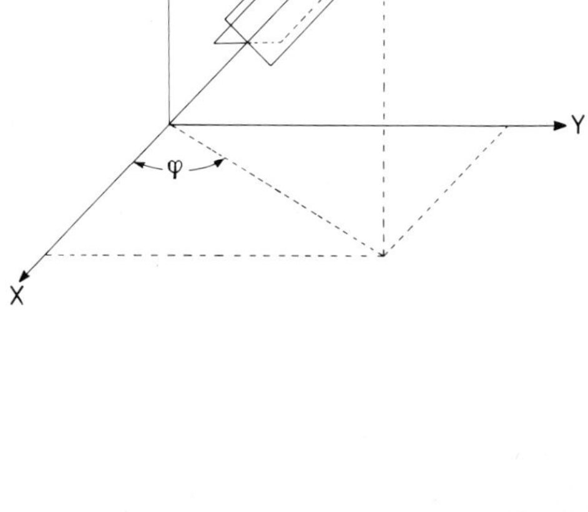

Fig. 27. Orientation of the molecular z axis. φ denotes a rotation around the ordinate Z and Ψ denotes a rotation of the molecule around its long molecular axis z

on S is identical for cholesteric and nematic phases and has already been discussed in 2.3.1.4.

While nematic phases can be described in the simplest case by consisting of uniaxially cylindrical molecules, for cholesteric phases at least a molecular biaxiality has to be considered, which quite naturally follows from the chirality of the molecules. This assumption is also the basis for the theoretical explanation of the twist of cholesteric phases, which has been given by Goossens and Vertogen [79,86]. Following these theories, the helical twist not only depends on S, which characterizes the parallel alignment of the long molecular axis z with respect to the director, but also on order parameters, which consider the rotation around any axis perpendicular to z (see Fig. 27). The cholesteric phase is then described by the following set of order parameters (average values) [79]

$$S = F_1 = \frac{3}{2}\left\langle \cos^2\theta - \frac{1}{3}\right\rangle \tag{15a}$$

$$D = F_2 = \langle \cos 2\psi \sin^2\theta \rangle \tag{15b}$$

$$F_3 = \langle \cos 2\varphi \sin^2\theta \rangle \tag{15c}$$

$$F_4 = \frac{1}{2}\langle \cos 2\varphi \cos 2\psi (1 + \cos^2\theta)\rangle \tag{15d}$$

which accounts for biaxiality on a molecular scale as well as on a phase biaxiality [87]. When only molecular biaxiality is considered, e.g. a hindered rotation of the molecules around the long molecular z-axis [86], F_3 and F_4 are zero and the cholesteric pitch, apart from molecular properties, only depends on the order parameter S and D of Eq. (15).

These theories are of importance with respect to the understanding of the helical pitch of cholesteric polymers. Our experiments have indicated, that the order parameter S is not influenced, when the rigid mesogenic moieties are linked via flexible spacers of different lengths to the polymer main chain. On the other hand, if we concentrate only on the rotation of the mesogenic molecules around the long molecular axis, there should be a strong difference between the monomers and the side chain polymers of different spacer length with respect to the order parameter D. While in the monomeric phase the motions of a molecule are only restricted by the (isotropic and) anisotropic interactions with the neighbouring molecules, for the polymers the rotation of the molecule around its long molecular axis is additionally determined by its covalent linkage to the polymer backbone. Steric effects will at least hinder this rotation, which will be alleviated with an increasing length of the flexible spacer.

Actually, these considerations are confirmed by experiments [82]. The systems investigated are shown in Table 8, No. 4, which are induced cholesteric polymer systems. The nematogenic host molecules of benzoic acid phenyl esters are linked via spacers of different length m (m = 3, 4, 5, 6) to the polymer backbone. The polymers are converted to polymers having a cholesteric phase by the chiral cholesteryl derivative, which is also linked to the polymer backbone (copolymer).

If we compare the different copolymers, they only differ in the length of the alkyl-chain of the flexible spacer, whereas the chemical constitution of the mesogenic moieties and of the polymer backbone remains unchanged (average degree of polymerization $\bar{r} = 95$). Therefore we can assume that the molecular properties, like polarizabilities, which could change the cholesteric pitch, also remain unchanged. In Fig. 28 the inverse wavelength of reflection $1/\lambda_R$ which is directly correlated to the pitch p by Eq. (14), is plotted versus the mole fraction of the chiral comonomer (the concentration is based on the monomer units). For $x_{ch} = 0$ ($1/\lambda_R$ is zero) the phase is nematic. With increasing concentration of the chiral comonomer the cholesteric twist and therefore $1/\lambda_R$ increases. If we refer to one defined concentration of the chiral derivative and compare the induced twist for the systems that only differ in the spacer length m, a strong difference in the induced twist is observed. The system having the shortest spacer m = 3 exhibits the strongest, the system with the longest spacer m = 6 exhibits the weakest induced twist. The lowest twist is observed for the corresponding l-l.c. mixture. These strong differences in the induced twist *cannot* be explained by any change of chemical constitution and can only be explained by the change in the order parameter D of Eq. (15b). With increasing length of the flexible spacer the hindered rotation of the mesogenic moieties around their long molecular axes is alleviated, causing a decrease of twist, which tends to the twist of the corresponding monomer system.

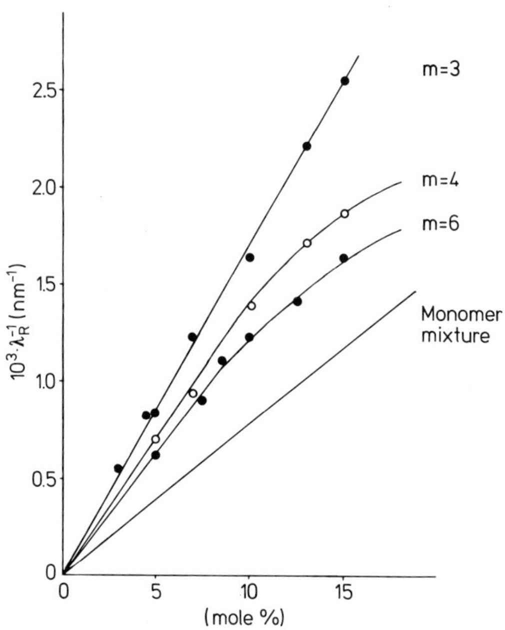

Fig. 28. Inverse wavelength λ_R^{-1} versus concentration in mole % of the chiral comonomer for induced cholesteric phases No. 4 of Table 8 at T* = 0.95.

Monomer mixture 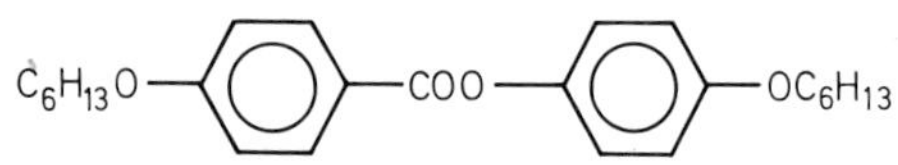

and $CH_2{=}CH{-}CH_2{-}COO$-cholesteryl

A second experiment supports these considerations [88]. For the induced cholesteric system No. 5 in Table 8 the chiral center of the chiral compound is not within the mesogenic moiety, like in case of the cholesteryl derivative No. 4 in Table 8, but is linked via a flexible chain to the rigid mesogenic moiety. Consequently a linkage of the mesogenic moiety to the polymer backbone does no longer effect rotational motions of the chiral center. Actually this is proved by the experiments, which are shown in Fig. 29. *No* difference is observed for the copolymer and polymer-monomer mixture and only a small change is observed with respect to the very similar l-l.c. system, which might be due to the change in S described in 2.3.2.1.

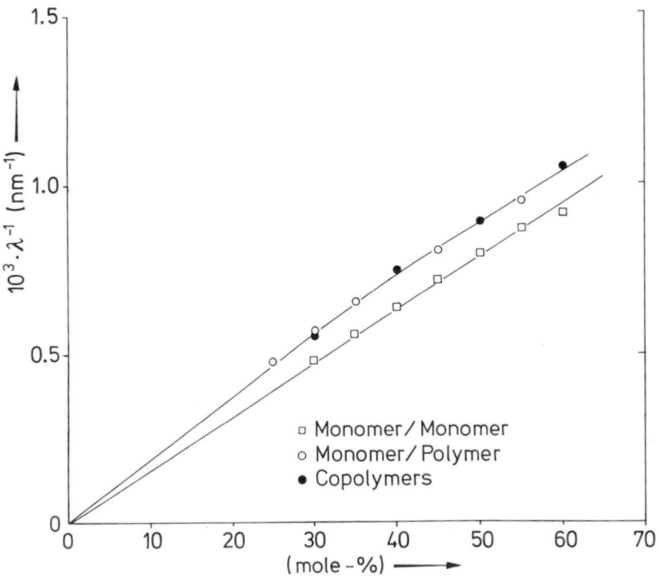

Fig. 29. Inverse wavelength of reflection λ_R^{-1} vs. concentration in mole % of the chiral component for induced cholesteric phases; ● = copolymer No. 5, Table 8; ○ = monomer-polymer mixture

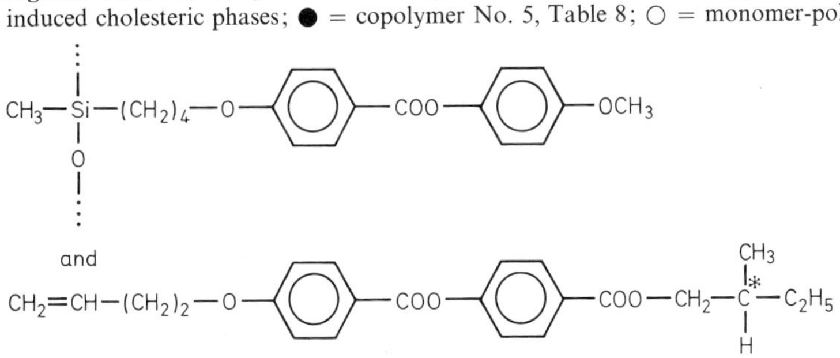

□ = monomer mixture

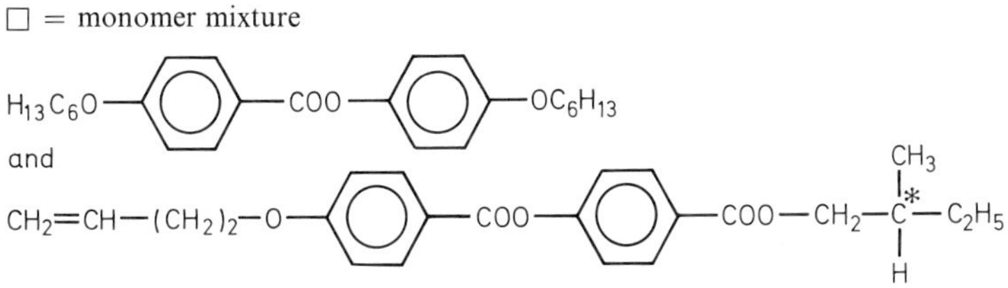

These experiments clearly confirm the theoretical concept of the helical twist of cholesteric phases developed by Goossens and Vertogen [79, 86], which is based on the hindered rotation of chiral molecules around their long molecular axis and the introduction of the order parameter D. Furthermore they support the concept of the flexible spacer, described in Chap. 2.1.

2.3.3 Smectic Polymers

In addition to the long range orientational order, in smectic phases, the molecules possess a long range positional order in one or two dimensions, causing the molecules to be arranged in planes with respect to their centers of gravity. Furthermore the molecules can be orthogonal or tilted with regard to their long molecular axis to the planes, causing the manifold polymorphism of smectic phases [3–6].

While for nematic polymers the statistical distribution of the centers of gravity of the mesogenic side chains is compatible with a more or less statistical main chain conformation, for smectic polymers a three dimensional coil conformation is no longer consistent with the layered structure of the mesogenic side chains. The backbone has to be restricted in its conformation, which will cause a more pronounced interaction between the main chain and the anisotropically ordered mesogenic side chains, compared to nematic and cholesteric polymers.

2.3.3.1 Chemical Constitution

Numerous experiments on homologous series of low molar mass l.c.'s have confirmed the principle that with increasing length of the substituents of the rigid mesogenic core, a smectic phase appears [3–6]. This is schematically presented in Fig. 30. Here it is presumed that within a homologous series the derivatives with short substituents of the length a and b are nematic. Increasing length of a and b favours the existence of a low temperature smectic phase and above a certain length only smectic l.c's appear. These experimental observations have been theoretically explained by McMillan [89] and Kobayashi [90] introducing an additional order parameter, which describes the arangement of mesogenic molecules within planes. Following their theories, the substituents of the rigid mesogenic groups act as spacers that keep the rigid cores within planes and which therefore determine the distance of the smectic layers.

In l.c. side chain polymers the rigid mesogenic groups normally also carry two substituents at their ends, where one substituent connects the mesogenic group via the flexible spacer to one monomer unit of the polymer backbone in case of homopolymers. The analysis of the phase behavior of homologous series have proved, that for polymers the same principles are valid as for the monomeric l.c.'s: Increasing length of the substituents of the mesogenic moieties favour the existence of smectic phases. For the polymers, however, we can differentiate between three different ways of lengthening the substituents (refer to Fig. 30):

i) the "free" substituent A,
ii) the spacer B and
iii) the length of the main chain segments C that connect the flexible spacers.

In Table 9a the phase transitions of two arbitrarily selected systems are compared, where the length of the "free" substituent A is varied. For the poly(methacrylate) No. 1 and the poly(siloxane) No. 3 nematic phases are observed which become smectic

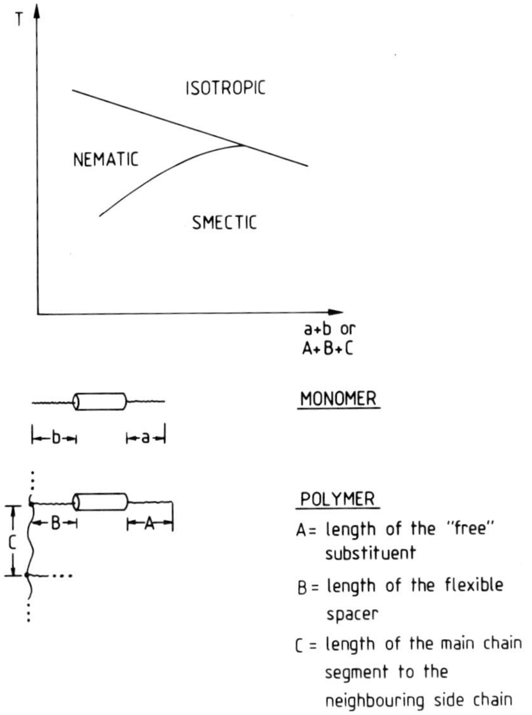

Fig. 30. Schematic phase behavior of l-l.c.'s and p-l.c.'s depending on the chain length a, b and A, B, C respectively of the substituents of the rigid mesogenic core

for polymer No. 2 and No. 4 having long substituents. Accordingly the typical low enthalpies of the phase transformation of the nematic polymers become larger for the smectic systems, which is well known from low molar mass l.c.'s. The same change from nematic to smectic can be observed when the length of the spacer is increased, as shown for two different systems in Table 9b. Polymers No. 1 and 2 have been synthesized by Zentel [91].

The third way to change the l.c. phase from nematic to smectic is the lengthening of the main chain segments C, which connect the side chains. For polymer No. 1 in Table 9c each monomer unit carries one mesogenic group, while for polymer No. 2 on average two monomer units have one mesogenic side chain. Polymer No. 1 is nematic, polymer No. 2 is smectic. This proves that the polymer main chain segments also act as substituents of the mesogenic group. Experiments on different systems confirm these results [92, 30].

With these three different examples it has been demonstrated that the systematics observed for the polymorphism of m-l.c.'s is also valid for the side chain polymers, provided that a flexible spacer connects the rigid mesogenic moieties to the polymer main chain. Deviations from this behavior are observed, when the mesogenic moieties are directly linked to the backbone. Under these conditions, normally no liquid crystalline behavior is to be expected, according to the model considerations mentioned in Chap. 2.1. Some examples, however, proved l.c. properties for such systems, which are characterized by two striking properties: Very high glass transition temperatures and only smectic structures even in case of short substituents

Table 9. Examples of smectic polymers by lengthening the segments A **(9a)**, B **(9b)** and C **(9c)** (for A, B and C refer to Fig. 30)

a)

NO	POLYMER	A	PHASE TRANSITIONS (K)	$\Delta\tilde{H}_{LC-i}$ (J/g)
1	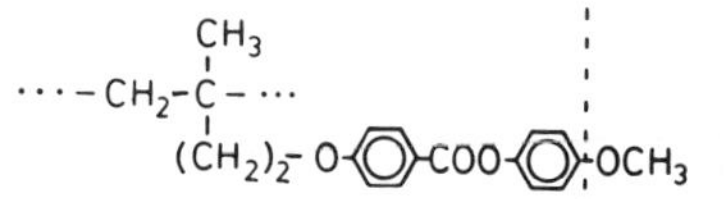	OCH_3	g 369 n 394 i	2.3
2		OC_6H_{13}	g 410 s 451 i	11.5
3	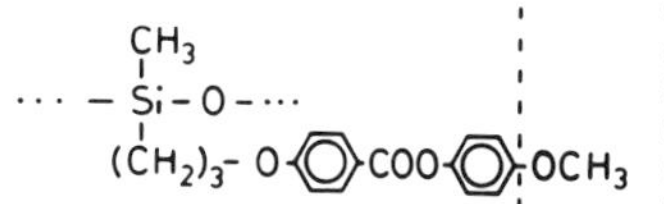	OCH_3	g 288 n 334 i	2.2
4		OC_6H_{13}	g 288 s 385 i	11.6

b)

NO	POLYMER	B	PHASE TRANSITIONS (K)	$\Delta\tilde{H}_{LC-i}$ (J/g)
1	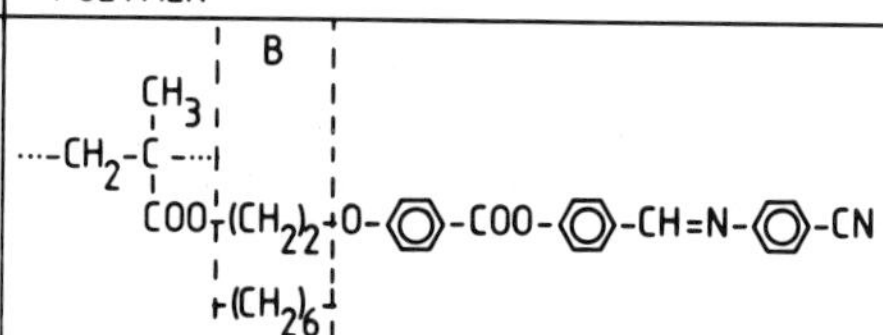	$(CH_2)_2$	g 361 n 580 i	–
2		$(CH_2)_6$	g 324 s 607 i	–
3	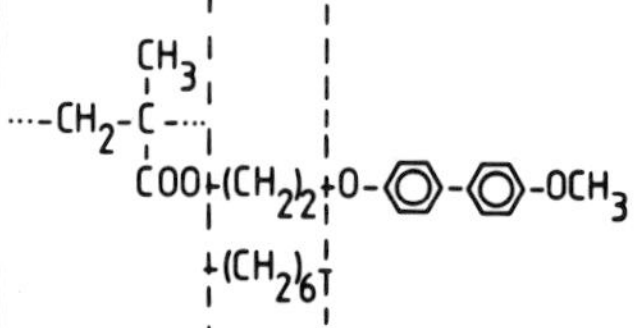	$(CH_2)_2$	g 393 n 425 i	2,8
4		$(CH_2)_6$	t_1 119 s 136 i	7,1

c)

NO	POLYMER	PHASE TRANSITIONS (K)
1	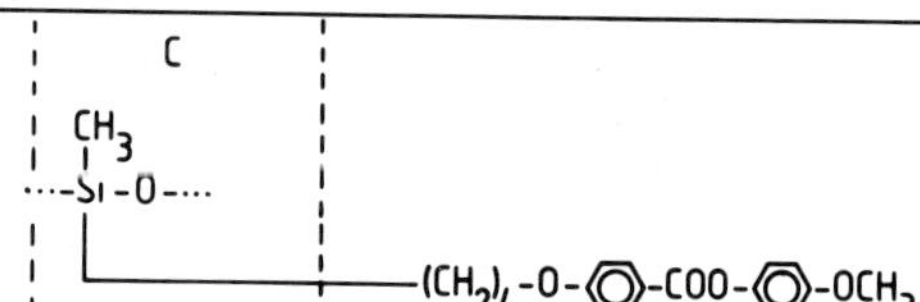	g 288 n 368
2	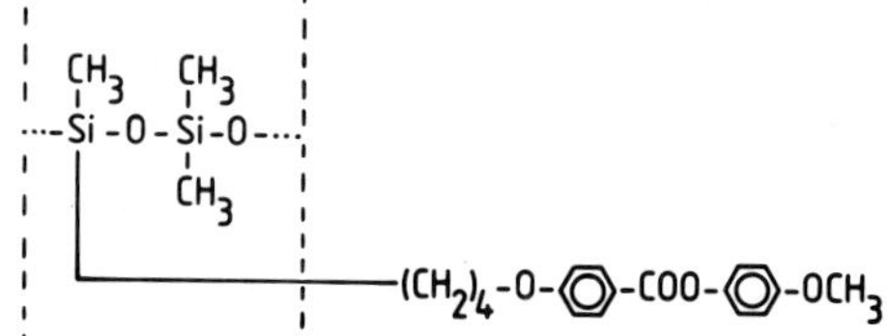	g 267 s 323

A. These anomalies can be understood by the restriction of packing of the mesogenic moieties, due to their direct linkage to the backbone. The rigid cores of the mesogenic groups stiffen the main chain, which becomes obvious because of the high glass transition temperatures. A nematic arrangement is not possible because of steric restrictions and therefore a smectic ordering remains.

2.3.3.2 Structure of Smectic Polymers

Numerous X-ray investigations on smectic l.c. side chain polymers have confirmed that the polymers exhibit a polymorphism similar to l-l.c.'s. For the packing of the mesogenic side chains, in principle the same arrangements have been proposed as known from monomeric systems. In most cases, however, only the packing of the mesogenic moieties has been considered, without making any definite proposals about the conformation of the polymer backbone. The first model of smectic polymers was proposed by Strzelecki [93], where the polymer backbone forms a two dimensional coil sandwiched between the smectic layers of the mesogenic side chains. Recently Mügge and Zugenmaier [94, 95] have deduced a model for smectic polymers from X-ray investigations of polymers in the crystalline and smectic state, which is schematically demonstrated in Fig. 31. Following their considerations, the polymer backbone forms a more or less ordered structure, from which the mesogenic groups stick out at an angle of 90 degree or near 90 degree, depending on packing effects and chemical constitution of the side chain. The macromolecules (backbone and mesogenic side chains) form elliptical bodies as schematically represented in Fig. 31, which can be shifted with respect to their long axes and which are not or only weakly correlated. It has to be emphasized that in contrast to the model of Strzelecki it is assumed that the backbone forms not twodimensional coil but an ordered main chain conformation.

In any case, both models have in common that owing to the positional ordering of the mesogenic side chains, the polymer backbone no longer exhibits a statistical three dimensional coil conformation. Therefore at the phase transformation isotropic to smectic or nematic (cholesteric) to smectic, in addition to the change of the anisotropic packing of the side chains, the main chain has to change its conformation, which must be consistent with the layered smectic structure. A direct interaction

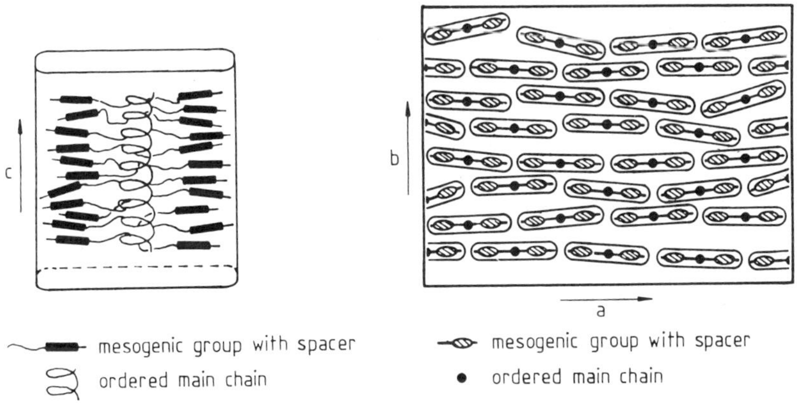

Fig. 31. Schematic model of the smectic structure of these liquid crystalline side chain polymers

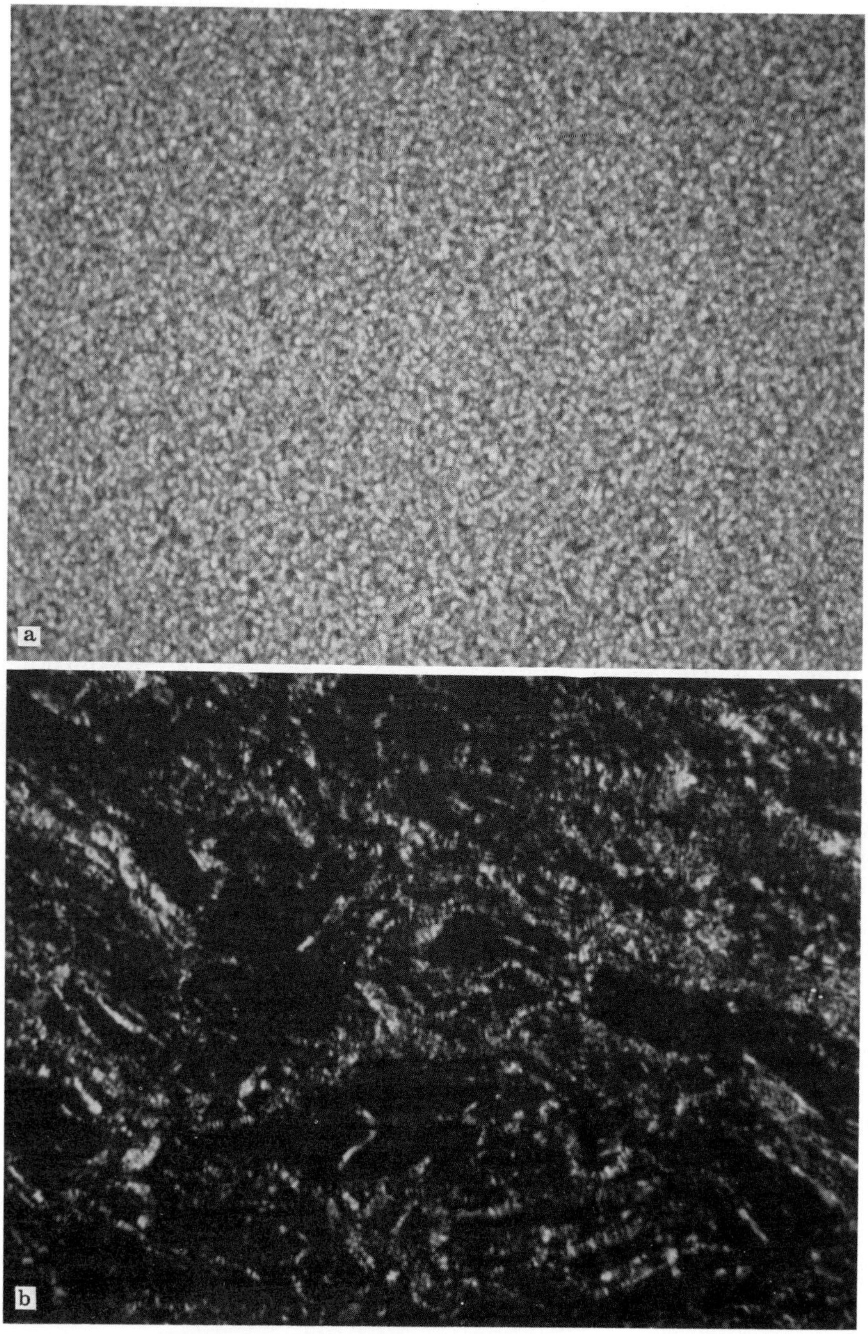

Fig. 32a. Micrograph of a freshly prepared smectic polymer (magnification 40); **b** S_A-texture of copolymer No. 4 of Table 8 (magnification 40)

of main chain and side chain results. Actually this is indicated by all polarizing microscopic investigations on smectic polymers. While for nematic and cholesteric polymers nearly always the typical textures can be observed after preparation of the material, for smectic polymers normally strongly disturbed textures are to be seen. A typical example is shown in Fig. 32a. Only after very long annealing periods, or in case of investigations at temperatures far above the glass transition of the polymer, smectic textures are observed [40, 96]. In Fig. 32b the homeotropic texture of the S_A polymer No. 4, Table 8, is shown, which could be observed after an annealing time of about one week. This indicates that in contrast to nematic systems the polymer backbone strongly influences the formation of macroscopic orientations, owing to its reorganization in the smectic state. The strong influence of the backbone on the formation of smectic clusters in cholesteric phases near the transformation cholesteric to smectic was already discussed in 2.3.2.2. and confirms these considerations. Detailed X-ray investigations on the packing of the side chains in smectic polymers are reviewed by Shibaev and Platé (this issue).

2.3.3.3 *Optical Properties*

Because of the difficulties in obtaining macroscopically oriented smectic polymers, up to now only little is known about the optical properties of smectic polymers. As mentioned above, polymer No. 4, Table 8, forms a homeotropic S_A texture, where the mesogenic groups are ordered orthogonally within planes. Measurements of the birefringence of this polymer [57] prove the optical uniaxial positive behavior with $n_{e,s} > n_{o,s}$. The temperature dependence of the refractive indices is shown in Fig. 22. At high temperatures the uniaxial negative cholesteric phase is observed for this polymer, which is discontinuously transformed into the uniaxial positive smectic A phase at $T^* = 0.91$. Under the condition of homeotropic orientation the values of birefringence refer to the principal polarizabilities with respect to the principal axes

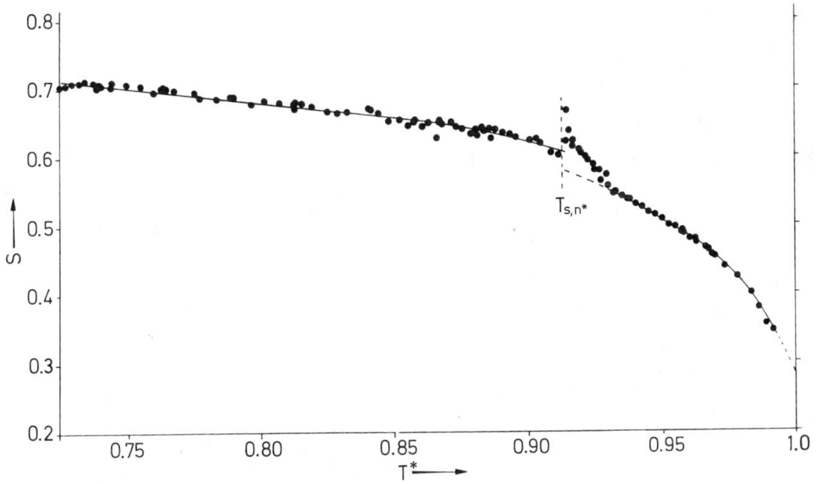

Fig. 33. The temperature dependence of the orientational order parameter S of the copolymer No. 4 of Table 8 in the cholesteric and smectic state. ($T^* = T_c/T_m$; T_c = clearing temperature, T_m = measuring temperature)

of the mesogenic side chains. The absolute value $\Delta n = n_{e,s} - n_{o,s}$ is of the same order as for l-l.c. S_A phases of similar chemical constitution [65]. Using the same method as described for nematic phases, we can roughly estimate the order parameter of the orientational order of the mesogenic groups. It has to be mentioned, however that using this method, the relation of Eq. (7) only correlates the refractive indices with molecular polarizabilities in the nematic state and need not necessarily be valid for smectics. The temperature dependence of S is shown in Fig. 33 for this system. At thc phase transformation cholesteric to smectic a small jump in S of about 5% indicates the first order transformation. In the smectic state, the absolute value of S as well as the temperature dependence of S is very similar to low molar mass S_A-phases.

From these measurements we can conclude, that at least for S_A-polymer phases the mesogenic groups exhibit the same packing as known for conventional thermotropic l.c.'s. But one must keep in mind that in this case the polymer backbone must participate in the anisotropic packing.

2.3.4 Glassy Polymers Having a Liquid Crystalline Structure

2.3.4.1 Phase Behavior

As already mentioned in Chap. 2.2. one of the most obvious features of the l.c. side chain polymers is their ability to become glassy. The glass transition can be observed by cooling nematic, cholesteric and smectic polymers depending on the chemical constitution of the system and is indicated e.g. by a bend in the V(T) curves as schematically shown in Fig. 8. Two questions are of interest which will be discussed in this chapter:

i) is the glass transition a transformation of second order and
ii) what is the influence of the glass transition on the l.c. phase behavior.

Whether the glass transition in case of liquid crystalline polymers can be described as a transformation of the second order has been generally checked for nematic and smectic side chain polymers [36-39] and will be illustrated for the polymethacrylate in Fig. 34. Here isobaric volume-temperature curves at different pressures are shown. The polymer has two first order transformations from isotropic to nematic and from nematic to smectic. The smectic state changes into the glassy state, which is indicated by the bend in the V(T) curves. In case of a second order transformation Eqs. (2a) and (2b)

$$(dP/dT)_{tr} = \Delta\alpha^*/\Delta\varkappa^* \tag{2a}$$

$$(dP/dT)_{tr} = \Delta c_p^*/T\,\Delta\alpha^* \tag{2b}$$

must hold. To check, whether these equations are applicable, the step values of the thermal specific expansion coefficient $\Delta\alpha^*$ and the isothermal specific compressibility $\Delta\varkappa^*$ have been determined as a function of pressure [37]. The change of the specific heat Δc_p^* at T_g could only be measured at 1 bar by DSC. The results for the polymer are shown in Fig. 35. Usually the pressure dependence of c_p^* is very small [97-99,119] and we can assume that Δc_p^* decreases with increasing pressure in a similar way as $\Delta\alpha^*$ and $\Delta\varkappa^*$. The line for the quotient $\Delta c_p^*/T\,\Delta\alpha^*$ is obtained with a pressure independent

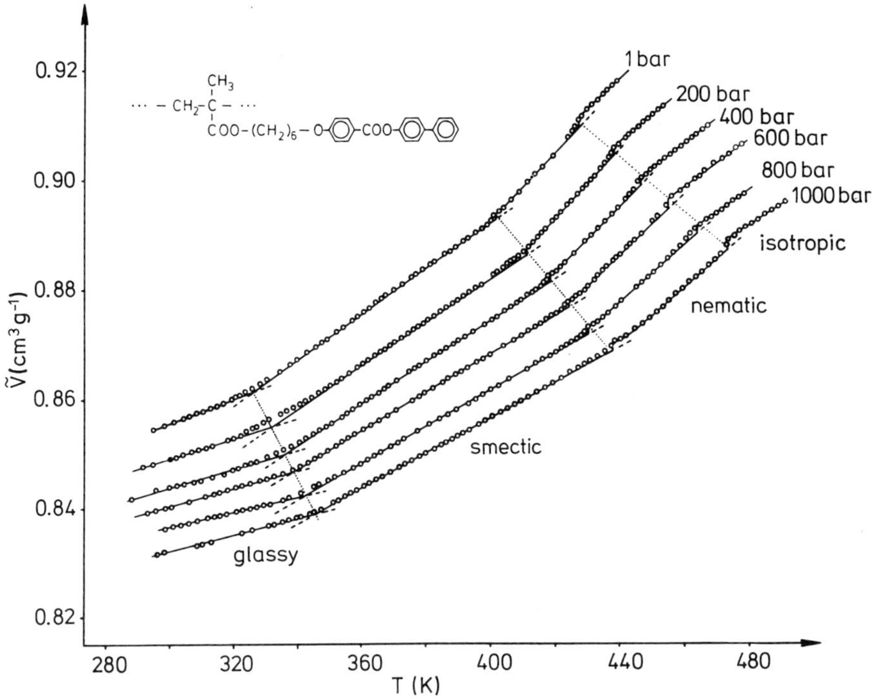

Fig. 34. Specific volume-temperature curves at different pressures for the poly(methylacrylate) (isobaric measurements)

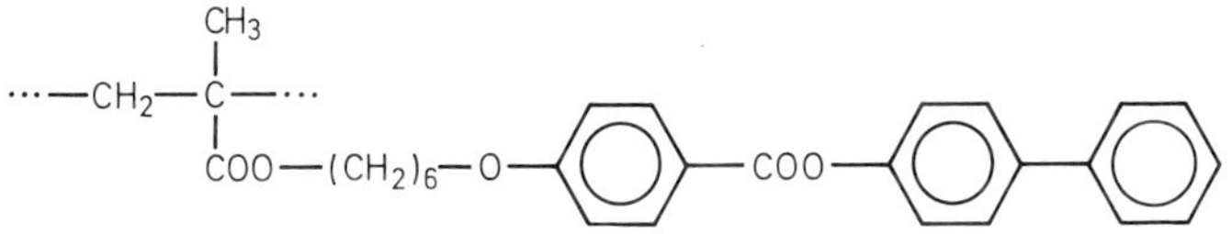

Δc_p^*. This curve is therefore the upper limit for the course of $\Delta c_p^*/T\,\Delta\alpha^*$ vs. P. The real values will be below this curve. At one bar the following relation is fulfilled:

$$(dP/dT)_{tr} \approx \Delta c_p^*/T\,\Delta\alpha^* > \Delta\alpha^*/\Delta\varkappa^*$$

The result confirms the results of other authors that at one bar, if at all, only the Ehrenfest Equation (2b) is valid within the experimental error, while the other equation is not valid. Therefore, the Ehrenfest equations are not applicable for the glass transition of the l.c. side chain polymers, because not *both* equations are verified. This means that the glass transition is not a transformation of the second order but a freezing-in process. In contrast to measurements on amorphous polystyrene [99], the value for $\Delta\alpha^*/\Delta\varkappa^*$ and $\Delta c_p^*/(T\,\Delta\alpha^*)$ differ strongly. This implies that for the description of the glassy state and the glass transition more than one internal order parameter ξ_i, which characterizes the internal order, is necessary [36]. This is not unreasonable because two different parts of the macromolecules having two different states of order freeze in together at T_g: the more or less coiled main chains and the anisotropically ordered mesogenic side chains.

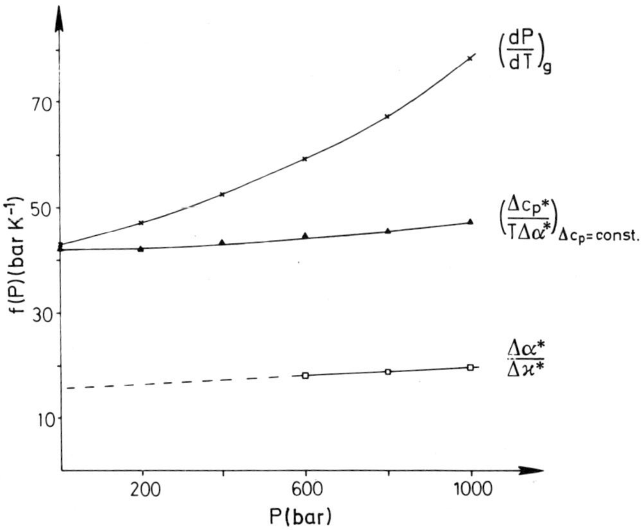

Fig. 35. Examination of the validity of the Ehrenfest equations for the glass transition of the poly(methylacrylate) shown in Fig. 34

The influence of the glass transition on l.c. phase behavior is the second question of interest. The glass transition of a polymer is — roughly described — determined by i) the primary structure of the backbone and ii) intermolecular interactions of the backbone segments. The primary structure describes the chemical constitution of the polymer main chain, which, on the other hand, directly influences its rigidity. Furthermore, the glass transition depends on the intermolecular interactions between the chain segments. For the l.c. state of the l.c. side chain polymers it is of interest to see whether the phase behavior is more affected by a change in the rigidity of the polymer main chain, or whether the intermolecular interactions between the chain segments having the mesogenic units are more important.

The influence of the glass transition on the extent of the l.c. phase, due to a change of the rigidity of the polymer backbone, is demonstrated for three examples in Table 3. All polymers have the same mesogenic side groups. These groups are linked to different polymer backbones, which differ in their flexibility. This is indicated with the glass transition temperature T_g. The poly(methycrylate) exhibits the highest T_g. T_g is lowered, if the poly(methacrylate) main chain is exchanged by the poly(acrylate), which has a more flexible main chain because of the missing methyl group. The lowest T_g is observed for the poly(siloxane). Looking at the phase transformation temperatures liquid crystalline to isotropic, $T_{n,i}$, we find out that they are strongly influenced by the change of T_g. Increasing flexibility of the main chain, which is indicated by the falling T_g, also lowers $T_{n,i}$. The same effect was observed by Wendorff et al. for l.c. poly(methacrylates) [40]. For these polymers the chemical structure of the main chain remained and only the tacticity was changed, which describes the stereochemical arrangement of the monomer units. A change in the tacticity shifts T_g, causing a shift of the phase transformation l.c. to isotropic in the same direction as well.

These examples indicate that the flexibility of the main chain, which is determined by its chemical constitution, directly influences the *extent* of the liquid crystalline state. With increasing rigidity of the main chain the motions of the mesogenic groups are more restricted. This can be noticed in an increase of the phase transformation temperature $T_{n,i}$.

As mentioned above, the flexibility of the polymer backbone is also determined by the interactions of the main chain segments. For conventional polymers, these *interactions* can be suppressed by adding low molar mass substances to the polymer. They cause a strong decrease of T_g (softening effect). Whether the softening effect also influences the liquid crystalline phase transformations can be deduced from the phase diagram Fig. 20 described in 2.3.1.6. For this mixture no severe distortion of the intermolecular interactions of the mesogenic groups can be assumed. Because of the very similar chemical structure, the enthalpy of mixing should be approximately zero or at least very small. Actually a strong shift of T_g is observed to lower temperatures with increasing amount of the low molar mass l.c., indicating the softening effect on the polymer system. However, in contrast to the behavior of the homopolymers mentioned in the last paragraph, the phase transformation liquid crystalline to isotropic does not show the strong shift as observed for the glass transition. The clearing temperature T_c depends on the composition as expected for nearly ideal mixtures of l.c. compounds of similar chemical constitution.

These examples clearly indicate that the extent of the l.c. phase of the polymer is mainly influenced by the flexibility of the polymer main chain, determined by its chemical constitution. The flexibility of the backbone also influences the mobility of the mesogenic side chains.

2.3.4.2 Optical Properties

The thermodynamic investigations have indicated that the glass transition is a freezing-in process. Consequently the anisotropic liquid crystalline orientation of the mesogenic side chain should also freeze in, yielding an anisotropic glass having a liquid crystalline *structure*. This process is of interest in view of the applicability of l.c. polymers.

A detailed insight into the freezing-in process is given by optical investigations. As described in 2.3.1.4 for nematics, 2.3.2.3 for cholesterics and 2.3.3.3 for smectics, the optical uniaxial character of the polymers in the liquid crystalline state has been proved by birefringence measurements and the state of order was calculated from these measurements. This method also provides information about the glassy state. For conventional l.c's it has been demonstrated, that the temperature dependence of $\log\left(\frac{\Delta\alpha}{\bar{\alpha}} S\right)$ yields a straight line in the l.c. state at low temperatures beginning some degrees below the phase transformation l.c. to isotropic, if plotted vs. the logarithm of the reduced temperature $(1 - T^*)$. This relation, which correlates the orientational order parameter S with the temperature, is also valid for the l.c. side chain polymers [57]. In Fig. 36, $\log\left(\frac{\Delta\alpha}{\bar{\alpha}} S\right)$ has been plotted vs. $-\log(1 - T^*)$ for a nematic polymer (Fig. 36a) and a smectic polymer (Fig. 36b). For both polymers the linear relation is obtained at high temperatures. At lower temperatures, however, a bend in the

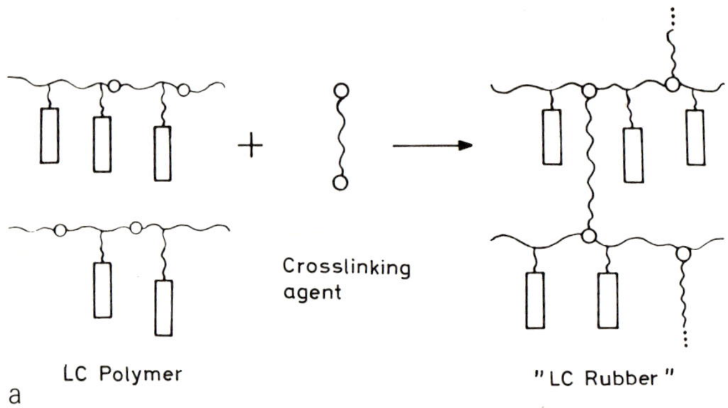

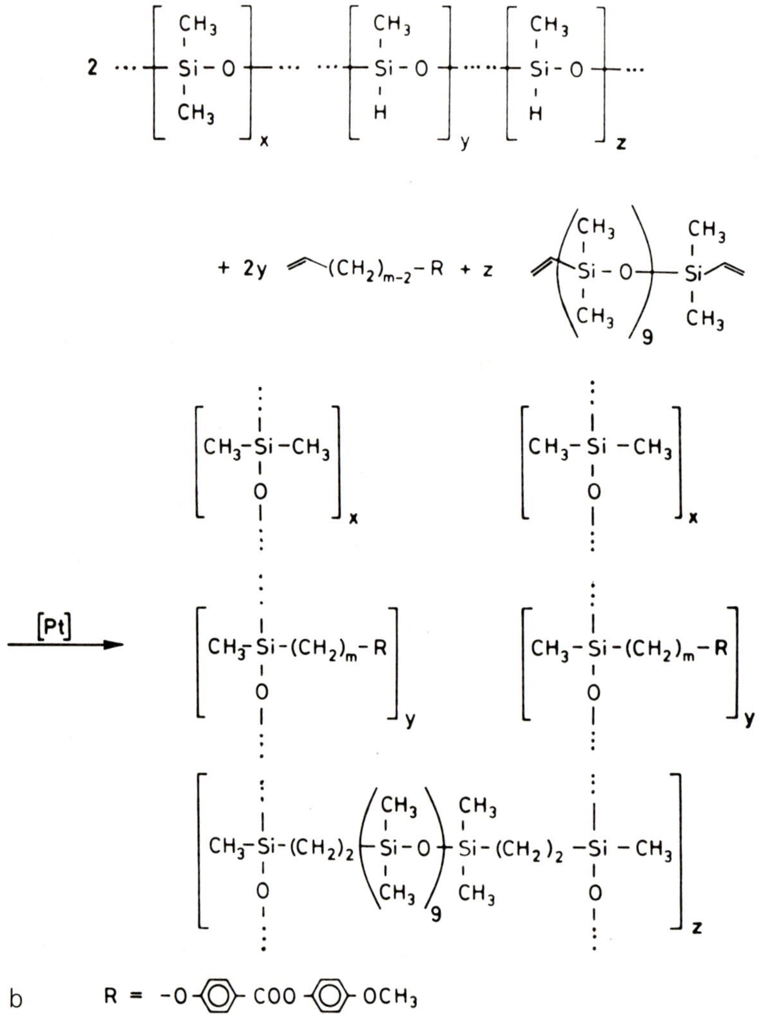

Fig. 37 a. Schematic synthesis route for l.c. elastomers; **b** Example for the synthesis of a l.c. elastomer with an additional crosslinking agent, the reactive centers have to be connected, forming the polymer network.

An example for a synthesis of a poly(siloxane) network is shown in Fig. 37b. In a one-step reaction the mesogenic moieties as well as the crosslinking agent are coupled via an addition reaction to the reactive linear poly(methylhydrogensiloxane) backbone [92]. Because of similar reactivity of the crosslinking agent and mesogenic molecules, a statistical, disordered addition to the backbone has to be expected.

The phase behavior of the systems can be obtained as discussed in detail for linear l.c. side chain polymers in Chap. 2. In Table 10 we will compare the phase

Table 10. Phase transition temperatures of linear and crosslinked l.c. side chain polymers

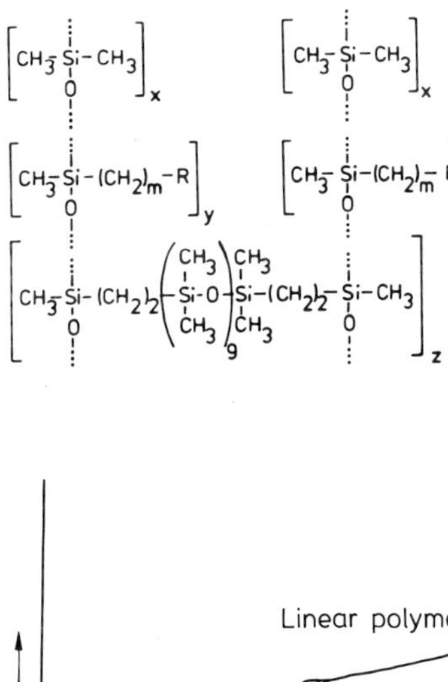

LINEAR POLYMERS

HOMOPOLYMERS	x	y	m	PHASE TRANSITION TEMPERATURES IN K
1 a	0	120	3	g 288 n 334 i
b	0	120	4	g 288 n 368 i
c	0	120	6	g 278 s 319 n 385 i
COPOLYMERS				
d	60	60	3	g 276 n 294 i
e	60	60	4	g 267 s 323 i
f	60	60	6	g 263 s 350 i

CROSSLINKED POLYMERS

HOMOPOLYMER	x	y	z	m	PHASE TRANSITION TEMPERATURES IN K
2 a	0	108	12	3	g 273 n 335 i
COPOLYMERS					
d	60	48	12	3	g 263 n 283 i
e	60	48	12	4	g 258 s 305 i
f	60	48	12	6	g 253 s 332 i

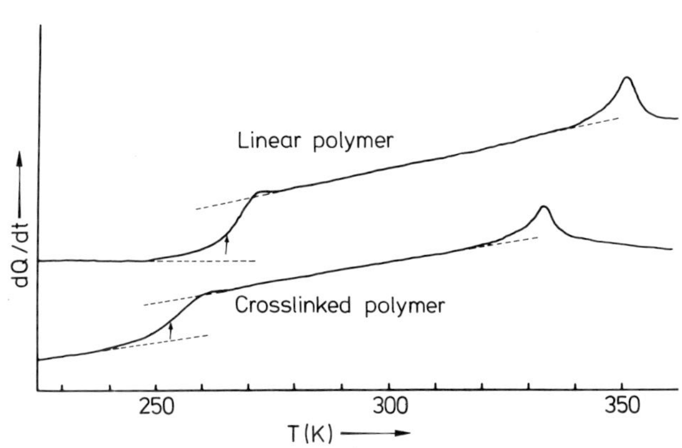

Fig. 38. DSC traces of the linear and crosslinked smectic polymer No. 1f and No. 2f of Table 10

transition temperatures of some arbitrarily chosen linear and crosslinked polymers. The linear polymers exhibit nematic and smectic phases depending on the length of the flexible spacer. Their phase transitions can be easily detected by DSC and the DSC-trace of the linear, smectic polymer No. 1 is shown in Fig. 38. The analogous cross-linked polymer, containing the flexible oligo(dimethylsiloxane) crosslinking agent, also shows the l.c. state: If we compare the DSC curve of the elastomer, which corresponds to the linear, smectic polymer No. 2f nearly the same DSC curve as in case of the linear polymer is obtained. Owing to the plasticizing effect of the highly flexible crosslinking agent, only the glass transition temperature and the phase transformation temperature isotropic to l.c. are shifted towards lower temperatures, but the extent of the l.c. phase remains constant. Additional X-ray analysis and optical investigations prove the l.c. state and therefore we can attribute the phase transitions listed in Table 10 to the polymer networks.

These examples prove that the considerations with respect to the realization of l.c. elastomers, as mentioned above, are justified. Therefore the crosslinking of linear l.c. side chain polymers provides a method to realize a new type of liquid crystalline material: form-retaining l.c. elastomers, that combine l.c. properties with rubber elasticity.

It has to be noted that this method is limited by the extent of the crosslinking density. As already mentioned, at the crosslinks the mobility of the chain segments is reduced and consequently also the mobility of the mesogenic groups in the neighbourhood of a crosslink. Therefore, with regard to linear polymers, a crosslink can be understood as a defect in the l.c. structure. Increasing crosslinking density consequently produces an increasing number of defects. Actually a high crosslinking density destroys the l.c. state. Duroplast [23] cannot be mate liquid crystalline.

2.4.2 Dynamic Mechanical and Optical Properties

Dynamic mechanical experiments, where the material is periodically strained, are common methods to characterize the visco-elastic behavior of elastomers by measuring the storage modulus G′ and loss modulus G″. G′ is a measure for the maximal, reversibly stored energy for a periodical deformation and G″ is proportional to the dissipated energy for the oscillation cycle. It is obvious to investigate, whether the l.c. state of the l.c. elastomers influences the dynamic mechanical properties and whether different modes of linking the mesogenic moieties to the backbone can be detected.

We will discuss some preliminary results, which have been performed recently [101]. In Fig. 39a the results for polymer No. 2d of Table 10 are shown, which were obtained by torsional vibration experiments. At low temperatures the step in the G′(T) curve and the maximum in the G″(T) curve indicate a β-relaxation process at about 120–130 K. Accordingly the glass transition is detected at about 260 K. At 277 K the nematic elastomer becomes isotropic. This phase transformation can be seen only by a very small step in G′ and G″ in the tail of glass transition region, which is shown in more detail in Fig. 39b. From these measurements we can conclude, that the visco-elastic properties are largely dominated by the properties of the polymer backbone: the change of the mesogenic side chains from isotropic to liquid crystalline acts only as a small disturbance and in principle the visco-elastic behavior of the elastomer

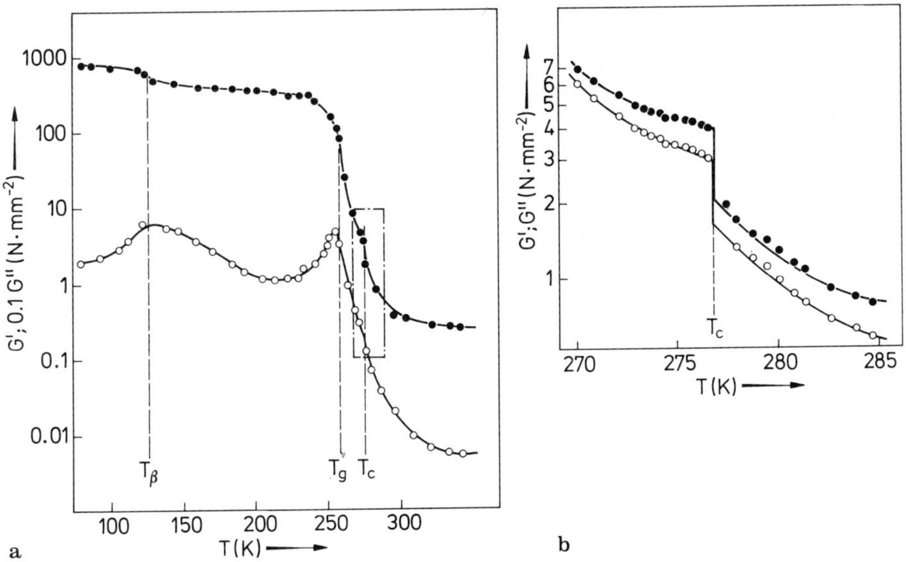

Fig. 39a. Temperature dependence of the storage modulus G′ (filled circles) and loss modulus G″ (Open Circles), measured at 0.5 Hz, for the l.c. elastomer No. 2d in Table 10; **b.** Detailed section of Fig. 39a

is only determined by the crosslinked siloxane backbone. This view is supported by the magnitude of G′ and G″, which are of the usual order found for common polymers, in the glassy state and in the rubbery region as well. On the other hand these results are consistent with the model considerations of Chap. 2.1, where we have assumed that the l.c. order of the side chains is more ore less independent of the conformation of the backbone and vice versa. Contrariwise, for l.c. *main-chain* elastomers strong effects of the l.c. order on the properties are expected, which are already theoretically predicted by de Gennes [102] and R. B. Meyer [103].

From the dynamic mechanical investigations we have derived a discontinuous jump of G′ and G″ at the phase transformation isotropic to l.c. Additional information about the mechanical properties of the elastomers can be obtained by measurements of the retractive force of a strained sample. In Fig. 40 the retractive force divided by the cross-sectional area of the unstrained sample at the corresponding temperature, σ^0 is measured at constant length of the sample as function of temperature. In the upper temperature range, $T > T_c$ (T_c is indicated by the dashed line), the typical behavior of rubbers is observed, where the (nominal) stress depends linearly on temperature. Because of the small elongation of the sample, however, a decrease of σ^0 with increasing temperature is observed for $\lambda < 1.1$. This indicates that the thermal expansion of the material predominates the retractive force due to entropy elasticity[a]. For $\lambda \cong 1.1$ the nominal stress σ^0 is independent on T, which is the so-called thermo-elastic inversion point. In contrast to this "normal" behavior of the l.c. elastomer

---

a) $\lambda = \frac{l_0}{l}$; l_0 unloaded sample length; l stretched sample length

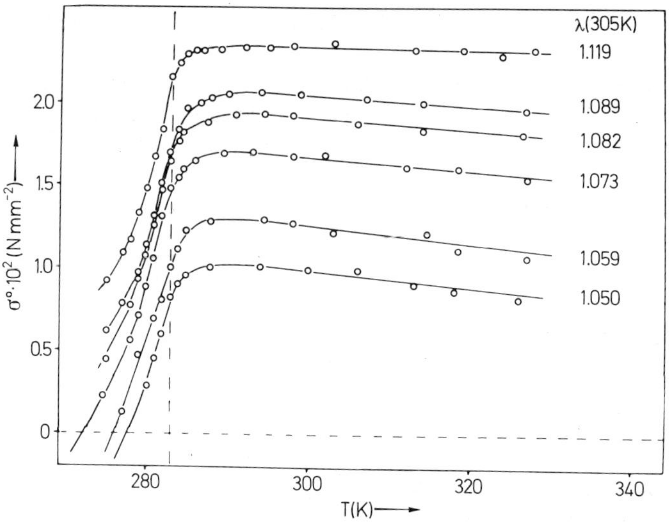

Fig. 40. Temperature dependence of the nominal stress at constant sample length l for the l.c. elastomer No. 2d in Table 10. Different stretching ratios $\lambda = \frac{l}{l_0}$ at 305 K are indicated. l_0 = unstretched sample length, at the given temperature, l = stretched sample length, constant for these experiments

above T_c, in the vicinity and below T_c a strong deviation from the normal thermoelastic behavior is observed, where the retractive force substantially decreases with falling temperature. Although the sample has been stretched at higher temperatures, the tensile force can change into a compressive force at lower temperatures. This means that the sample, which is a thin rubber strip for these measurements, elongates by an amount, which is larger than the length obtained by the initial stretching ratio applied at higher temperatures. This is an absolutely unusual and surprising behavior in the rubbery region and is due to the anisotropic ordering of the mesogenic side chains below T_c.

In the undeformed state above T_c, the network is isotropic, the statistical segments being randomly distributed. When the sample becomes stretched, the statistical segments in the chains of the network have an anisotropic distribution, which essentially does not influence the isotropic arrangement of the mesogenic groups. Cooling below T_c, under these conditions the elastomer becomes now l.c. As the network is already deformed uniaxially, the mesogenic groups macroscopically also adapt an orientation, due to the anisotropy of the network. A uniform macroscopically ordered liquid crystal is the consequence. If now the initially applied stretching force is released (still below T_c), the sample does not adopt its initial length obtained in the isotropic state, but remains at some intermediate length between the initial length in the isotropic state and the length of the deformed sample below T_c. Because of the anisotropic order of the mesogenic groups, the network itself remains also to some extent in the macroscopic dimension anisotropic. This behavior can also be detected by direct measurements of the dimension of the unloaded sample. An increase of the length l_0 of the unloaded sample below T_c, owing to the macroscopic orientation of the mesogenic groups, on the other hand is the reason for the strong drop of the retractive force, if the sample length is kept constant.

The macroscopical, uniform orientation of the l.c. side chains of the elastomers in the l.c. state can be qualitatively nicely observed by a simple experiment shown in Fig. 41. In Fig. 41a the unstretched elastomer in the l.c. state behaves like a monomeric l.c. melt, where because of thermal fluctuations the optical axis changes its direction, causing a strong light scattering. The sample is turbid. If the network becomes uniaxially deformed by stretching the sample (Fig. 41 b) the mesogenic groups also become macroscopically ordered. This is indicated by the uniform transparent sample. The markings, affixed behind the elastomer, become clearly recognizable. This experiment impressively demonstrates the combination of properties of liquid crystal and elastomer, resulting in a material with new characteristics. While l-l.c. can be macroscopically oriented by applying electric or magnetic fields, this can be performed by mechanical stress for the l.c. elastomers.

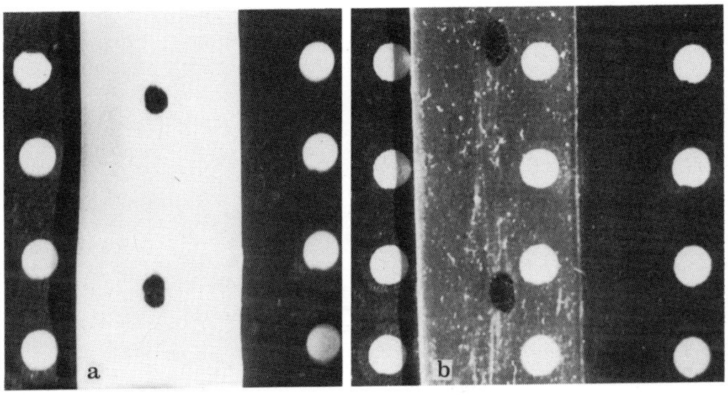

Fig. 41a and b. Elastomer in the l.c. state; **a** unstretched, macroscopically disordered and turbid; **b** stretched, macroscopically ordered and transparent. The samples are layered on a perforated black background

As mentioned before, in the isotropic phase the network becomes anisotropic by uniaxial stress. This implies, that even in the isotropic phase the mesogenic side chains will be influenced by the anisotropic matrix of the network, and in analogy to the Kerr effect or Cotton-Mouton effect by applying electric or magnetic fields on isotropic phases, an optical response is to be expected. Following the simple statistical theory of Kuhn and Grün [104)], the stress optical coefficient C, which is the birefringence of an uniaxially stretched elastomer divided by the applied true stress σ (σ = stress per area of the deformed sample), should solely depend on the mean refractive index $\bar{n}$ and the optical anisotropy $\Delta\alpha$ of the statistical segment:

$$C = \frac{2}{45\,kT} \frac{(\bar{n}^2 + 2)^2}{\bar{n}} \Delta\alpha \tag{16}$$

k = Boltzmann constant

$\bar{n} = \frac{n_{\parallel} + 2n_{\perp}}{3}$, where $n_{\parallel}$ and $_{\perp}$ are the refraction indices parallel and perpendicular to the optical axis.

If the product CT is plotted versus T a nearly constant value should be obtained, which is essentially fulfilled for common rubbers like natural rubber (NR) or poly(ethylene) (PE). This is shown in Fig. 42. At high temperatures, far above the phase transformation isotropic to nematic, the nematic elastomer No. 2a from Table 10 behaves very similarly to conventional material, having approximately a temperature independent CT. On lowering the temperature, however, a strong deviation from a constant CT(T) sets in about 30 K above T_c, where CT increases exponentially with T. This strong deviation from Eq. 16 can be easily understood by a partial orientation of the mesogenic groups in the pretransformational region, owing to the anisotropic network; a phenomenon analogous to the well known effect of l-l.c. and linear polymers in electric or magnetic fields [65]. These measurements on the other hand clearly demonstrate the direkt influence of the anisotropy of the network on the mesogenic side chains and vice versa.

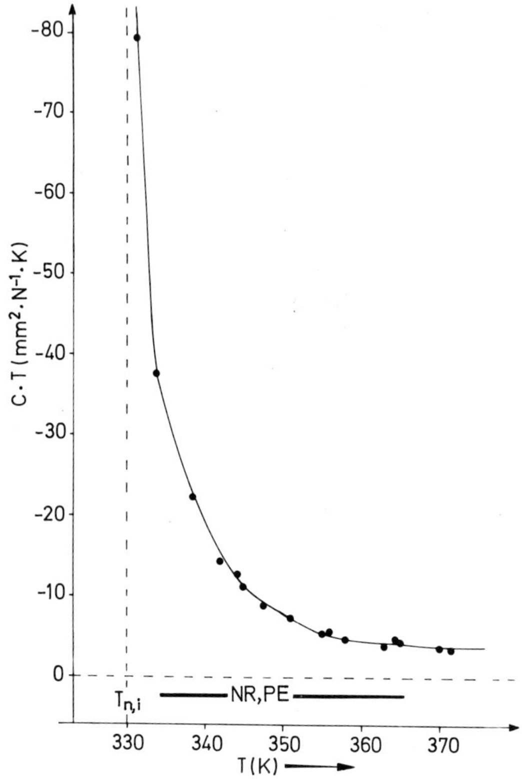

Fig. 42. Temperature dependence of the product of stress optical coefficient and temperature CT for natural rubber (NR), poly(ethylene) (PE) and the l.c. elastomer No. 2a from Table 10

These experimental results give only a first insight into the fascinating properties of these very new l.c. elastomers. Further detailed studies have to come to a more profound understanding of the interactions of polymer networks and l.c. order of the mesogenic side chains. The exceptional photoelastic and thermoelastic properties promise to open up a new scientific and technological field of interest.

3 Amphiphilic L.C. Side Chain Polymers

Amphiphilic molecules, consisting of a hydrophobic and hydrophilic part as schematically presented in Fig. 1, associate in aqueous solutions. Beginning at very low concentrations of the amphiphiles, the originally molecular disperse dissolved molecules form mainly globular micelles above the critical micelle concentration (c.m.c.). With increasing concentration the globular micelles grow to rod like or planar micelles. The different micelles can further associate to isotropic three dimensional networks [117] or liquid crystalline phases at high concentrations. In Fig. 43a schematic phase diagram shows the l.c. phases, which in principle can be observed. At lower concentrations rod like micelles of indefinite length become hexagonally ordered in the "hexagonal l.c. phase". With increasing concentration planar micelles of indefinite size become ordered in the "lamellar phase". At very high concentrations inverse rod like micelles form an "inverse hexagonal phase", where the water molecules are within the center of the micelles. These common types of l.c. phases can be separated by cubic phases, or cubic phases can be found before as well as behind the phase region of the hexagonal and lamellar phases [118].

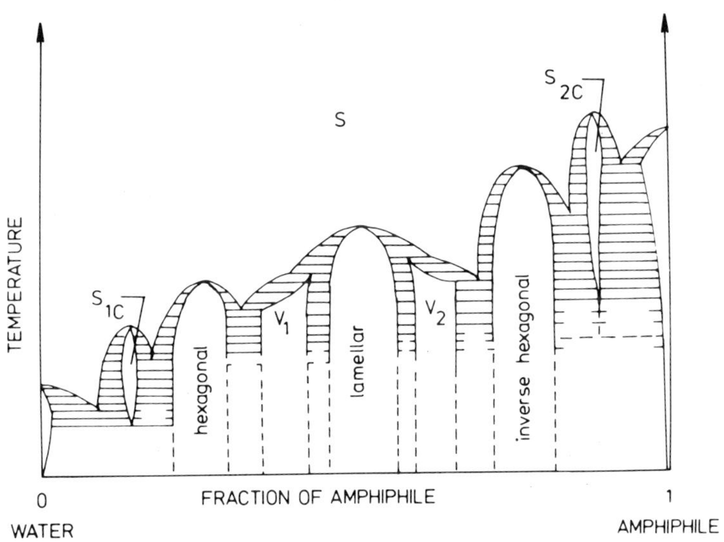

Fig. 43. Hypothetic diagram of liquid crystalline phases formed by amphiphilic molecules in solution [111] S_{1c}, V_1, V_2, S_{2c} = cubic phases

Although numerous amphiphilic monomers have been polymerized and a great number of amphiphilic polymers have been investigated, only some hints in the literature indicate mesomorphic behavior of concentrated polymer solutions [105, 106, 121]. Nearly all measurements were carried out at low concentrations, in order to characterize the properties of these "polysoaps" in view of applications as surfactants. For these applications high polymer concentrations are, of course, not of interest.

The change of lyotropic l.c. behavior of a monomer in the hexagonal phase by polymerization has been described for the first time by Friberg et al. [107, 108]. A change from the hexagonal monomer phase to the lamellar phase of the polymer was observed and carefully identified. Complete phase diagrams of monomer and polymer, however, were not compared.

3.1 Model Considerations and Synthesis

According to the model considerations on non amphiphilic l.c. side chain polymers, the linkage of amphiphilic l.c.'s to a polymer backbone restricts their translational and rotational motions and neighboring molecules are strongly correlated via the polymer main chain. Presuming the chemical constitution of the systems in principle does not prevent a micellar association, characteristic changes in the extent and structure of the lyotropic l.c. phases are to be expected. With respect to the "stability" of the l.c. phase an extension of the l.c. phase region is to be foreseen in analogy to the non amphiphilic systems.

The phase structure of the p-l.c.'s should be directly influenced by the kind of the linkage of the amphiphilic molecules forming the macromolecule. Two possibilities exist (Fig. 44): the l-l.c.'s are connected:

i) at their hydrophobic ends (Type A) or
ii) at their hydrophilic ends (Type B).

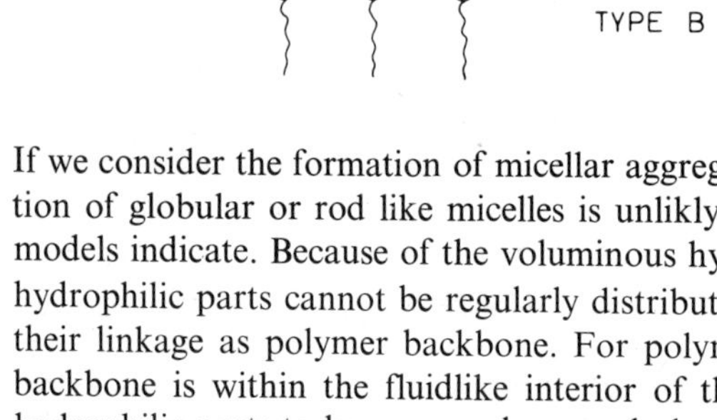

Fig. 44. Amphiphilic side chain polymers; m.-l.c. connected via the hydrophobic part (Type A); m-l.c. connected via the hydrophilic part (Type B) to the polymer backbone

If we consider the formation of micellar aggregates in aqueous solutions, the formation of globular or rod like micelles is unlikly for polymers type B as space filling models indicate. Because of the voluminous hydrophobic parts of the molecules the hydrophilic parts cannot be regularly distributed on the shell of the micelle, due to their linkage as polymer backbone. For polymers type A, on the other hand, the backbone is within the fluidlike interior of the micelle and does not prevent the hydrophilic parts to be more or less regularly arranged at the outside of the micelle. Contrariwise, inverse rod like micelles might easily be formed by type B and hardly formed by polymer type A. If we consider the formation of micellar aggregation in non aqueous solutions, of course the reversed situation exists. While the position of the backbone might be of importance for hexagonal phases, lamellar phases can be built up by both types of polymers without any disturbing effect of the backbone.

Starting with amphiphilic m-l.c's for the synthesis of polymers, the well known methods can be applied. For the monomeric amphiphiles a broad variety exists in view of ionic or non-ionic surfactants. Up to now, however, only for the three polymer systems listed in Table 11 l.c. behavior has been established.

Table 11. Examples of amphiphilic l.c. side chain polymers

NO	MONOMER		POLYMER	TYPE	LIT
1	$CH_2{=}CH-(CH_2)_8-COO^-\ Na^+$	→	$\cdots-CH_2-CH(-(CH_2)_8-COO^-\ Na^+)-\cdots$	A	108
2	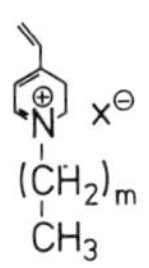	→	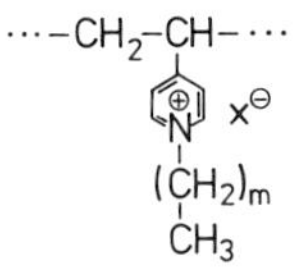	B	109
3	$\cdots-Si(CH_3)(H)-O-\cdots$ + $CH_2{=}CH-(CH_2)_8-COO-(CH_2-CH_2-O)_x-CH_3$	→	$\cdots-Si(CH_3)(-CH_2-CH_2-(CH_2)_8-COO-(CH_2-CH_2-O)_x-CH_3)-O-\cdots$	A	110,111

3.2 Linear Polymers

3.2.1 Phase Behavior in Aqueous Solutions

The first detailed investigations on the phase behavior of a p-l.c. compared with the chemically very similar low molar mass system have been reported recently [110]. The system investigated is shown in Table 11, No. 3. In order to determine the influence of the linkage of the monomeric surfactant to the siloxane backbone, we will compare the phase behavior of the monomer consisting of the hydrophobic 10-undecenoic acid esterified with the hydrophilic octa-ethylenglycol-monomethyl-ether (x = 8) with that of the corresponding polysiloxane.

In Fig. 45 the phase diagram of the m-l.c. in aqueous solution is shown, which equals diagrams of well known surfactants [5]. Because of a dehydration of the hydrophilic ethylen glycol chain, at high temperatures a broad miscibility gap, is found, having a lower consolute point at 14.1 % of m-l.c. in water. Besides the normal diagram of binary mixtures forming an eutectic system, in the concentration range of 49 % to 70 % of the m-l.c. a hexagonal phase is observed, which can be easily identified by the characteristic texture shown in Fig. 46. Two important data with respect to the extent of the hexagonal phase should be pointed out. The maximum of the

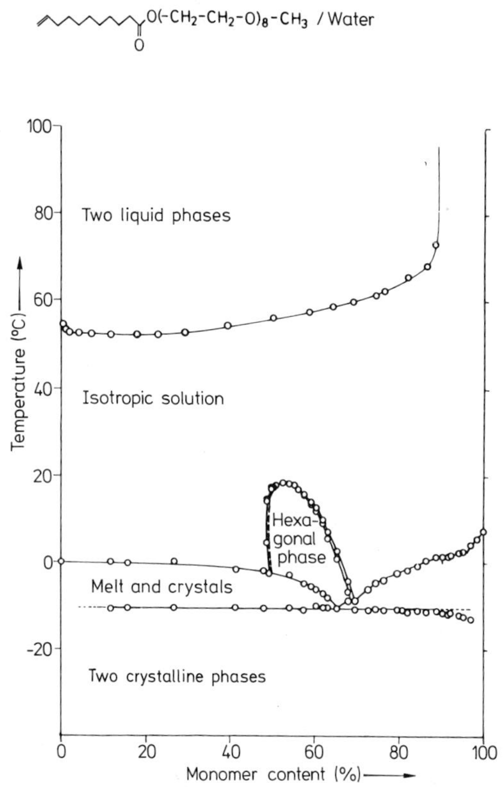

Fig. 45. Phase diagram of the system: water-monomeric surfactant (see Table 11, No. 3)

Fig. 46. Fan-like texture of the m-l.c. (Table 11, No. 3) in the hexagonal phase (Magnification 180 ×)

coexistence line of the hexagonal phase and the isotropic solution is observed at a monomer content of 53%, which corresponds to a tripolyhydrate of the surfactant (for each oxygen atom of the glycol chain three water molecules are available) [120]. The temperature of the maximum lies at 19 °C.

In the following we will discuss the corresponding polymer: By addition reaction to a poly(hydrogenmethylsiloxane) ($\bar{r} = 95$), the m-l.c. is attached to the backbone

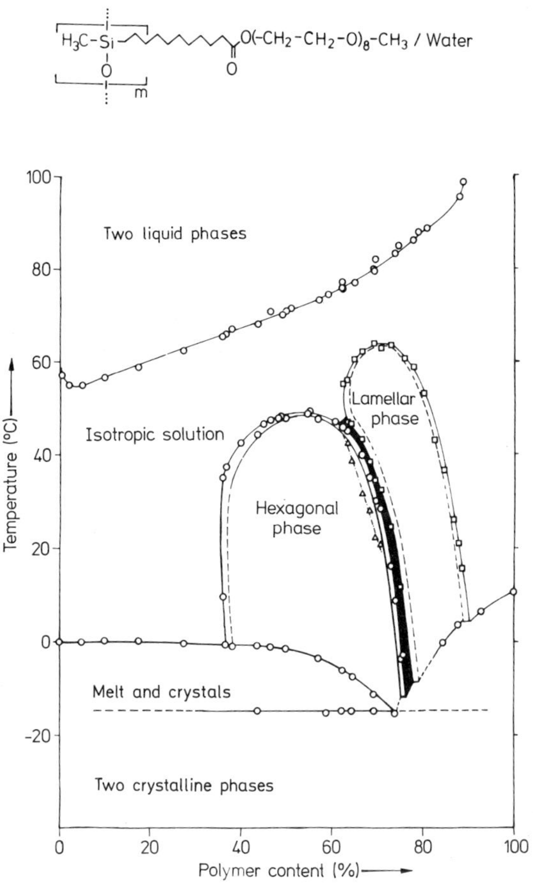

Fig. 47. Phase diagram of the system: water-polymeric surfactant (see Table 11, No. 3)

at the hydrophobic end, forming an amphiphilic l.c. side chain polymer of type A. In this case, the formation of rod like micelles in aqueous solutions should not be affected by the backbone according to the model considerations mentioned before. The phase diagram of the polymer/water system is shown in Fig. 47. It is obvious, that the extent of the l.c. phase region is strongly enlarged, compared to the monomer/water system: anisotropic phases exist from 36% to 90%. The maxima of the coexisting lines l.c. — isotropic lie at 55% polymer (49 °C) and at 72% polymer (64 °C). The maximum temperature, where a liquid crystalline phase exists, is shifted by 45 °C to higher temperatures compared with the monomer/water system. As for the m-l.c.'s the maxima in the phase equilibrium curves correspond to a tripolyhydrate and additionally to a 1,5-polyhydrate, respectively, with respect to the monomer unit of the polymer.

A second important aspect is the structure of the l.c. phases. At low concentrations the polymer/water system exhibits in analogy to the monomer/water system a hexagonal phase, showing the typical fan like texture (Fig. 48). Additionally at higher concentrations a lamellar phase exists, which is separated from the hexagonal phase by a cubic phase of small extent (black area in Fig. 47). With this finding, the

Fig. 48. Fan-like texture of the p-l.c. (Table 11, No. 3) in the hexagonal phase (Magnification 180×)

succession of the l.c. phases known from conventional amphiphilies, is analogously formed by the amphiphilic p-l.c.

Besides the l.c. phases, the phase diagram of the p-l.c./water is very similar to the diagram of the m-l.c./water. The broad miscibility gap of the polymer/water system shows a lower critical consolute point, which is shifted to lower concentrations (3.2% of polymer). This is consistent with experiments and theory on the position of miscibility gaps in polymer solutions [112].

From these very first measurements we have to conclude the following essential aspects:

i) The textures of the polymeric mesophases are similar to the textures of l-l.c.'s. Therefore we can conclude that the arrangement of the amphiphilic side chains of the polymer/water system corresponds to the structure of the low molar mass systems. In case of polymers of type A, hexagonal micelles in aqueous solutions are formed where the backbone is arranged within the micelle. Similar considerations are valid for the lamellar phases.

ii) The extent of the l.c. phases of the polymeric system is enlarged compared with the l.-l.c. systems. This is analogous to the systematics observed for non amphiphilic l.c. side chain polymers described in Chap. 2. Therefore this behavior can be explained by the restriction of motions of the amphiphilic units, when they are attached to a polymer backbone.

For the systems investigated, however, each monomer unit is lengthened by one siloxane main chain segment, which might also contribute to an extension of the l.c.

phase region [115,116]. Therefore it cannot be definitely excluded that the lengthening owing to the siloxane backbone shifts the phase transformations at least to the same extent. On the other hand, a lengthening of the hydrophobic part of amphiphilic molecules normally lowers the lower critical consolution point, which is not observed for the system investigated. This supports the argument that the extension of the l.c. state for the polymer system is mainly due to the restriction of motions of the amphiphilic molecules because of their linkage as side chains to the polymer backbone.

Further systematic investigations have to clarify the effect of the polymer main chain on the structure and the extent of the l.c. state of the attached side chain amphiphiles and will open a new field of experimental and theoretical interest.

References

1. H. Kelker: Mol. Cryst. Liq. Cryst. *21*, 1 (1979)
2. F. Reinitzer: Monatsh. *9*, 421 (1888)
3. G. W. Gray: in: "Polymer Liquid Crystals", eds. A. Ciferri, W. R. Krigbaum and R. B. Meyer: Academic Press, NY (1982) chap. 1 and literature cited herein
4. G. W. Gray: in: "The Molecular Physics of Liquid Crystals", eds. G. R. Luckhurst, G. W. Gray: Academic Press, NY (1979) chap. 1
5. G. W. Gray, P. A. Winsor: in: "Liquid Crystals and Plastic Crystals" eds. G. W. Gray, P. A. Winsor: Vol. 1, Ellis Horwood, Chichester England (1974)
6. G. W. Gray: "Advances in Liquid Crystals" ed. G. H. Brown: Vol. 2, Academic Press, N.Y., p. 1 (1976)
7. J. Billard: in: "Liquid Crystals of One- and Two Dimensional Order", eds. W. Helfrich, G. Heppke: Springer-Verlag, Berlin. p. 383 (1980)
8. P. Ekwall: in: "Advances in Liquid Crystals" ed. G. H. Brown: Vol. 1, Academic Press, NY, p. 1 (1975)
9. S. Friberg, K. Larsson: in: "Advances in Liquid Crystals", ed. G. H. Brown: Vol. 2, Academic Press N.Y., p. 173 (1976)
10. R. G. Laughlin: in: "Advances in Liquid Crystals" ed. G. H. Brown: Vol. 3, Academic Press, N.Y., p. 42 (1978)
11. L. J. Yu, A. Saupe: Phys. Rev. Lett. *45*, 1000 (1980)
12. W. Maier, A. Saupe: Z. Naturforsch., *13a*, 564 (1958)
13. P. J. Flory: Adv. Pol. Sci., *59*, 1 (1984)
14. L. P. Yu, E. T. Samulski: in: "Ordered Fluids and Liquid Crystals", Vol. 4, eds. Johnson, Porter: Plenum Press, in press
15. B. Gallot: in: "Liquid Crystalline Order in Polymers", ed. A. Blumstein: Academic Press, N.Y., p. 192 (1978) and literature cited herein
16. E. Perplies, H. Ringsdorf, J. H. Wendorff: Ber. Bunsenges. Phys. Chem. *9*, 921 (1974)
17. Y. Tanaka, Yukikagaku Kyokashi: *34*, 2 (1976) and references cited herein
18. Y. Tanaka, H. Hitotsuyanagi et al.: Makromol. Chem. *117*, 3035 (1976)
19. Y. Tanaka, F. Yamaguchi et al.: J. Polym. Sci., Polymer Chem. Ed., *16*, 1027 (1978)
20. Y. Tanaka, H. Tsuchiya: Journal de Physique, C3, *4*, Tome 40, C3-41 (1979)
21. A. Blumstein, E. C. Hau: in: "Liquid Crystalline Order in Polymers", ed. A. Blumstein: Academic Press, N.Y., p. 150 (1978)
22. V. P. Shibaev, N. A. Plate: Polym. Sci. USSR (English Translation) *19*, 1065 (1978)
23. H. J. Lorkorwski, F. Reuther: Plaste Kautsch. *2*, 81 (1976)
24. H. Finkelmann, H. Ringsdorf, J. H. Wendorff: Makromol. Chem. *179*, 273 (1978)
25. A. Dubault, C. Casagrande, M. Veyssie: Mol. Cryst. Liq. Cryst, *41*, 239 (1978)
26. N. A. Platé, V. P. Shibaev: in: "Grebneobraznye polimeri i zhidkie Kristally (Comblike Polymers and Liquid crystals), Khimiya, Moscow, 1980

27. D. Demus, H. Demus, H. Zaschke: „Flüssige Kristalle in Tabellen", VEB Dtsch. Verlag Grundstoffind., Leipzig (1976)
28. D. Hisgen, K. Kreuder, H. Ringsdorf: Abstracts of "13. Freiburger Arbeitstagung Flüssigkristalle", Freiburg FRG (1983)
29. H. Finkelmann, G. Rehage: Makromol. Chem. Rapid Commun. *1*, 31 (1980)
30. H. Ringsdorf, A. Schneller: British Polymer J. *13*, 43 (1981)
31. H. Finkelmann, M. Happ, M. Portugall, H. Ringsdorf: Makromol. Chem. *179*, 2541 (1978)
32. V. V. Sinitzyn, R. V. Talroze, V. P. Shibaev, N. A. Platé: Abstracts of the 4. Int. Liquid Crystal Conference of Sozialist Countries, Tbilisi, USSSR, 213 (1981)
33. R. V. Talroze, S. G. Kostromin, V. P. Shibaev, N. A. Platé, H. Kresse, K. Sauer, D. Demus: Makromol. Chem. Rapid Commun. *2*, 305 (1981)
34. J. H. Gibbs, E. A. Di Marzio: J. Chem. Phys. *28*, 373 (1958)
35. G. Adam, J. H. Gibbs: J. Chem. Phys., *43*, 139 (1965)
36. G. Rehage: J. Macromol. Sci.-Phys., *B18(3)*, 423 (1980)
37. J. Frenzel: PhD-Thesis Clausthal (1981)
38. J. Frenzel, G. Rehage: Makromol. Chem., Rapid Commun., *1*, 129 (1980)
39. J. Frenzel, G. Rehage: Makromol. Chem.
40. B. Holm, J. H. Wendorff, M. Portugall, H. Ringsdorf: Colloid and Polymer Sci., *259*, 875 (1981)
41. G. Rehage, J. Frenzel: Brit. Polym. J., *14*, 173 (1982)
42. H. Finkelmann, B. Lühmann, G. Rehage, H. Stevens: in: "Liquid Crystals and Ordered Fluids", eds. Johnson and Porter, Plenum Press, Vol. 4, in press
43. H. Stevens, H. Finkelmann, G. Rehage: Macromolecules, in press
44. P. H. Hermans, P. Platzek: Kolloid Z. *88*, 65 (1939)
45. H. Finkelmann, H. Ringsdorf, W. Siol and J. H. Wendorff: in: "Mesomorphic Order in Polymers", ACS Symposium Series *74*, p. 22 (1978)
46. J. H. Wendorff, H. Finkelmann, H. Ringsdorf: in: "Mesomorphic Order in Polymers", ACS Symposium Series *74*, p. 12 (1978)
47. J. H. Wendorff: in: "Liquid Crystalline Order in Polymers", ed. A. Blumstein: Academic Press, N.Y. (1978)
48. H. Finkelmann, D. Day: Makromol. Chem. *180*, 2269 (1979)
49. H. Finkelmann, J. H. Wendorff: ACS preprints, ACS-meeting, Washington D.C. (1983)
50. F. Cser, K. Nitrai, G. Hardy: Advances in Liquid Crystal Research and Applications, ed. L. Bata: Pergamon Press, Oxford, p. 845 (1980)
51. D. Demus, L. Richter: "Textures of Liquid Crystals", Verlag Chemie, Weinheim (1978)
52. H. Kelker, U. G. Wirzing: Mol. Cryst. Liq. Cryst. Lett., *49*, 175 (1979)
53. A. Saupe, W. Maier: Z. Naturforschg. *16a*, 816 (1961)
54. M. F. Vuks: Optics and Spectroscop., *20*, 361 (1966)
55. S. Chandrasekhar, N. V. Madhusudana: J. de Physique, *C4*, 24 (1969)
56. J. M. Daniels, P. E. Cladis et al.: J. Applied Phys. *53*, 6127 (1982)
57. H. Finkelmann, H. Benthack, G. Rehage: Journal de chimie physique, *80*, 163 (1983)
58. H. Finkelmann, unpublished results
59. H. Geib, B. Hisgen, U. Pschorn, H. Ringsdorf, H. W. Spiess, J. Am. Chem. Soc. *104*, 917 (1982)
60. Ch. Boeffel, B. Hisgen, U. Pschorn, H. Ringsdorf, H. W. Spiess: to be published in a special issue of the Israel Journal of Chemistry on NMR Spectroscopy in Liquid Cystalline Systems, ed. Z. Luz
61. K. H. Wassmer, E. Ohmes, G. Kothe: Makromol. Chem., Rapid Commun. *3*, 282 (1982)
62. U. Müller, H. Stegemeyer: Ber. Bunsenges. Phys. Chem. *77*, 20 (1973)
63. J. Haller, H. A. Huggins, H. R. Lilienthal, T. R. McGuire: J. Phys. Chem., *77*, 950 (1973)
64. J. Haller: Prog. Solid State Chem., *10*, 103 (1975)
65. H. Kelker, R. Hatz: "Handbook of Liquid Crystals", Verlag Chemie, Weinheim (1980)
66. H. Finkelmann, U. Kiechle, G. Rehage: Mol. Cryst. Liq. Cryst. *92*, 49 (1983)
67. H. Finkelmann, D. Naegele, H. Ringsdorf: Makromol. Chemie *180*, 803 (1979)
68. G. Meier: Abstracts of "3. Arbeitstagung Flüssigkristalle", Freiburg, FRG (1973)
69. H. Ringsdorf, R. Zentel: Makromol. Chem. *183*, 1245 (1982)
70. H. Pranoto, W. Haase: to be published
71. C. Casagrande, M. Veyssie, C. Weill, H. Finkelmann: Mol. Cryst. Liq. Cryst. Lett. (1983) in press

72. H. Arnold, H. Sackmann: Z. Phys. Chem. *213*, 137 (1960); *213*, 145 (1960)
73. E. C. H. Hsu, J. F. Johnson: Mol. Cryst. Liq. Cryst. *20*, 177 (1973); *25*, 145 (1974)
74. D. Demus, C. Fietkau, R. Schubert, H. Kehlen: Mol. Cryst. Liq. Cryst. *25*, 215 (1974)
75. H. Finkelmann, H. J. Kock, G. Rehage: Mol. Cryst. Liq. Cryst. *89*, 23 (1982)
76. H. Ringsdorf, H. W. Schmidt, A. Schneller: Makromol. Chem., Rapid Commun. *3*, 745 (1982)
77. C. Casagrande, M. Veyssie, H. Finkelmann: J. Physique, Letters *43* L671 (1982)
78. W. J. A. Goossens: Mol. Cryst. Liq. Cryst., *12*, 237 (1970)
79. W. J. A. Goossens: J. Phys. Colloq. (Orsay, Fr.) *40*, 158 (1979)
80. H. Finkelmann, G. Rehage: Makromol. Chem., Rapid Commun. *3*, 859 (1982)
81. H. Finkelmann, J. Koldehoff, H. Ringsdorf: Angew. Chem. Int. Ed. Engl. *17*, 935 (1978)
82. H. Finkelmann, G. Rehage: Makromol. Chem., Rapid Commun. *1*, 733 (1980)
83. V. P. Shibaev, H. Finkelmann, A. V. Kharitonov, M. Portugall, N. A. Platé, H. Ringsdorf: Vysokomol. Soedin, *A23*, 919 (1982)
84. H. De Vries: Acta Crystallogr. *4*, 219 (1950)
85. P. Pollmann, H. Stegemeyer: Ber. Bunsenges. Phys. Chem. *78*, 843 (1974)
86. B. W. van der Meer, G. Vertogen: Phys. Lett. *A59*, 279 (1976)
87. J. P. Straley: Phys. Rev. *A10*, 1801 (1974)
88. H. Finkelmann: Phil. Trans. R. Soc. Lond. A, *309*, 105 (1983) in press
89. W. L. McMillan: Phys. Rev. *A4*, 1238 (1971) and Phys. Rev. *A6*, 936 (1972)
90. K. K. Kobayashi: Phys. Letters, *3117*, 125 (1970)
91. R. Zentel: Diplomarbeit, Mainz, FRG (1980)
92. H. Finkelmann, H. J. Kock, G. Rehage: Makromol. Chem., Rapid Commun. *2*, 317 (1981)
93. L. Strzelecki, L. Liebert: Bull. Soc. Chim. Fr., 597 (1973)
94. J. Mügge: Diplomarbeit, Clausthal, FRG, 1982
95. P. Zugenmaier, J. Mügge: Makromol. Chem., Rapid Commun. *5*, 11 (1984)
96. S. G. Kostromin, V. V. Sinitzyn, R. V. Talroze, V. P. Shibaev, N. A. Platé: Makromol. Chem., Rapid Commun., *3*, 809 (1982)
97. P. Andersson, G. Bäckström: High Temperatures, High Pressures, *4*, 101 (1972)
98. G. Gee: Contemp. Phys. *11*, 313 (1970)
99. H. J. Oels, G. Rehage: Macromolecules, *10*, 1036 (1977)
100. G. Rehage: Ber. Bunsenges. Phys. Chem. *81*, 969 (1977)
101. W. Oppermann, K. Braatz, H. Finkelmann, W. Gleim, H. J. Kock, G. Rehage: Rheologica Acta, *21*, No. 4/5, 423 (1982)
102. P. G. de Gennes: in: "Polymer Liquid Crystals" eds. A. Ciferri, W. R. Krigbaum, R. B. Meyer: Academic Press, N.Y., chap. 5 (1982)
103. R. B. Meyer: in: "Polymer Liquid Crystals" eds. A. Ciferri, W. R. Krigbaum, R. B. Meyer: Academic Press, N.Y., chap. 6 (1982)
104. W. Kuhn, F. Grün: Kolloid Z. *101*, 248 (1942)
105. A. Mathis, A. Schmitt, A. Skoulios, R. Varoqui: Europ. Polym. J. *15*, 255 (1979)
106. A. Mathis, A. Skoulios, R. Varoqui, A. Schmitt: Europ. Polym. J. *10*, 1011 (1974)
107. S. E. Friberg, R. Thundathil, J. O. Stoffer: Science, *205*, 607 (1979)
108. R. Thundatil, J. O. Stoffer, S. E. Friberg: J. Polym. Sci., Polymer Chem. Ed., *18*, 2629 (1980)
109. U. P. Strauss, N. L. Gershfeld, E. H. Crook: J. Phys. Chem. *60*, 577 (1956)
110. H. Finkelmann, B. Lühmann, G. Rehage: Colloid and Polymer Sci., *260*, 56 (1982)
111. L. E. Scriven: in: "Micellization, Solubilisation and Microemulsions", ed. L. L. Mittal: Plenum Press, N.Y., Vol. II, pp. 877 (1977)
112. P. J. Flory: Principles of Polymer Chemistry, Cornell University Press, Ithaca and London (1953)
113. P. Dias: C.R. Acad. Sci., Ser. A, *282*, 71 (1976)
114. P. Piéransky, F. Brochard, E. Guyon: J. Phys. *34*, 35 (1973)
115. J. S. Clunie, J. F. Goodman, P. C. Symons: Trans. Faraday Soc. *65*, 287 (1969)
116. J. S. Clunie, J. M. Corkill, J. F. Goodman, P. C. Symons, J. R. Tate: Trans. Faraday Soc. *63*, 2839 (1967)

117. H. Rehage: PhD-thesis Bayreuth (1982)
118. G. J. T. Tiddy: Physics Reports, *57*, 1 (1980)
119. H. J. Oels, G. Rehage: Macromolecules, *10*, 1036 (1977)
120. R. Heusch: Ber. Bunsenges. Phys. Chem. *83*, 834 (1979)
121. J. Herz, F. Reiss-Hussen, V. Luzatti, A. Skoulios: Patent, F. 1,295,525, June 8, 1962, Appl. Apr. 27, 1961

M. Gordon (Editor)

Received July 28, 1983

Thermotropic Liquid-Crystalline Polymers with Mesogenic Side Groups

Valery P. Shibaev, Nicolai A. Platé
Department of Chemistry, Moscow State University, Moscow 119899, USSR

The article covers synthesis, structure and properties of thermotropic liquid-crystalline (LC) polymers with mesogenic side groups. Approaches towards the synthesis of such systems and the conditions for their realization in the LC state are presented, as well as the data revealing the relationship between the molecular structure of an LC polymer and the type of mesophase formed. Specific features of thermotropic LC polymers and copolymers of nematic, smectic and cholesteric types are considered.

The possibility to affect the structure of an LC polymer by the influence of electric and magnetic fields is demonstrated. The kinetics and the mechanism of structural rearrangements are discussed.

The initial steps of mesophase formation in dilute solutions of polymers are examined.

Advances in Polymer Science 60/61

List of Abbreviations and Main Symbols

LC	— Liquid Crystalline
K	— Crystalline State
N, S, Ch	— Nematic, Smectic and Cholesteric Phases, respectively
Chol	— Cholesterol group
PChMA-n and PChMO-n	— Cholesterol esters of poly-N-methacryloyl-ω-aminocarboxylic and ω-oxycarboxylic acids, respectively
A-n and Pa-n	— n-Alkyl Acrylates and Poly-n-Alkylacrylates
MA-n and PMA-n	— n-Alkyl Methacrylates and Poly-n-Alkyl Methacrylates
IMM	— Intramolecular Mobility
E_f, E_A and E_{or}	— Activation energies of the viscous flow, dielectric relaxation process and orientational process in an electric field
$E^b_{d.p}$ and $E^s_{d.p}$	— Activation energies of the dipole polarization in bulk and solution
$\tau^b_{d.p}$ and $\tau^s_{d.p}$	— Relaxation times of the dipole polarization in bulk and solution
f_R	— Dielectric relaxation frequency.

1 Introduction

Thermotropic liquid-crystalline polymers belong to a relatively new class of liquid-crystalline compounds. Indeed, if lyotropic polymeric liquid crystals (such as, for instance, solutions of synthetic polypeptides) have been well-known and are under investigation already for quite a long time (see Refs. in [1]), the first attempts to synthesize thermotropic polymeric liquid crystals date only to the beginning of the 70-ies of our century [2-10]. It is in this period, that on the background of vital interest and extensive practical utilization of low-molecular liquid crystals, publications revealing various approaches towards synthesis of thermotropic polymers LC systems begin to appear and mesomorphic polymers become the object of intensive attention of scientists, working in the field of polymer science [11-17].

The study of this type of polymers is of interest in its own right, which is inspired by the need to clarify the nature and specific features of LC state of macromolecular compounds.

On the other hand, the interest towards this field is accounted for by the possibility to create polymeric systems, combining the unique properties of low-molecular liquid crystals and high molecular compounds, making it feasible to produce films, fibers and coatings with extraordinary features. It is well-known that the utilization of low-molecular thermotropic liquid crystals requirs special hermetic protective shells (electrooptical cells, microcapsules etc.), which maintain their shape and protect LC compounds from external influences. In the case of thermotropic LC polymers there is no need for such sandwich-like constructions, because the properties of low-molecular liquid crystals and of polymeric body are combined in a single individual material. This reveals essentially new perspectives for their application.

The study of thermotropic, as well as of lyotropic LC polymers is directly linked to a series of practical tasks, regarding the construction of polymeric materials with set properties. For instance, making use of anisotropy of the LC state in processing (particularly in moulding) of polymeric materials discloses impressive prospects for the production of so called high modulus fibers and films [18-25].

At present at least three types of thermotropic LC polymers may be identified — these are:

1) the melts of some linear crystallizable polymers;
2) polymers with mesogenic groups incorporated within the backbone; and
3) polymers with mesogenic side groups.

The first two types of polymers are reviewed in the article by McIntire included in this volume [26].

This review deals with LC polymers containing mesogenic groups in the side chains of macromolecules. Having no pretence to cover the abundant literature related to thermotropic LC polymers, it seemed reasonable to deal with the most important topics associated with synthesis of nematic, smectic and cholesteric liquid crystals, the peculiarities of their structure and properties, and to discuss structural-optical transformations induced in these systems by electric and magnetic fields. Some aspects of this topic are also discussed in the reviews by Rehage and Finkelmann [27], and Hardy [28]. Here we shall pay relatively more attention to the results of Soviet researchers working in the field.

2 Synthesis of Liquid-Crystalline Polymers with Mesogenic Side Groups

The general approach towards the synthesis of thermotropic LC polymers is confined to "chemical binding" of polymer chains with the molecules of low-molecular liquid crystals (with mesogenic groups, to be more precise), which may be incorporated either within the main chains or within the side chains of macromolecules [2–16, 27, 28].

The latter involves the synthesis of monomers containing LC (mesogenic) groups, with the subsequent homopolymerization or copolymerization with mesogenic or non-mesogenic compounds (Fig. 1), or the attachment of low-molecular crystal

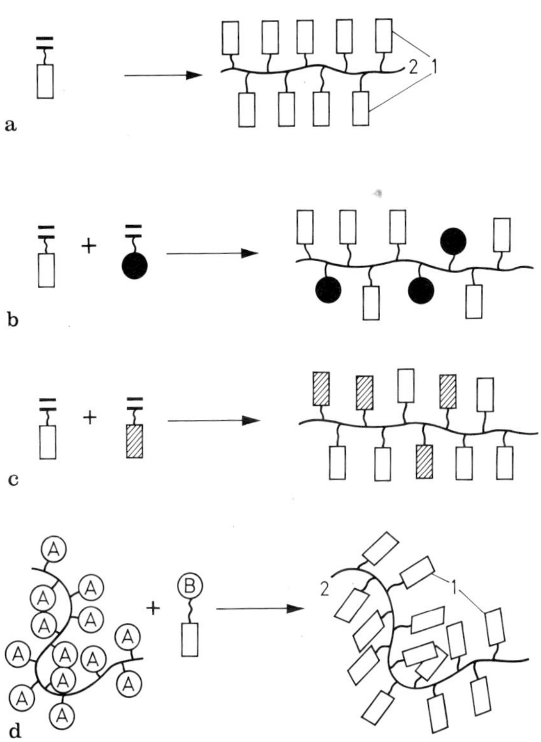

Fig. 1a–d. Synthesis of LC polymers with mesogenic side groups: **(a)** — homopolymerization; **(b)** copolymerization of mesogenic and nonmesogenic monomers; **(c)** copolymerization of mesogenic monomers; **(d)** polymer-analogous reaction; 1 — mesogenic groups; 2 — main chain; A, B — functional groups

molecules to a polymeric backbone via polymer-analogous reactions. The second pathway is chosen much more rarely and there are few LC polymers synthesized by this method. The examples of the reactions used for binding of mesogenic compounds to polymers are [29–31].

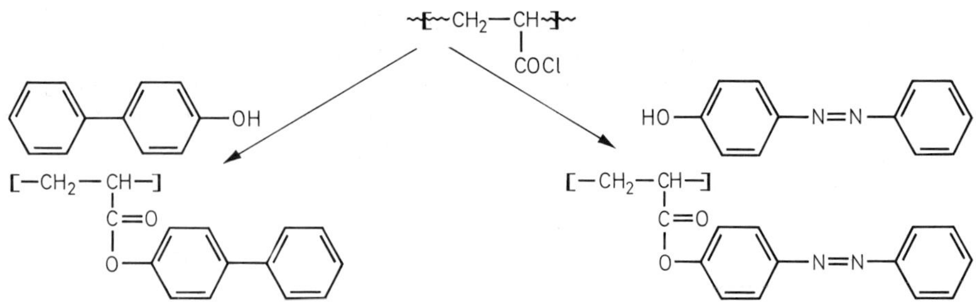

Poly - p - biphenylacrylate

Poly - p - acryloyloxiazobenzene

A modified method was proposed in [32] and consisted in that an original mesogenic olefin monomer was used as a mesogenic compound:

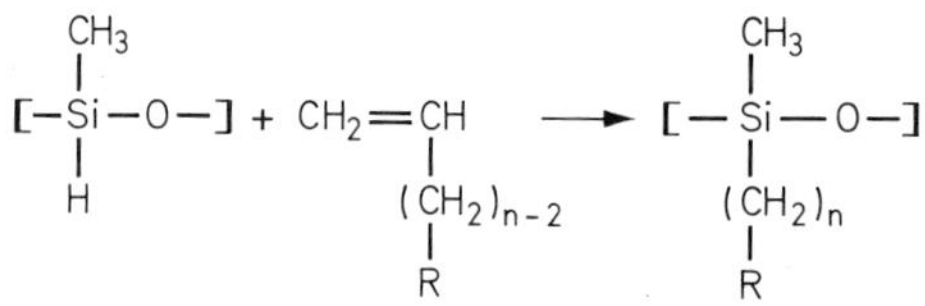

where R is a mesogenic group

What is important, is the essential role of the length of side chain (so called spacer), linking a mesogenic group to the backbone, in the realization of the LC state. The direct attachment of a mesogenic group to the backbone (without the spacer-group) does not always lead to LC polymer. This is accounted for by steric hindrance imposed by the main chain on the packing of mesogenic groups. As a result, most of the polymers of this type are amorphous. Examples of this kind of polymers are discussed in detail in our book [12].

A synthetic pathway that appeared to be most convenient and promising for the synthesis of LC polymers was proposed and proved in Moscow State University in 1973 [6,8-10] and involved the use of so called comb-like polymers containing long paraffinic fragments in each monomer unit [12,33] (Fig. 2). Macromolecules of comb-like polymers are constituted of two types of structural units — the main chains and the side chains. Their behaviour is mutually dependent as they are chemically linked and at the same time both parts are sufficiently independent because the side chains are long enough.

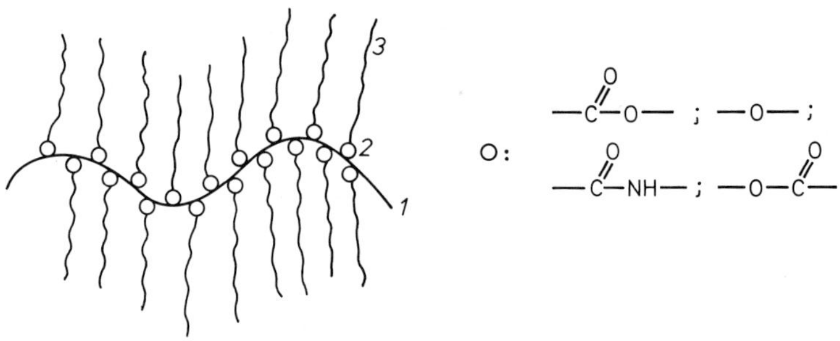

Fig. 2. Scheme of a comb-like macromolecule: *1* — main chain; *2* — attachment bridge (chemical junction); *3* — n-aliphatic side chain

The autonomy of side groups in comb-like polymers, exhibited in their ability to form layer structures in melts, and even to crystallize independently of the main chain configuration [34, 35], provides the basis for the synthesis of LC polymers on the basis of comb-like polymers [6–12]. The attachment of mesogenic groups to the side branches of comb-like polymers efficiently reduces steric restrictions brought upon the packing of mesogenic groups by the main chain, when compared to polymers with mesogenic groups bonded directly to the backbone.

Thus, the "remoteness" of mesogenic groups from the backbone provided by a polymethylene spacer secures them sufficient autonomy from the main chain. On the other hand, the fact that mesogenic groups are chemically linked with the main chain of the macromolecule assists their cooperative interaction. This is why comb-like polymers have come to be accepted as convenient "matrices" for constructing LC polymers. Already a few hundred liquid-crystalline polymers with various mesogenic side groups have been synthesized.

Among the multitude of liquid-crystalline polymers, those of acrylic and methacrylic series, containing various types of widespread fragments of low-molecular

Table 1. Some types of liquid-crystalline polymers with mesogenic side groups

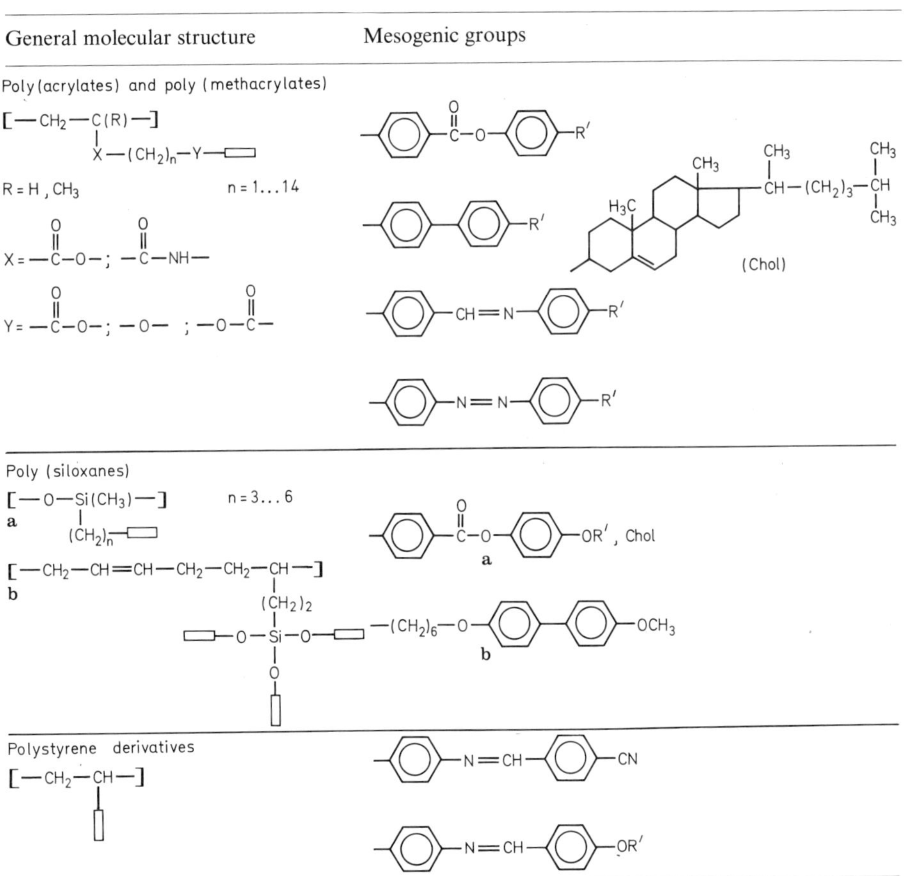

▭ Mesogenic group R' = alkyl or alkyloxy group Chol = cholesteryl group

Table 2. Copolymers of cholesterol-containing monomer (ChM-11) with n-alkyl acrylates (A-n) and n-alkyl methacrylates (MA-n) [36]

$CH_2{=}C(CH_3)$–$OC{-}NH{-}(CH_2)_{11}C({=}O){-}OChol$ + $CH_2{=}C(R)$–$OC{-}OC_nH_{2n+1}$ R = H (A-n), R = CH_3 (MA-n)

(ChM-11)

Copolymer, mol.-% of ChM-11	T_g, °C	T_{cl}, °C
Copolymers of ChM-11 with A-4		
100	120	180
42	65	160
37	60	140
17	20	100
Copolymers of ChM-11 with MA-4		
90	115	180
67	105	170
40	85	160
Copolymers of ChM-11 with MA-10		
75	90	180
58	70	170
25	20	no LC phase
Copolymer ChM-11 with A-16		
45	45	100
Copolymers of ChM-11 with MA-22		
75	70	no LC phase
50	40	no LC phase

liquid crystals (Schiff's bases, cyanobiphenyl groups, esters of alkoxybenzoic acids, cholesterol esters etc.) occupy the most prominent place. Some types of LC polymers are given in Table 1. Recently a number of copolymers of mesogenic monomers with alkylacrylates and methacrylates was synthesized (Table 2) [36,37]; organometallic compounds, such as linear and crosslinked polysiloxanes displaying LC properties were obtained [38,39].

Worthy of attention are the attempts to produce LC polymers on the basis of inorganic polymers: those are polyphosphazenes with mesogenic side groups (cholesterol) although the first results to have been published were not promising [40]. A broad class of heterocyclic compounds could have probably contributed to the synthesis of new systems. The synthetic possibilities of this approach are quite evidently far from being exhausted.

3 Peculiarities of Thermotropic Liquid-Crystalline Polymers Related to Their Macromolecular Nature

One of the main peculiarities of thermotropic LC polymers is related to the high molecular mass of polymers themselves, which implies high viscosity of polymeric mesophases [41] exceeding the viscosity of corresponding low-molecular liquid crystals [42,43] by two orders of magnitude. This should be the cause for the slowing down of all structural rearrangements influenced in thermotropic LC polymers by external fields. Thermotropic polymers were thus viewed somewhat sceptically by researchers working in the field of low-molecular liquid crystals. It must be mentioned in advance that LC polymers form an independent new class of compounds and materials; they are not to be evaluated only in a way analogous to common liquid crystals.

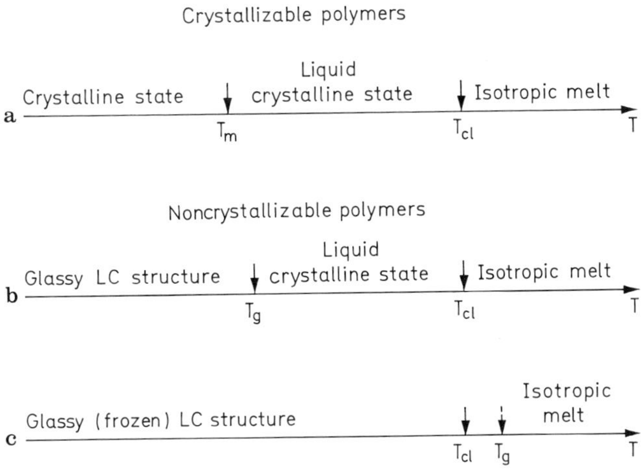

Fig. 3. Relationship between glass temperature T_g, melting temperature T_m, and clearing point T_{cl} for LC polymers

Let us consider some aspects of thermal behaviour of LC polymers [45] (Fig. 3). In case of crystallizable polymers, which are mainly those containing mesogenic groups in the main chain, the LC state is observed from above the melting temperature (T_m); and up to the clearing temperature (T_{cl}), the melt displays anisotropy and may flow. The polymer thus behaves alike low molecular liquid crystals (Fig. 3a), the viscosity of the former being, however, essentially higher.

A different situation is observed for non-crystallizable polymers, which include the vast majority of polymers with mesogenic side groups (Fig. 3b, c). In this case the low temperature limit for the existence of LC state is the glass transition temperature T_g (and not the melting temperature as for crystallizable polymers), above which the so-called segmental mobility, originating from the lability of distinct macromolecular segments, is exhibited. As a rule this temperature is lower than the clearing temperature and in a $T_g - T_{cl}$ interval the polymer either in the form of elastomer or in the form of viscous melt is in a LC state (Fig. 3b).

In contrast to low-molecular liquid crystals, which usually crystallize on cooling, polymers with mesogenic groups being cooled down undergo a glass transition. A liquid-crystalline structure characteristic of the mesophase is then preserved in a glassy state. Below the T_g the LC structure may thus be frozen and we actually deal with a glassy polymer having LC structure. This constitutes one of the most interesting peculiarities of LC polymers, which makes it possible, by making use of "flowable" LC phase and by cooling the polymer below T_g, to vitrify and fix LC structures with intrinsic anisotropy of mechanical, optical, electric and other properties in a solid material. If up to T_{cl} the polymer does not soften on heating, it actually implies that its hypothetic T_g is higher than T_{cl}, and the polymer is in a glassy state with a "frozen-in" LC structure (Fig. 3c). Most LC polymers with mesogenic groups attached directly to the backbone belong exactly to glassy LC compounds.

The possibility for the existence of mesophase in a rubbery state [36,46], typical only for macromolecular compounds with their natural ability to display big reversible deformations, reveals interesting prospects from the viewpoint of creation of new types of liquid-crystalline materials in the form of elastic films, as well as for development of the theory of viscoelastic behaviour of such unusual elastomers.

The significant feature of LC polymers in comparison with low-molecular liquid crystals consists in the broadening of the temperature range for the existence of mesophase. This is easy to see when comparing transition temperatures for low-molecular and polymeric LC compounds with identical mesogenic groups [47,48] (Fig. 4).

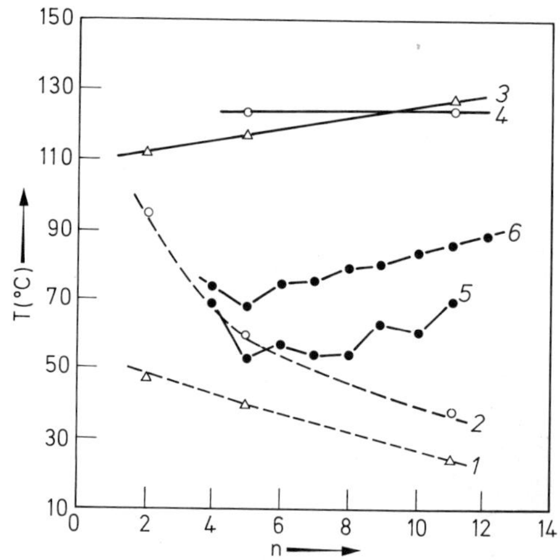

Fig. 4. Glass transition temperatures (1, 2), clearing points (3, 4, 6) and melting temperatures (5) vs number (n) of carbon atoms in the aliphatic substituent [36,47,50]:

1, 3 — LC acrylic polymers:

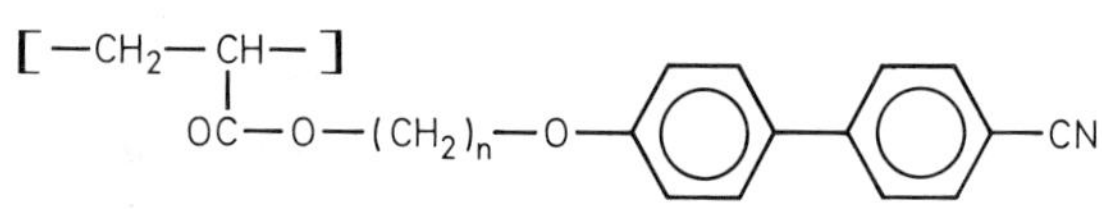

2, 4 — LC methacrylic polymers:

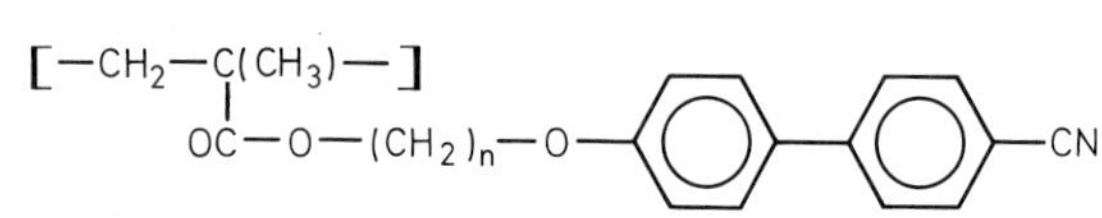

5, 6 — alkoxycyanobiphenyls:

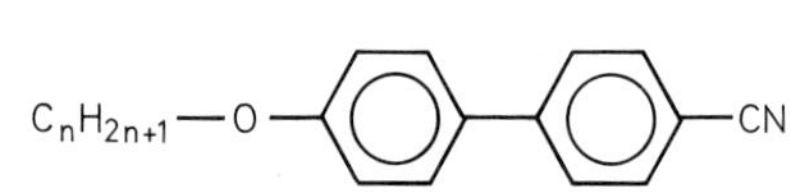

Besides, it is possible by copolymerization of one and the same mesogenic monomer with non-mesogenic comonomers to vary the type and temperature range of the mesophase [36,37] (Table 2). It is seen from the table that using alkylacrylates with alkyl groups of different length (A-n) as comonomers and varying the ratio of components it is possible to shift the transition temperatures of a LC phase.

As the formation of LC phase in comb-like polymers is predetermined by the interaction of mesogenic groups, it would have seemed, that the temperature range of LC state for such systems should not depend on the length of the main chain, i.e. on the degree of polymerization (DP). However, studies [44,49] on the dependence of T_{cl} on DP carried out for some polyacrylic and polymethacrylic derivatives of cyanobiphenyl, as well as for polyparabiphenylacrylate [51] (Fig. 5), have shown that

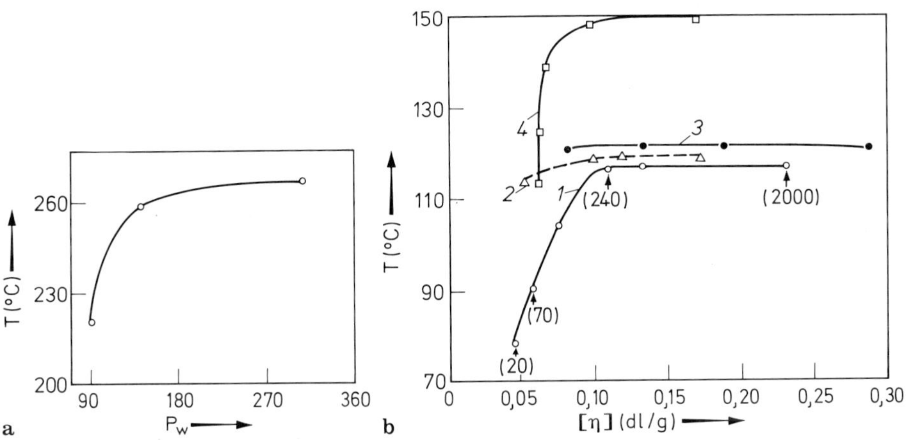

Fig. 5a and b. Clearing points of poly(*p*-biphenyl acrylate) **(a)** and biphenyl-cyano-containing polymers

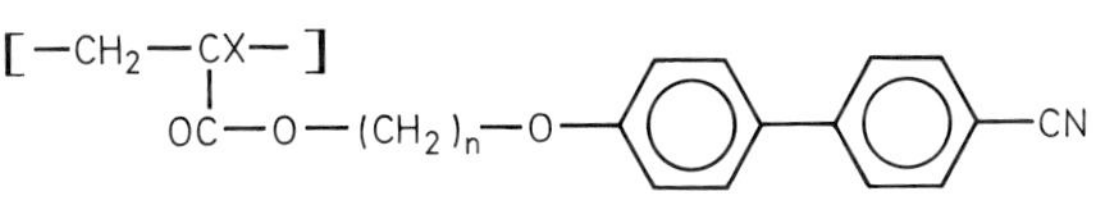

as a function of the degree of polymerization P_w and intrinsic viscosity [η] of their solutions in 1,2-dichloroethane: 1 — X = H, n = 5; 2 — X = CH_3, n = 5; 3 — X = CH_3, n = 11; 4 — X = H, n = 11 (in parentheses: P_w for some fractions of polymer 1) [44,49,51]

there is a definite critical length of a polymer chain, starting from which the clearing temperature really does not depend on DP [49,51][1].

The analysis of enthalpy and entropy changes on melting of mesophases has shown that in the low-molecular region the LC phase is probably formed mainly due to intermolecular contacts of the side groups, which gives the dependence of T_{cl} on DP. Exceeding the critical DP value leads evidently to the predominance of intramolecular contacts and to a smaller number of defects, which are unavoidable when the mesophase is formed only via intermolecular contacts of mesogenic groups.

1 Curves 2–4 in Fig. 5 give the dependence of T_{cl} on intrinsic viscosity of polymer solutions, which is in a first approximation proportional to the degree of polymerization.

Attempts were made [12–15] to disclose any relationship between the ability of low-molecular compounds and of polymers based on them to exist in a LC state. The analysis of data referring to the synthesis of LC polymers, leads to the following conclusions:

a) LC polymers can be synthesized, as a rule, from mesogenic compounds (i.e. monomers containing mesogenic groups), but it is not necessary for a monomer to form a mesophase to be transformed into a LC polymer. There exists a vast number of mesogenic monomers that do not form mesophases but their polymerization leads to mesomorphic LC polymers.

b) If the monomer is not mesogenic, the polymer, as a rule, does not display LC properties. In the majority of cases this is actually true; there exist exceptions, however, relating for example, to biphenyl derivatives [51–53]. Acyl derivatives of p-oxybiphenyl (I) and of its partially hydrogenated analog (II)

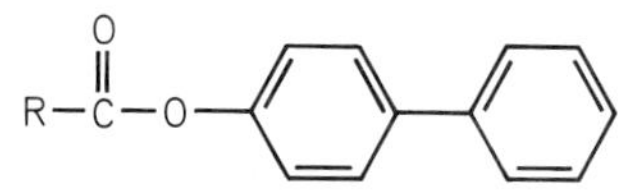

I

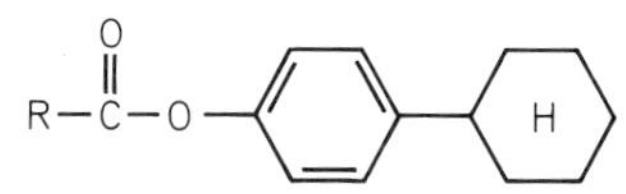

II

for instance, are not mesogenic (because they lack 4′ substituents in benzene and cyclohexane rings) and do not form a mesophase. At the same time, their polymeric analogs, carrying the same substituents in the side groups, — poly-para-biphenyl-acrylate (III) and poly-para-cyclohexylphenylacrylate (IV)

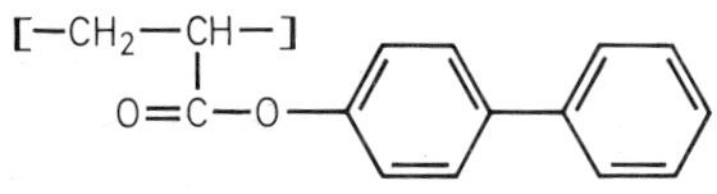

III

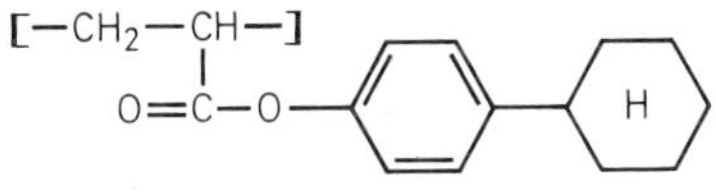

IV

may exist in a LC state of smectic type with clearing temperatures lying in the interval of 260–285 °C (for polymer III) and 205 °C (for polymer IV). The reasons for mesomorphism of these polymers are not yet clear; it may be suggested that the involvement of rigid biphenyl fragments along the backbone into a single ensemble enhances their dispersive interaction in a melt, which leads to the formation of mesophase.

The incorporation of rigid fragments of low-molecular compounds within a macromolecule may thus lead to manifestation of LC properties by polymers even if the respective low-molecular compounds do not form a LC phase by themselves. The possibility to induce mesomorphism in organic compounds is of significant interest as extending the line of LC compounds and exploiting non-mesogenic low-molecular compounds as a raw material to make polymeric LC compounds.

4 The Principles of Formation and Some Properties of Smectic, Nematic and Cholesteric Mesophases of Liquid-Crystalline Polymers

As is known, the question of how to predict the type of mesophase on the basis of the molecular structure of liquid crystals alone is not yet solved in spite of its vital importance for low-molecular crystals. Researchers rely only on some empirical rules [54]. For LC polymers the problem is even more complicated because of significantly smaller number of systems investigated up to date and their more complicated structures. However, some approaches how to predict the type and, particularly, limiting thermodynamic conditions of mesophase formation in thermotropic LC polymers are already being developed. Among theoretical studies of this kind the works of American scientists [55] should be noted, as well as calculation of thermodynamic conditions of smectic phase formation in comblike polymers [56], of nematic phase formation in linear polymers with macromolecules consisting of alternating rigid and flexible fragments [57–59] and, finally, investigation on theoretical computations for cholesteric LC polymers [60] — all of the latter carried out in the Soviet Union.

4.1 Smectic Liquid-Crystalline Polymers

4.1.1 Some Theoretical Approaches

Studies of LC thermotropic polymers with mesogenic side groups do convincingly show that these polymers are most liable to smectic type of ordering. The tendency to predominantly layer type of packing is probably predetermined by the fact that all mesogenic side groups are united into a single ensemble by the main chain. And it is exactly this type of mesogenic group ordering that is most easy to treat theoretically [56]. Let us consider thermodynamic conditions of smectic mesophase formation in melts of comb-like polymers, containing mesogenic groups [56]. Schematic presentation of the comb-like macromolecule is given in Fig. 6a. Its structure may be simulated on the tetrahedral lattice with x_1, x_2 and x_3 expressed in "steps along the lattice" units, and proportional to the C-C bond length. Assuming that smectic layers are formed only by mesogenic groups, and the main chains together with spacers (except for mesogenic groups themselves) are located in isotropic (amorphous) phase between the layers, it is then according to [61], possible to compute the difference in free energies of isotropic and smectic phases and to obtain the condition for phase equilibrium. The latter is given by the stipulation (1) [56]

$$x_1 = \frac{1}{1 + \varepsilon/kT} \ln \frac{z - 1}{P(x_2, x_3)} \tag{1}$$

where ε — is the energetic factor, corresponding to the energetic gain per lattice cell of the smectic layer formation; k is the Boltzmann constant; z is the coordination number of the lattice, which is 4; $P(x_2, x_3)$ is the probability that in a random walk along the lattice, which has begun at the boundary of an isotropic layer, we will return in $x_2 + 2x_3$ steps to the same or the opposite boundary of this layer. In calculation of $P(x_2, x_3)$, performed with a computer, several cases, corresponding to flexible and semiflexible x_2 and x_3 chains, were considered. Rigidity was expressed as the probability for the occurence on the lattice of trans- and gauche-isomers. With x_1 greater than the value calculated according to formula (1), the smectic phase is thermodynamically stable, in the opposite case — the isotropic phase. The computations resulted in the drawing of phase diagrams demonstrating the relationship between basic molecular parameters of polymeric chains x_1, x_2 and x_3 (at $\varepsilon = 0$).

Table 3. LC smectic polymers with indefinite type of smectic mesophase

Polymer No.	Structure of monomer unit	T_g, °C	T_{cl}, °C	ΔH_{cl}, J/g	Ref.
A. Polymers containing phenyl, diphenylbenzoate and diphenyl groups					
1.	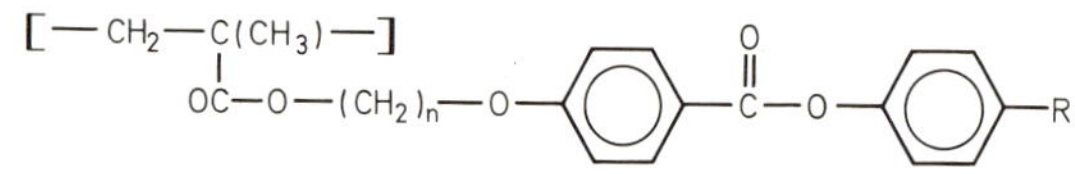				
1.1. n = 2,	R = $-C_6H_{13}$	100	156	—	[79]
1.2. n = 2,	$-OC_3H_7$	120	129	9.2	[79,80]
1.3. n = 2,	$-OC_6H_{13}$	100	140	11.3	[79,80]
1.4. n = 3,	$-OC_6H_{13}$	100	120	6.8	[81]
1.5. n = 3,	$-C_6H_{13}$	100	120	—	[81]
1.6. n = 6,	$-C_6H_{13}$	60	90	—	[81]
1.7. n = 6,	$-OC_6H_{13}$	60	115	15.5	[79]
2.	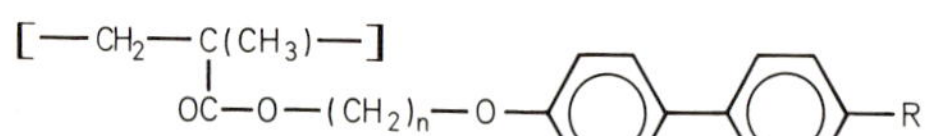				
2.1. n = 6,	R = $-OC_5H_{11}$	80	129	16.8	[82]
2.2. n = 6,	$-OC_6H_{13}$	—	159	22.7	[82]
3.[a]	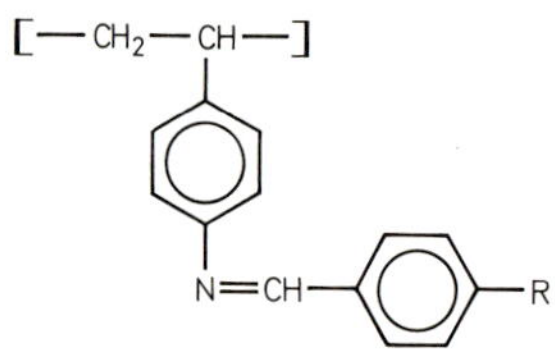				
3.1.[a]	R = $-OC_4H_9$	55	—	—	[83-85]
3.2.[a]	$-OC_6H_{13}$	—	—	—	[83]
B. Cholesterol containing Polymers					
3.	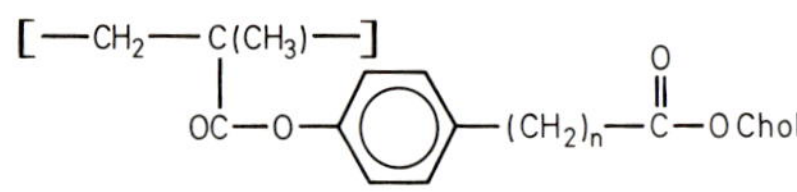				
3.1. n = 2		—	—	—	[80]
3.2. n = 6		100	182	6.7	[80]
3.3. n = 12		100	168	4.2	[80]
4.	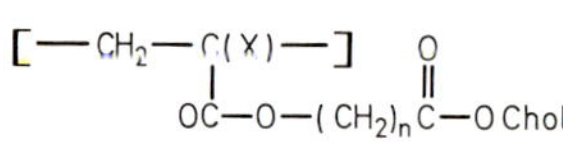				
4.1.	X = CH_3, n = 5	85	210	2.1	[86,87]
4.2.	X = CH_3, n = 10[b]	60	124; 158	0.4; 1.4	[86,87]
4.3.	X = CH_3, n = 14[b]	50	54; 153	4.2; 6.3	[86,87]
4.4.	X = H, n = 5	55	218	4.2	[86,87]
4.5.	X = H, n = 10	35	148	3.2	[86,87]

Table 3 (continued)

Polymer No.	Structure of monomer unit	T_g, °C	T_{cl}, °C	ΔH_{cl}, J/g	Ref.
C. Poly(siloxanes)					
5.	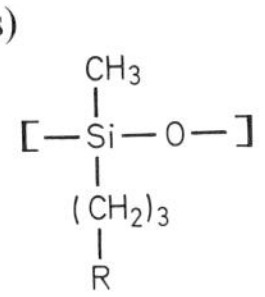				
5.1.	R = —Chol	45	115	2.7	32)
5.2.	$—OC_6H_{13}$	15	85	11.6	32)
5.3.	—CN	20	61	1.9	32)

[a] Authors of the papers [83–85] define this type of mesophase as an "intermediate" between S and N phases;
[b] There exist two smectic phases

Table 4. LC polymers with supposed S_A structure

Polymer No.	Structure of monomer unit	T_g, °C	T_{cl}, °C	ΔH_{cl} (J/g)	Ref.
A. Polymers containing Schiff-base derivatives					
1.	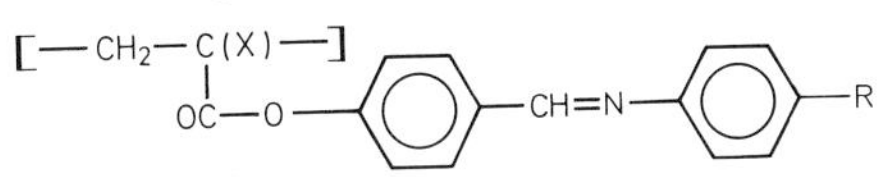				
1.1.	X = H, R = $—OC_2H_5$	180	310	—	7, 88)
1.2.	$—CH_3$, $—OC_2H_5$	—	310	—	79, 80)
1.3.	$—CH_3$, —COOH	154	—	—	84)
2.	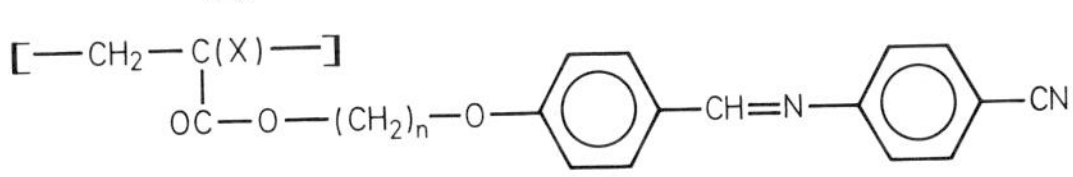				
2.1.	X = H, n = 11	10	169	8.6	65, 89)
2.2.	$—CH_3$, n = 6	35	125	4.2	65, 89)
2.3.	$—CH_3$, n = 11	25	155	7.3	65, 89)
B. Polymers with diphenyl groups					
1.	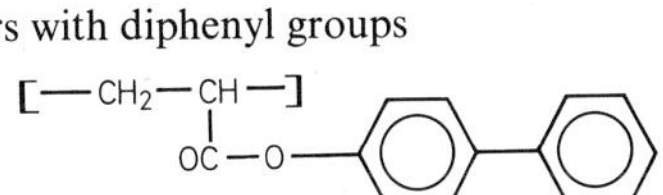	110	285	21.4	51–53)
2.	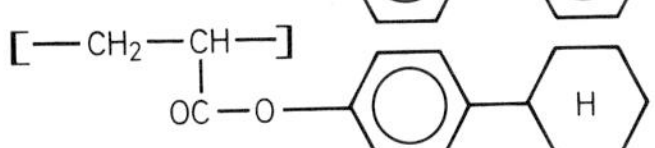	101	205	17.9	51–53)
3.	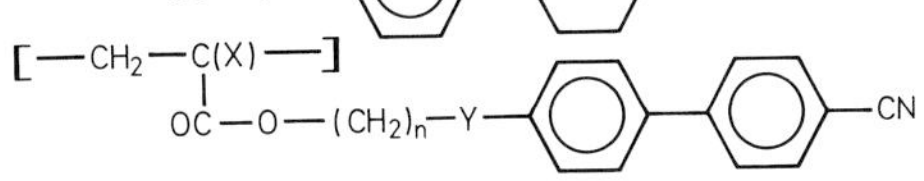				
3.1.	Y = —O—, X = CH_3, n = 5	60	121	6.7	47, 62)
3.2.	Y = —O—, X = CH_3, n = 11	40	121	8.8	47, 62)
3.3.	Y = $—CH_2—$, X = CH_3, n = 10	30	81	7.1	47, 62)
3.4.	Y = —COO—, X = CH_3, n = 10	45	93	4.2	47, 62)
3.5.	Y = —O—, X = H, n = 11	25	145	6.3	47, 62)

Table 4 (continued)

Polymer No.	Structure of monomer unit		T_g, °C	T_{cl}, °C	ΔH_{cl} (J/g)	Ref.
C. Polymers with phenyl and diphenylbenzoate groups						
1.	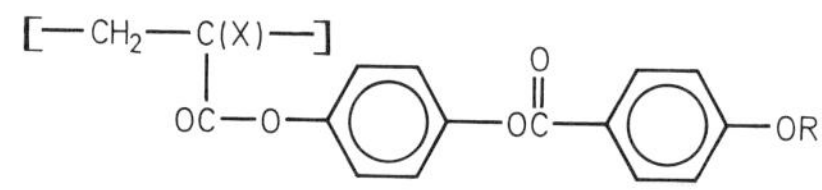					
1.1.	$X = CH_3$,	$R = n\text{-}C_4H_9$	180	270	27.8	
1.2.	$X = CH_3$,	$n\text{-}C_9H_{19}$	140	220	7.2	
1.3.	$X = CH_3$,	$n\text{-}C_{12}H_{25}$	130	215	7.7	[76,77,78,90]
1.4.	$X = CH_3$,	$n\text{-}C_{16}H_{33}$	160	220	8.5	
1.5.	$X = H$,	$n\text{-}C_{16}H_{33}$	100	200	—	
D. Cholesterol containing polymers						
1.	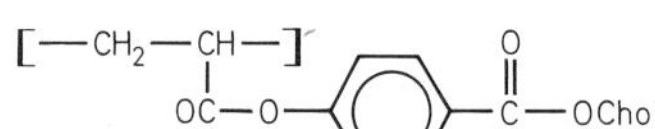		—	—	—	[84]
2.	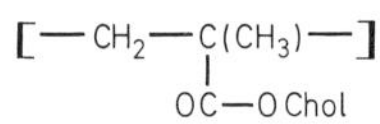		160	—	—	[84]
3.	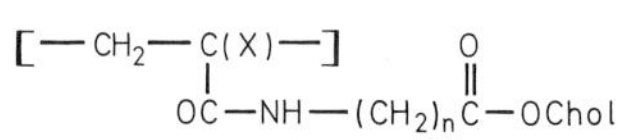					
3.1.	$X = CH_3$,	$n = 2$	180	—		
3.2.	$X = CH_3$,	$n = 5$	130	220		
3.3.	$X = CH_3$,	$n = 6$	130	215		
3.4.	$X = CH_3$,	$n = 8$	130	200		
3.5.	$X = CH_3$,	$n = 10$	125	185		[91,92]
3.6.	$X = CH_3$,	$n = 11$	120	180	3.3	
3.7.	$X = H$,	$n = 2$	190	150		
3.8.	$X = H$,	$n = 5$	165	220		
3.9.	$X = H$,	$n = 11$	110	185		

polymers (polymers A.2.1, A.2.2, B.3.2–B.3.5 in Table 4). The widespread structural type for these polymers is two-layer packing with overlapping of aromatic groups (Fig. 11a), the smectic layer thickness being approximately one and a half of the length d_1 of the side branch. Such setting of mesogenic groups is evidently stabilized by the interaction of constant dipoles induced by nitrile groups. A one-layer packing with complete overlapping of antiparallel-oriented side groups, which ensures their densest packing, corresponds to the second structural type, when the period d_2 is observed.

4.1.2.2 S_B-Mesophase and Other Phase Structure Types with Translationally Ordered Arrangement of the Side Groups in Layers

In S_B mesophase the side groups form layers, the axes of the side chains being perpendicular to the layers. The dictinctive feature of S_B mesophase is that the side

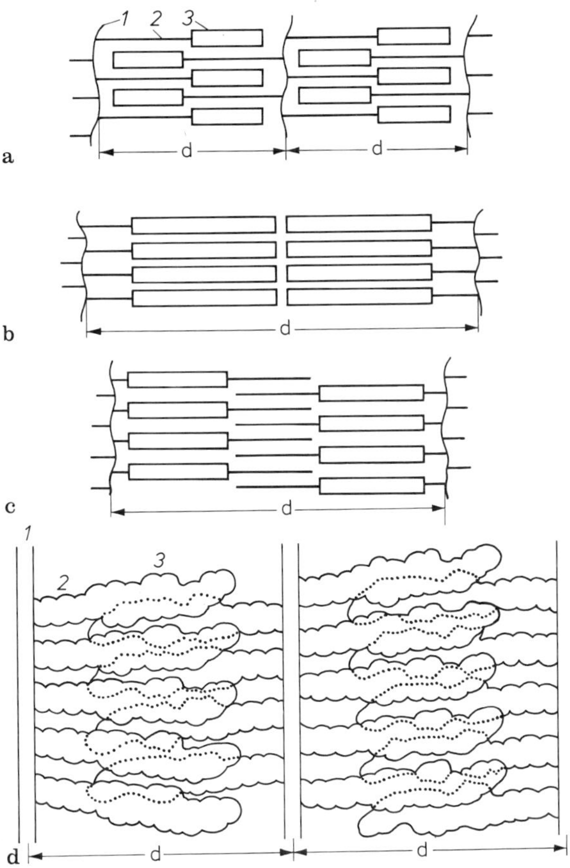

Fig. 10a–d. Schemes of side chains packing of macromolecules in S_A-mesophase of orientated LC polymers; **(a)** one-layer packing; **(b)** two-layer packing; **(c)** packing with overlapping of alkyl "tails"; **(d)** packing with partial overlapping of mesogenic fragments (side chains are designated by van-der-Waals radii): 1 — main chain; 2 — spacer; 3 — mesogenic group (Side groups of neighbouring macromocules (in the cases a, c and d) are arranged in different planes parallel to the plane of the figure)

branches are more ordered as compared to S_A mesophase and, as a rule, they form an hexagonal type of packing that approaches the structure of crystalline compounds.

That is why here we describe two groups of polymers–those, forming an "ordered" smectic phase (of S_B type) and polymers with mesogenic groups, for which the formation of crystalline structure is proposed (Table 5). Sometimes it is difficult to distinguish between crystalline and LC states of ordered smectic phases. Thus, the interpretation of data on crystalline phases of some polymers, listed in Table 5, is also possible from the viewpoint of smectic polymorphism.

As is seen from Table 5, there are quite a few polymers, that demonstrate an S_B mesophase type. X-ray patterns of such polymers, along with a series of small angle reflexes, display, as a rule, a sharp and intensive reflex at wide scattering angles (Fig. 12). For LC polymers of this type there exist two possibilities of packing: with

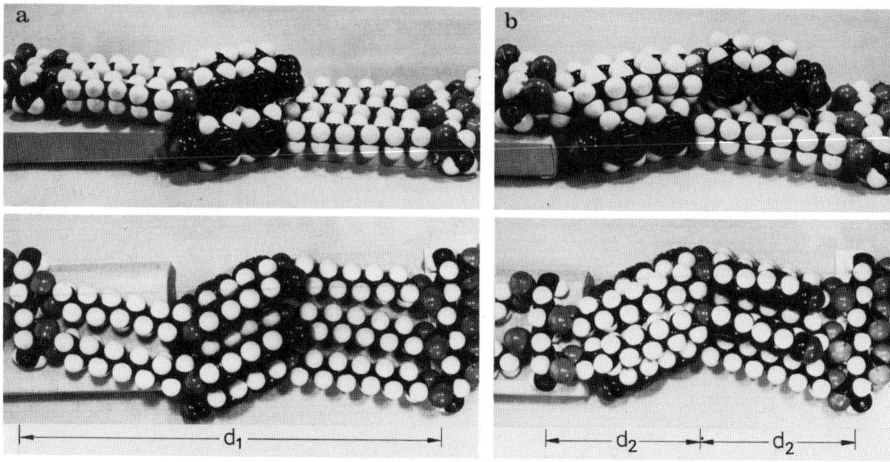

Fig. 11. Molecular models of layer packings of macromolecules in S_A mesophase with d_1 **(a)** and d_2 **(b)** spacings for polymer B.3.5 (Table 4)

Table 5. LC polymers with side groups translationally-ordered in layers

Polymer No.	Structure of monomer unit	T_g, °C	Phase transitions	ΔH, (J/g)	Number and type of reflexes at wide angles	Supposed type of structure	Ref.
1	2	3	4	5	6	7	8
1.	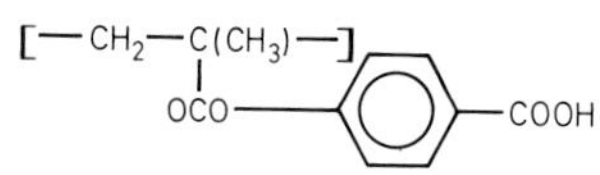	110	—	—	1, sharp	S_B K	94) 84,95)
2.	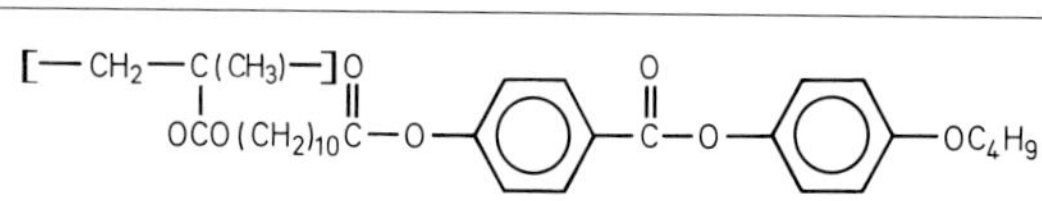	55	M_1130I	11.3	1, sharp	S_B	90)
3.	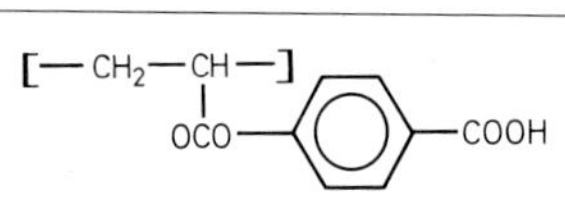	—	M_1252I	—	3, sharp	S_E K	94) 84)
4.	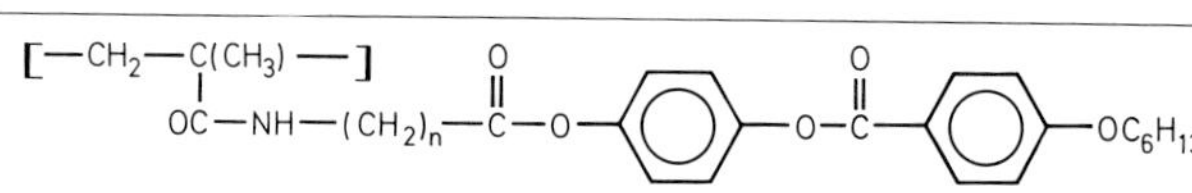						
4.1.	n = 5	—	M_1139M_2183I	13.0; 1.3	1, sharp 1, diffuse	K S	97–99)
4.2.	n = 11	—	M_173M_2160I	2.5; 8.4	1, sharp 1, diffuse	K S	97–99)

Table 5 (continued)

Polymer No.	Structure of monomer unit	T_g, °C	Phase transitions	ΔH, (J/g)	Number and type of reflexes at wide angles	Supposed type of structure	Ref.
5.	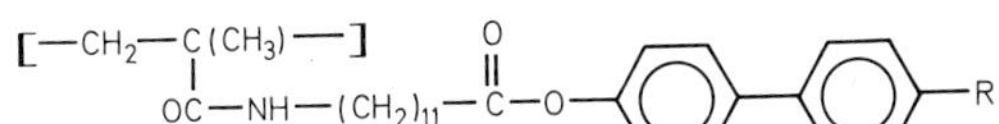						
5.1.	R = H	—	$M_1 75 I$	6.3	1, sharp	K	100)
5.2.	R = OC_6H_{13}	—	$M_1 143 M_2 160 I$	9.2 4.9	1, sharp	K S	100)
6.	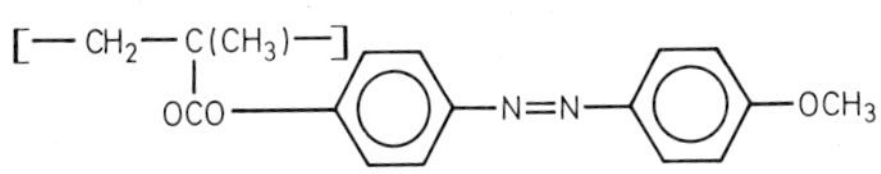	100	$M_1 270 I$	—	4, sharp	K	101)
7.	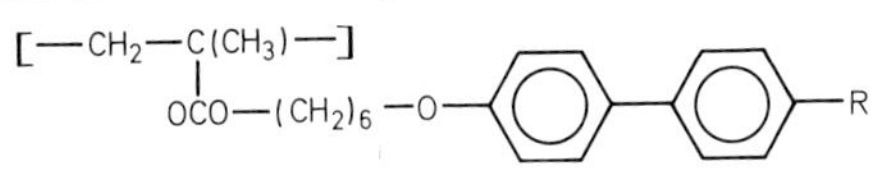						
7.1.	R = OCH_3	—	$M_1 119 M_2 136 I$	12.6; 7.1	—	K S	82)
	R = OCH_3	—	$M_1 117 N(?) 131 I$	32; 7.4	3, sharp	K N(?)	102)
7.2.	R = OC_2H_5	—	$M_1 134 M_2 162 I$	10.5; 10.9	—	K S	82)
8.	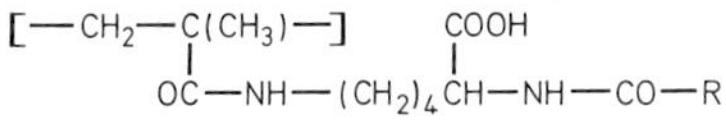						
8.1.	R = n-$C_{15}H_{31}$	—	$M_1 32 M_2$	—	3, sharp	K	103–104)
8.2.	n-$C_{17}H_{35}$	—	$M_1 34 M_2$	20.9	3, sharp	K	
8.3.	n-$C_{21}H_{43}$	—	$M_1 56 M_2$	49.0	3, sharp	K	
9.	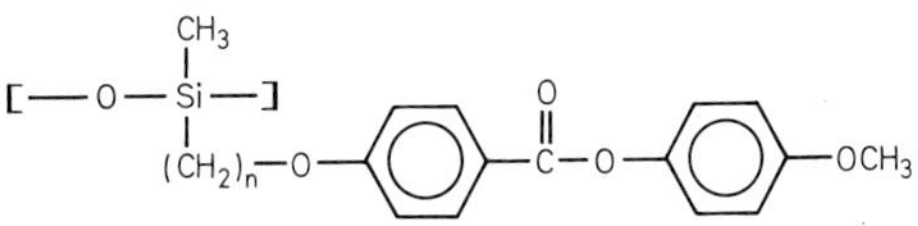						
9.1.	n = 5	—	$M_1 87 N_2 115 I$	—	—	K	39, 105)
9.2.	n = 6	—	$M_1 52 N_2 112 I$	—	—	K	

[a] M_1 and M_2 — mesophases with translationally-ordered and disordered arrangement of the side groups in the layers

orientation of side groups perpendicular to a layer plane, — a true S_B phase (polymer 2 in Table 5); and with a tilted orientation of side groups — a S_{Bt} phase [94]. According to the classification accepted for low-molecular liquid crystals, the latter should be ascribed an S_I mesophase type [96].

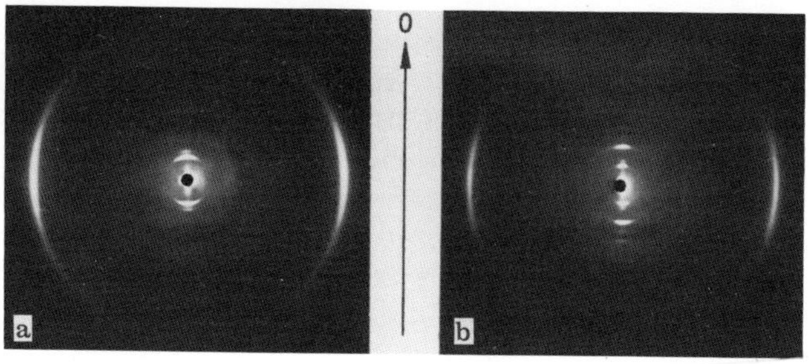

Fig. 12a and b. X-ray diagrams of oriented LC smectic polymers in S_B-mesophase [65, 66, 89]: **a)** polymer 2 (Table 6); **b)** polymer 5 (Table 6) (0 is texture axis)

The second group involves polymers with three-dimensional ordering of side branches (e.g., those forming M_1-phase)(Table 5). On X-ray patterns of these polymers 3–4 narrow reflexes at wide angles are observed. As a rule, the authors define this type of structure as crystalline, or ascribe a smectic type of structure, characteristic for ordered smectics in S_E or S_H phases. The heats of transition from anisotropic state to isotropic melt are usually small and do not exceed the heats of transition "smectic liquid crystal — isotropic melt". The similarity of structural parameters of three-dimensionally "ordered" smectics and that of crystalline polymers of the type here considered, make their correct identification quite a difficult task.

For polymeric smectic mesophases polymorphism is just as plausible as for low-molecular smectics. Examples of polymorphism of the type $S_B \rightleftarrows S_A$, discovered for polymeric smectics containing Schiff bases as mesogenic groups, are listed in Table 6 [66, 75].

Table 6. Polymorphism (S_B and S_A phases) of LC polymers with Schiff base fragments in side chains [66, 75]

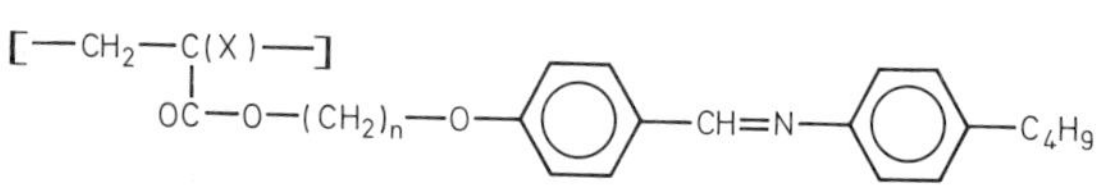

X	n	$S_B \rightarrow S_A$ phase transition		$S_A \rightarrow I$ phase transition	
		T_1, °C	ΔH_1 (J/g)	T_{cl}, °C	ΔH_{cl} (J/g)
H	3	70	—	165	2.5
H	6	76	4.2	115	8.4
H	11	90	6.7	149	11.3
CH_3	6	77	3.3	115	7.9
CH_3	11	86	5.0	140	12.5

Table 7. LC polymers with supposed S_C structure

Polymer No.	Structure of monomer unit	T_g, °C	T_{cl}, °C	Type of packing; angle of tilt of the side groups in layer	Ref.
1.	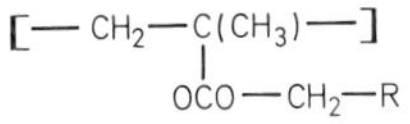				
1.1	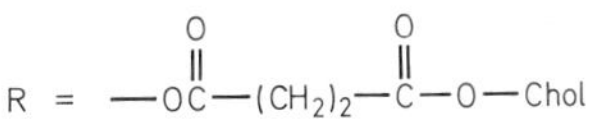	—	—	S_C	[72]
1.2.	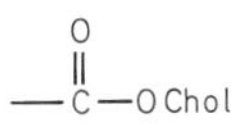	—	—	S_C	
1.3.	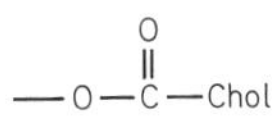	—	—	S_C	
2.	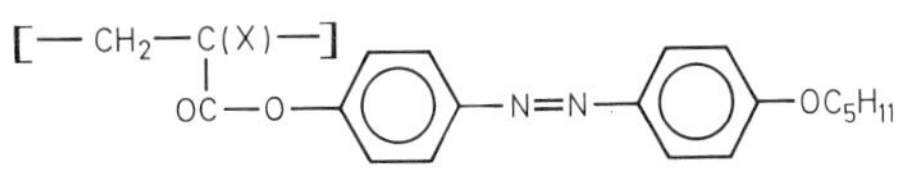				
2.1.	X = H	75	>270	Two layer packing, 36°	[101, 106, 107]
2.2.	$X = CH_3$	135	>270	Two layer packing, 50°	
3.1.	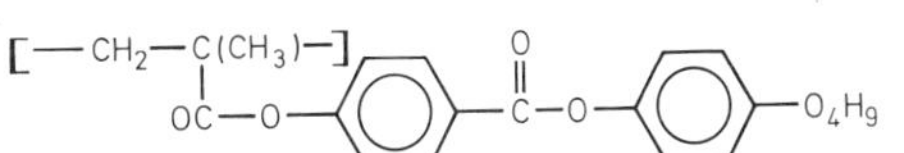	140	260	"fir-tree" structure	[76, 77]
3.2.	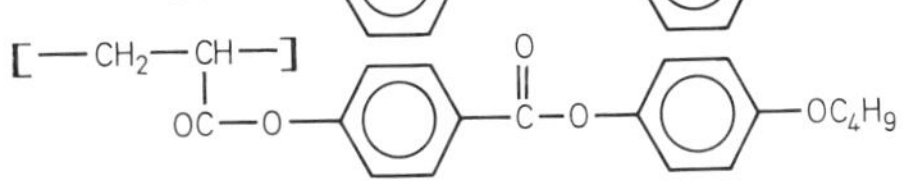	100	240	"fir-tree" structure	

4.1.2.3 S_C-Mesophase

This type of structure, characterized by tilted orientation of side groups in respect to smectic layers, one can find very rarely for LC polymers. An S_C phase was suggested in the works of Hungarian researchers [72], but no detailed structural data on macromolecular packing was reported. The ascription of polymers to this type of mesophase was based on the compatibility data for cholesterol-containing polymers (polymers 1.1–1.3, Table 7) mixed with an S_C phase of low-molecular liquid crystal terephthalyl-bis-butylaniline. It is known, however, that the principle of LC phase identification according to their compatibility, valid for low-molecular liquid crystals, is not always applicable to systems involving LC-polymer and low-molecular liquid crystals. For instance a tilted phase was suggested for polymers 2.1–2.2 (Table 7) in [101, 105, 107]. The proposition was based on X-ray data on the thickness of the smectic layer, which appeared to be smaller than the value calculated for two-layer S_A structure. It is however possible to suppose as an alternative (oriented samples were not investigated in this work), the formation of S_A mesophase with a packing typical for polymeric liquid crystals and characterized by overlapping of paraffinic "tails".

The most convincing and unambiguous criterion for the existence of S_C phase can be taken from X-ray investigations of oriented samples only. X-ray patterns of this type are characterized by four-point splitting of small angle reflexes; and this was observed for polymers 3.1–3.2 (Table 7). The formation of S_C phase from S_A via a phase transition occuring on cooling of polymers (Table 8) was reported in [62,75]. $S_A \rightarrow S_C$ transition, which was first reported for polymeric liquid crystals in [75], is revealed on an X-ray pattern exactly as a four-splitting of a small angle reflex

Table 8. Polymorphism (S_C and S_A phases) of LC polyacrylates with CN-containing mesogenic groups [62,75]

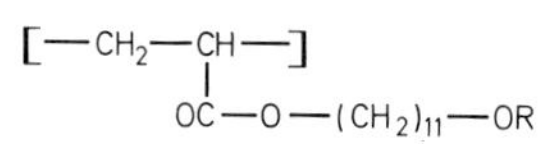

R	$S_C \rightarrow S_A$ phase transition		$S_A \rightarrow I$ phase transition	
	T_1, °C	ΔH_1 (J/g)	T_{cl}, °C	ΔH_{cl} (J/g)
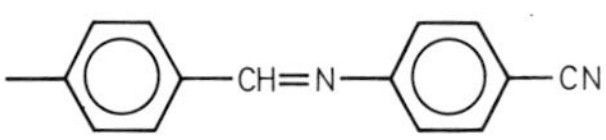	31	~0.9	169	8.3
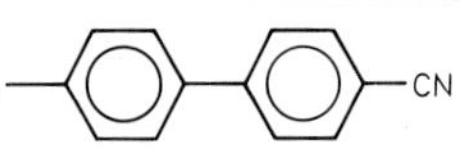	30	—	145	6.3

(see Fig. 13a, b) and is accompanied by a change of long period d_1. The schemes of macromolecular packing in S_A and S_C phases are given on Fig. 13c, d. An $S_A \rightarrow S_C$ transition is also characterized by an endothermic peak on the DTA curve (see values for transition enthalpy in Table 8) and by a significant change of optical texture (Fig. 14). Above the $S_A \rightarrow S_C$ transition temperature, both polymers display a fan texture of S_A mesophase (Fig. 14a). At temperatures, corresponding to the beginning of the transition, at first circular bands appear on the fans (Fig. 14b), which are then completely destroyed (Fig. 14c). It is important that a dense one-layer packing (Fig. 10a), coexisting with a two-layer packing with mesogenic group overlapping (Fig. 13) is unaffected by an $S_A \rightarrow S_C$ transition; i.e. two smectic types of packing actually coexist — one with a one-layer packing of side groups (S_A) and the other with a packing of side groups (S_C) tilted to the layer plane two-layer. Coexistence of the two types of packing was also reported for polymers 3.1–3.2 (Table 7) [76,77].

Apparently, coexistence of at least two (and probably even more) types of packing constitutes one of the peculiarities of LC polymers with mesogenic side groups. This feature, probably, is exclusive to polymers, as no similar phenomena have been observed in low-molecular liquid crystals.

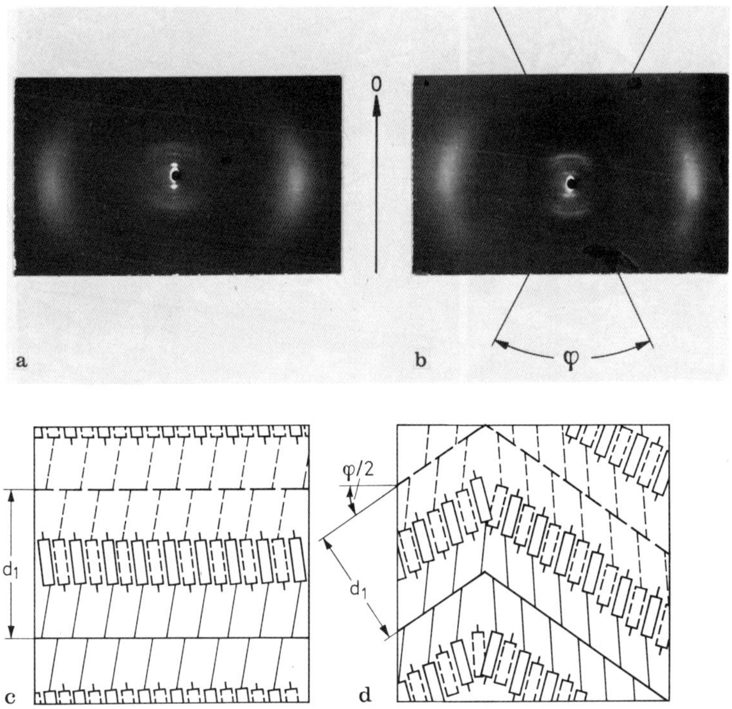

Fig. 13a–d. X-ray diagrams **(a, b)** of oriented LC polymer 2 (Table 8) and schemes of molecular packing **(c, d)** in S_A (a, c) and S_C (b, d) mesophases (0 — is texture axis)

4.1.2.4 Some General Remarks on the Structure of Smectic Polymers

A specific feature of all polymeric smectics, independently of the mesophase type, is that the regions of LC ordering are limited, while low-molecular liquid crystals display macroscopic ordering. This follows from the analysis of absolute intensities of small angle X-ray diffraction performed recently [108,109] (see polymers C.1.1–C.1.5; Table 4).

Using the scheme of macromolecular packing of a LC polymer and the values of the electron density distribution (Fig. 15a, b) the authors calculated one-dimensional correlation functions $\gamma_1(x)$ obtained by Fourier-transform of X-ray scattering intensity curves. Figure 15c shows a one-dimensional correlation function $\gamma_1(x)$ for two polymers with identical

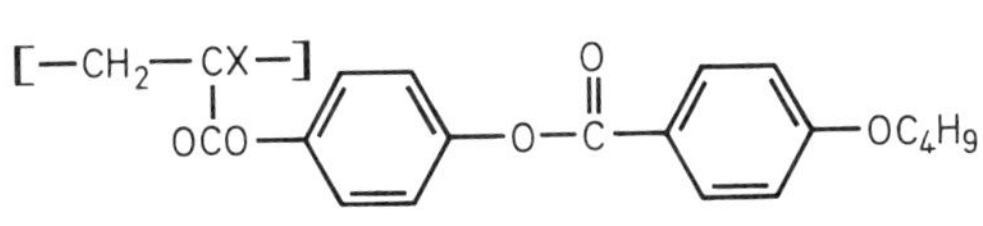

V : X = H
VI : X = CH3

Table 9 (continued)

Polymer No.	Structure of monomer unit			T_g, °C	T^a_{cl}, °C	ΔH_{cl} (J/g)	Method[b] of determination	Ref.
2.5.	X = H ,	n = 6 ,	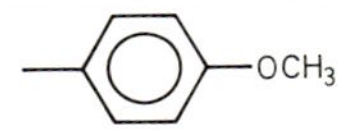	100	S121 N271	—	DSC	[119]
2.6.	X = H ,	n = 6 ,	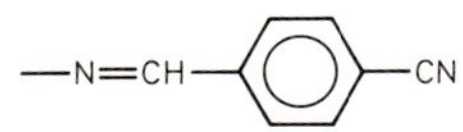	56	K110 S138 N257	—	DSC	[120]
				35	N211	—	DSC	[119]
3.			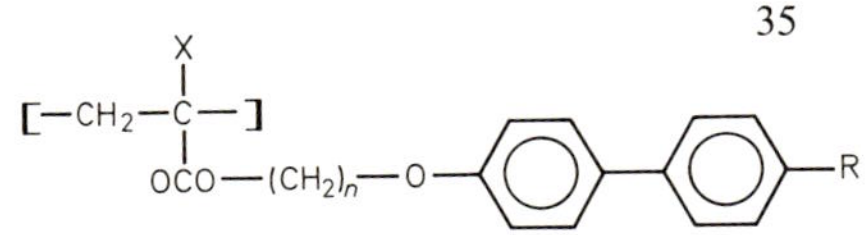					
3.1.	X = CH_3 ,	n = 2 ,	R = OCH_3	120	152	2.8	DSC	
3.2.[e]	X = CH_3 ,	n = 6 ,	OCH_3	—	K117 N(?)	32; 7.4	DSC, X-ray, Opt.	
				—	K119 S136	12.6; 7.1	DSC	[82]
3.3.	X = H ,	n = 2 ,	R = CN	50	112	0.4	DSC, X-ray, Opt.	[62, 121]
3.4.[f]	X = H ,	n = 5 ,	CN	40	120	1.7	DSC, X-ray, Opt.	[62, 121]
4.			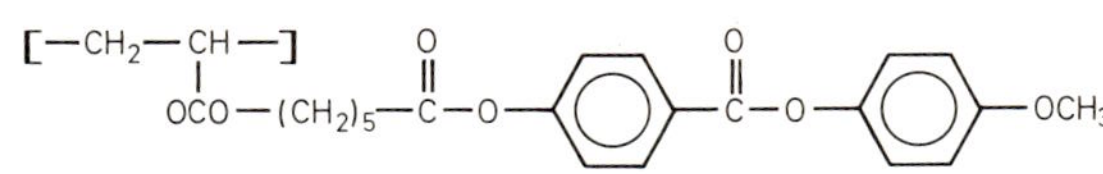	15	105	2.1	DSC, X-ray, Opt.	[123]

C. Poly(siloxanes)

5.

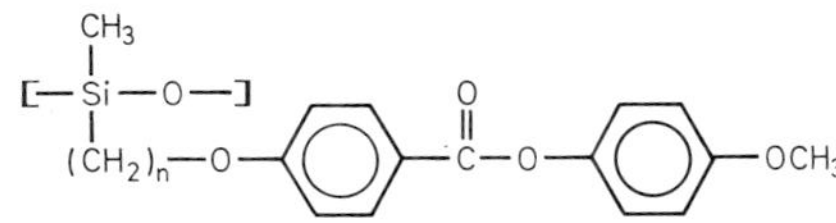

5.1.	n = 3	15	61	2.2	DSC	32, 105)
5.2.	n = 4	15	95	—	DSC	39)
5.3.	n = 5	—	K87 N115	—	DSC	105)
5.4.	n = 6	—	K52 N112	—	DSC	39, 105)

[a] T_{cl} corresponds to the melting temperature of the nematic mesophase; if there exist another type of structure, corresponding symbols are added;
[b] X-ray — X-ray analysis; DSC — Differential scanning calorimetry; Opt. — Optical microscopy; Mix. — Investigation of mixing of LC polymers with low molecular liquid crystals;
[c] crosslinked polymer;
[d] polymers have layered structure according to X-ray analysis;
[e] polymer forms fan-shaped texture;
[f] there are two sharp reflexes on X-ray diagrams at small angles

Thus, polymers with mesogenic groups in side chains form structural mesophases of the same types as low-molecular liquid crystals. This makes it possible to apply traditional mesophase classification for the description of the structure of LC polymers. At the same time, the structure of some of comb-like polymers (see Table 5) considered as crystalline, may probably be treated as one of highly-ordered smectic mesophases (S_H or S_I), whose study is only started [74].

4.2 Nematic Liquid-Crystalline Polymers

LC polymers of nematic type are mainly linear polymers, containing mesogenic groups in their main chains. Amongst LC polymers with mesogenic side groups nematic type of ordering occurs more seldom than the smectic one. As a rule, nematic structure occurs for homopolymers with short end substituents in mesogenic groups (CH_3, OCH_3, CN) and with a spacer of two to six methylene units long (Table 3). The exclusions are a few polymers, lacking a flexible link (e.g. polymer A.1 in Table 9), and crosslinked polymers, produced by polymerization of divinyl derivatives (polymer 2, Table 3). As regards crosslinked polymers, owing to the lack of data on the crosslink densities, and the slowness of the ordering processes towards the formation of equilibrium structures, their structure may not be considered to be unambiguously proved as nematic. Polymers 3.1–3.4 in Table 9 were assigned by Hungarian investigators to the nematic type according to the data on miscibility with low-molecular nematics. On the other hand, small angle reflexes are present on the X-ray patterns of these polymers. The difference between these results was explained by Cser [70] as resultant of the specific structure of nematic polymers which will be discussed later.

It is seen from Table 9, that nematic structure is really not typical for polymers without spacer groups and more often occurs among comb-like LC polymers. However, when identifying nematic polymeric mesophases the situation is somewhat more complicated than in case of polymeric smectics. The assignment of mesophase to the nematic type is mainly based on the results of calorimetric studies (see Table 9, methods of investigation). Structural studies are usually restricted to recording the absence of small angle reflexes on X-ray patterns of unoriented samples. Low enthalpy of transition from anisotropic to isotropic state (Table 9; its values are close to the respective values for low-molecular liquid crystals) and the absence of layer reflexes, seem to indicate a one-dimensional type of ordering. However, these data are, apparently, insufficient for the complete description of the structure of nematic polymers.

Similarly with low-molecular nematics is manifested in that the nematic polymers may form equally well schlieren texture, typical for low-molecular nematics (Fig. 18a) (polymers B.3.3–B.3.4, Table 9). The enthalpy of transition from LC state to isotropic melt is also close to that for low-molecular nematics. At the same time, there also exist definite structural differences. X-ray patterns of the same polymers, even in unoriented state, display certain elements of structural ordering in the arrangement of side branches (a weak diffuse halo at small angles), which could indicate a sibotactic nematic type of ordering. These differences are most distinct for oriented polymer films. As an example Fig. 18b, c, present X-ray patterns of unoriented and oriented samples of one and the same nematic polymer [121, 124]. In fact two sharp small angle

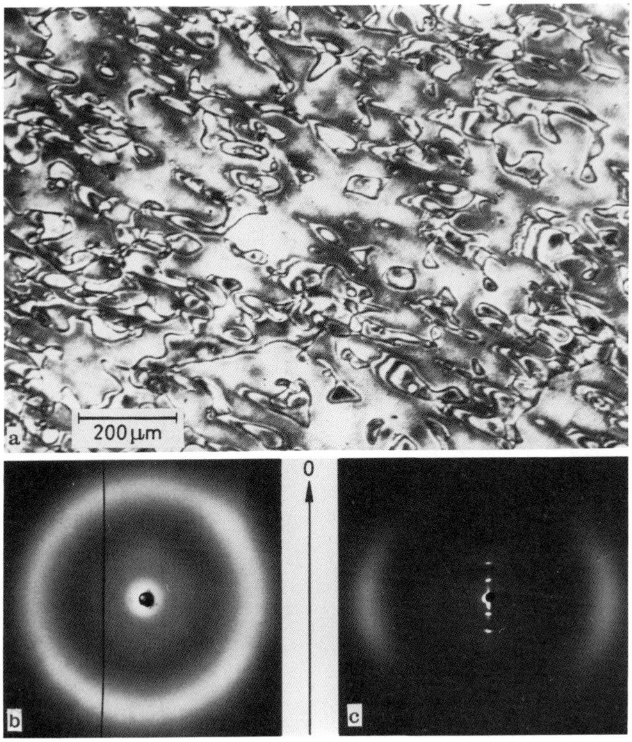

Fig. 18a–c. Schlieren texture **(a)** and X-ray diagrams of unoriented **(b)** and oriented **(c)** samples of polymer B.3.4 (Table 9) (0 is texture axis)

reflexes which are observed on X-ray diagramms indicate perfect layer ordering of side branches in oriented state rather than a sibotactic packing.

On the other hand, polymer B.3.2 (Table 9), which according to the X-ray analysis, does not exhibit a layer structure, displays fan texture, typical for low-molecular and polymeric smectics. It might be suggested, that this polymer also possesses some elements of layer ordering, which are different from those known for low-molecular sibotactic nematics.

It could be concluded that there exists a more close structural analogy between nematic and smectic phases in LC polymers than for low-molecular liquid crystals.

Are the observed peculiarities specific only for the polymers listed, or are they common for the whole class of polymeric nematics? Today, the experimental results are too few to permit an unambigous conclusion. Nevertheless, for comb-like polymers for those that do not contain mesogenic groups, a certain tendency exists to the formation of layer structures even in isotropic melts [33]. It may thus be suggested that for comb-like polymers with mesogenic groups as well, the polymeric nature of such liquid crystals may cause certain enhancement in ordering of nematic structures, formed by mesogenic groups. Structural data on polymer A.4 (Table 9) also confirm this [65, 89]. It appeared that for oligomeric fractions of this polymer the formation of typical for low-molecular nematics "schlieren" texture is common. Increasing the degree of polymerization leads to "degeneration" of this type of texture and to the formation of structures resembling to a greater extent the fan textures of smectic phases.

Another specific feature of polymeric nematics was disclosed when investigating oriented films.

Such polymers adopt, when affected by a mechanical field, an optically uniaxial homeotropic structure; polymers B.1.2, B.1.7, B.1.8 (Table 8) have positive birefringence; polymers B.1.1, B.1.8. (Table 9) have negative birefringence, which has not been reported to our knowledge, for low-molecular nematic liquid crystals. Although the authors do not comment on the cause for the observed phenomenon, the fact in itself is sufficiently uncommon.

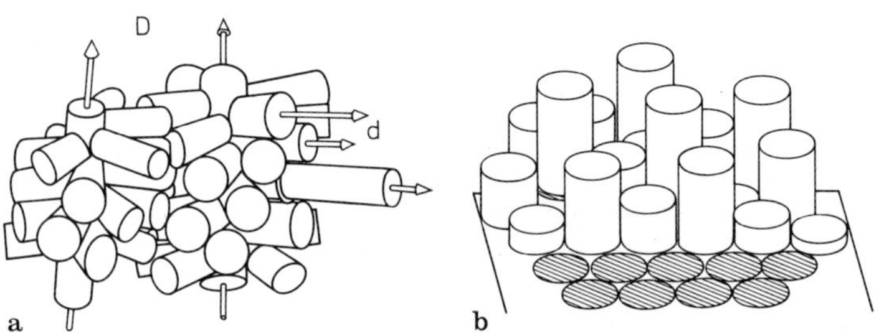

Fig. 19a and b. Schematic model of cylindrical macromolecules **(a)** and their semihexagonal packing **(b)** for nematic LC polymers [70]

The only general statement that might be made is that the specific features of the structure of polymeric nematics are provided for by the ordering effect of the main chain on the packing of mesogenic groups in LC phase.

Because of rather "poor" X-ray patterns of nematic LC polymers, displaying some elements of layer ordering, most authors confine themselves to the constatation of only orientational one-dimensional order in these polymers. The exception is probably only the work of Hungarian scientists (Cser in particular[70]), who develop quite different views on the structure of nematic LC polymers. In a model be proposed [70] (Fig. 19a, b) the macromolecule has an overall shape of a cylinder with a diameter D, where the side chains are located perpendicularly to the axis of the cylinder (Fig. 19a). The pseudohexagonal packing of these cylinders (as it is shown in Fig. 19b) in the author's opinion is in good agreement with the spatial packing density calculated on the basis of density measurements. As a criterion for the assignment of a polymer to either a smectic or a nematic type the author [70] suggests the orientability in electromagnetic field. The structure of a polymer is to be nominated smectic, when orientation of the side groups is parallel to the orientating field, and nematic, when the side group orientation is perpendicular to the field. The direction of the main axes of the rigid polymer is to be considered as the dorector. However, recent investigations on the structure and behaviour of LC polymers in an electric field (see Chap. 5) indicate that the considered type of macromolecular packing does not seem to be sufficiently justified.

Probably, the majority of LC nematic homopolymers exhibit certain elements of layer ordering in the arrangement of mesogenic groups, which is more or less close to a cibotactic or smectic type. Further structural investigations are necessary to work out a special classification and model considerations for this group of LC polymers. It is doubtless also necessary, when investigating mesomorphic polymers, to make use of successful techniques of low-molecular liquid crystals research, such as the construction of radial distribution functions, cylindrical functions of interatomic distances and the method of optical masks. These are only beginning to be applied to LC polymers [78–98, 110, 125, 126].

4.3 Comparison of Properties of Smectic and Nematic Liquid-Crystalline Polymers

In a majority of works on LC polymers, the main attention was paid to the synthesis and structural studies of such polymers. Significantly less information is available on physical properties of LC polymers, especially, when compared to low-molecular liquid crystals. In this chapter some rheological and dielectric properties of polymeric liquid crystals, characteristics of their dynamic properties and intramolecular mobility, are considered.

4.3.1 Homopolymers

4.3.1.1 Rheological Properties

The essential concern for the investigation of rheological properties of LC polymers is stimulated by the recently discovered ability of these polymers to orient in electric and magnetic fields (see Chap. 5). The understanding of the mechanism and the study of the kinetics of structural rearrangements, which are evidently impossible without displacement of macromolecular fragments, implies a detailed study of rheological behaviour. First data on rheological behaviour of these polymers were obtained in a study of shear flow for LC polymers of smectic (polymer XI) and nematic (polymer XII) types, using polymers having the same side groups and closely similar degrees of polymerization, but differing in the structure of their main chains (see Table 4 polymer B.3.1 and Table 9 polymer B.3.4) [41]

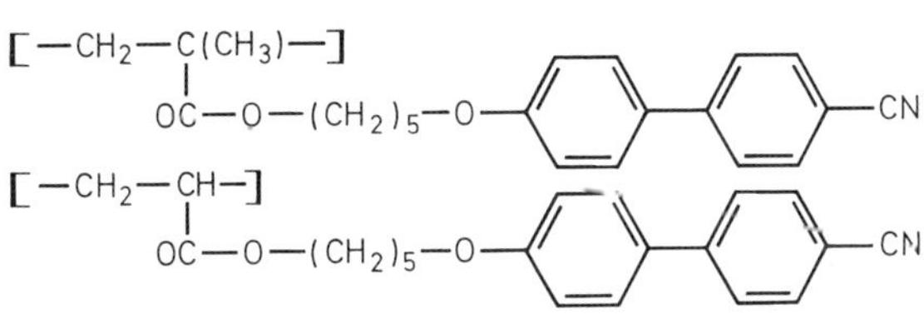

XI

XII

Analysis of flow curves of these polymers has shown that for a nematic polymer XII in a LC state steady flow is observed in a broad temperature interval up to the glass transition temperature. A smectic polymer XI flows only in a very narrow temperature interval (118–121 °C) close to the T_{cl}. The difference in rheological behaviour of these polymers is most nearly disclosed when considering temperature dependences of their melt viscosities at various shear rates (Fig. 20).

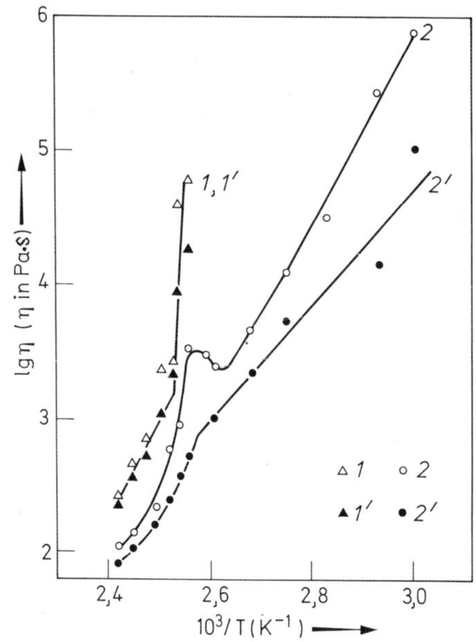

Fig. 20. Temperature dependence of the melt viscosity for polymer XI (1, 1′) and polymer XII (2, 2′) at lg γ = $\bar{3}$,96 (1, 2) and $\bar{1}$,16 (1′, 2′) [41]

What attracts attention in the first place, is a big absolute value for the viscosity of polymeric liquid crystals. For instance, already at T_{cl} (Fig. 20) the viscosity of polymers (10^3 Pa · s) is 1–2 orders higher than that of S_A phase and 4–5 orders higher than that of a nematic phase of low-molecular liquid crystals.

For a nematic polymer in a transition region from LC to isotropic state, maximal viscosity is observed at low shear rates $\dot{\gamma}$. For a smectic polymer in the same temperature range only a break in the curve is observed on a lg η — 1/T plot. This difference is apparently determined by the same reasons that control the difference in rheological behaviour of low-molecular nematics and smectics [126]. A "polymeric character" of liquid crystals is revealed in higher values of the activation energy (E_f) of viscous flow in a mesophase, e.g., E_f for a smectic polymer is 10^3 kJ/mole, for a nematic polymer[3]: 80–140 kJ/mole.

The fusion of LC phases above T_{cl} causes a sharp change in the character of flow and the values of E_f for nematic and smectic polymers become closer. In an isotropic phase E_f for a polymeric smectic (~140 kJ/mole) is only twice as large as E_f for a polymeric nematic (70–80 kJ/mole). In other words, the transition from LC phase to isotropic melt, accompanied by the "liberation" of mesogenic groups from the mesophase levels the differences in the character of flow of smectic and nematic polymers. The differences in E_f for isotropic phase are determined only by the differences in chemical nature of the main chain of smectic and nematic polymers. The values of E_f, in this case, are close to the E_f values for poly(butylmethacrylate) and poly(butylacrylate), respectively, which are structurally similar to polymers XI and XII except that they do not contain mesogenic groups.

3 The interval obtained for E_f is accounted for by the dependence of E_f on shear rate.

The study of rheological properties of LC polymers thus reveals a certain similarity of their behaviour with that of low-molecular liquid crystals. On the other hand there exists a serious distinction: absolute values of viscosity and flow activation energies are much higher for polymeric LC. This is accounted for by the contribution of polymeric main chains to hydrodynamic friction. Further investigations of the rheology and dynamics of polymeric LC systems (the latter being in its infancy [127]) are definitely interesting from the viewpoint of theoretical aspects of viscoelastic behaviour of such extraordinary systems.

4.3.1.2 Molecular Mobility in Solid State

Elucidation of the nature of molecular mobility in LC polymers is inseparable from clarification of the mechanisms of motions of distinct macromolecular fragments. Insufficiency of experimental data does not to date allow to comprehend these problems completely. Some aspects of these will be considered in Chap. 5, dealing with electrooptical phenomena in LC polymers.

We turn to the relaxation processes observed in smectic polymers with different attachment of mesogenic groups to the macromolecular backbone and compare dielectric behaviour of smectic and nematic polymers having identical mesogenic groups but different main chain structure.

Quantitative comparison of dipole relaxation characteristics of polymers listed in Table 10, shows that at temperatures T_r (i.e. lower than T_g) a relaxation process,

Table 10. Temperature of transitions, T_r, T_g and T_{cl}, activation energy $E^b_{d.p.}$ and relaxation time $\tau^s_{d.p.}$ of dipole polarization of smectic cholesterol-containing polymers [128, 129]

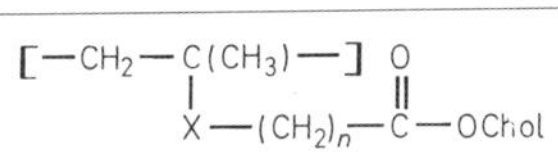

X	n	Temperature of transitions and activation energies (in bulk): T_r, °C	$E^b_{d.p.}$, (kJ/mole)	T_g, °C	T_{cl}, °C
CONH	11	50	360	120	180
COO	14	37	190	50	150

Relaxation times of dipole polarization in solution (Toluene, 40 °C) [131–134]

X	n	$\tau^s_{d.p.}$ (ns)
CONH	11	540
	6	150
COO	10	18
	5	3

corresponding to the mobility of kinetic units including the cholesterol radical with adjoining ester group, is observed. The process is characterized by a high activation energy, its value for polymer with amide spacer being higher than for polymer with ester group. Relaxation times of dipole polarization $\tau^{s}_{d.p.}$ for polymers with CO-NH groups are more than an order higher than the respective values of $\tau^{s}_{d.p.}$ for polymers with ester bonds. Together with structural investigations these results demonstrate that the presence of amide groups enabling hydrogen bond formation causes additional retardations for cholesterol group arrangement. As a result, in cholesterol-containing polymers with amide groups the LC state is achieved only under certain conditions of sample preparation, while for cholesterol-containing polymers with ester linkages LC structure is realized in all cases [135, 136]. It is worth noting, that dielectric method and NMR investigations detected for some LC polymers several relaxation processes preceding, as a rule, structural transitions of smectic-smectic, smectic-isotropic melt types [128]. It might be supposed that mobility of groups participating in "repacking" of macromolecular fragments is a requirement for structural transitions in LC polymers.

A detailed comparative study of dielectric behaviour of smectic and nematic polymers was carried out for polymers of acrylic and methacrylic series, containing identical cyanbiphenyl groups (polymers XI and XII) [137, 138]. The difference in structural organization of these polymers consists in a more perfect layer packing of smectic polymer XI (see Chaps. 4.1 and 4.2) with antiparallel orientation of CN-dipoles. This shifts the relaxation process of CN-dipole reorientation to a low frequency region compared to nematic polymer XII. Identification of Arrhenius plots for dielectric relaxation frequencies f_R shows that for a smectic polymer the value of f_R is a couple of orders lower than for a nematic polymer (Fig. 21). Though the values

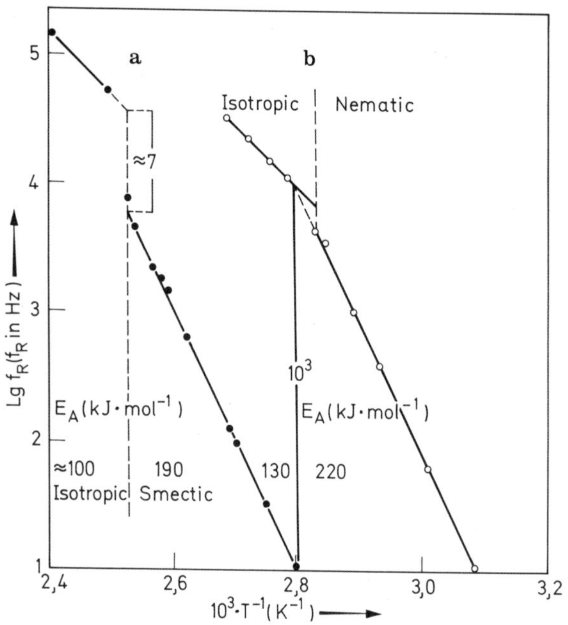

Fig. 21a and b. Relaxation frequencies of smectic polymer XI **(a)** and nematic polymer XII, **(b)** as functions of the reciprocal temperature. (Values of E_A correspond to activation energies of the dielectric relaxation process in different phase states)

for activation energies for both polymers are close to one another (as in the isotropic, so in the LC phase) (see Fig. 21), a more substantial change in f_R on "mesophase-isotropic melt" transition for a smectic polymer reflects apparently the greater change in structural order for a "smectic-isotropic melt" transition than for a "nematic-isotropic melt" one.

Complex molecular motion of a few polar groups in side branches significantly complicates exact assignment of the observed relaxation processes to the motion of distinct macromolecular fragments. Overlapping and combination of separate processes often takes place.

The elucidation of the mechanism of relaxation processes might be achieved only on the basis of strict analysis and juxtaposition of structural, thermodynamic and kinetic results with all molecular parameters of LC polymers taken into consideration.

4.3.2 Copolymers

Copolymerization appeared to be a rather fruitful method of producing new LC systems. All of the LC copolymers known to date may be divided into two groups:

1) copolymers, obtained by copolymerization of two monomers, one of which does not contain a mesogenic group;

2) copolymers, obtained by copolymerization of two mesogenic monomers (see Chap. 4.4).

Unfortunately, there are practically no systematic investigations that allow to survey the changes in physical parameters of copolymers for a broad interval of copolymer compositions. Moreover, in a majority of works published the copolymer composition was not specially determined but was accepted to be equal to the initial monomer mixture composition. This complicates the establishment of correlations between copolymer composition and properties. Nevertheless some conclusions for the first group of copolymers, as well as for the second can be made.

4.3.2.1 Copolymers of Mesogenic and Non-mesogenic Monomers

The study of polymers of mesogenic monomers with monomers, that do not contain mesogenic fragments has lead to the establishment of concentration limits for mesophase existence. With copolymers of cholesterol-containing monomers with alkylacrylates (A-n) and alkylmethacrylates (MA-n) taken as examples it was shown that the concentration region of mesophase existence essentially depends on the length of alkyl substituent C_nH_{2n+1} in a nonmesogenic monomer (Table 2) [26]. For instance, copolymers of butylacrylate (A-4) or butylmethacrylate (MA-4) display mesomorphism at mesogenic component content of less than 20 mole-%, while copolymers with decylmethacrylate (MA-10) at 25 mole-% mesogenic component content still fail to form a mesophase. An increase in alkyl substituent length (MA-22) is not accompanied by mesophase formation even at 75% mesogenic component content. In this case, long paraffinic branches screen mesogenic groups preventing their interaction.

As regards the minimum concentration of mesogenic component in LC copolymer, it is known at present to be 9 mole-%. This was shown for a copolymer with a polysiloxane backbone [118]:

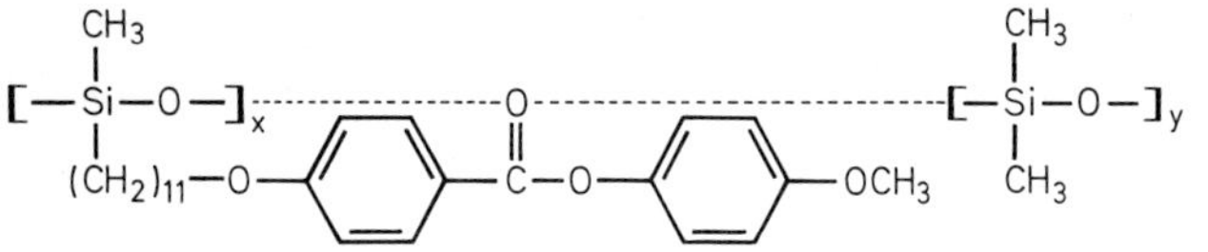

The copolymer displays two glass transition temperatures at —114° and —57 °C and forms a smectic mesophase with T_{cl} = —15 °C.

One of the important merits of the method of copolymerization of mesogenic and nonmesogenic monomers consists in the possibility of varying the temperature range of mesophase existence (Table 2). The main underlying principle is that the glass transition temperature changes gradually between the glass transition temperatures of the respective homopolymers as the copolymer composition is changed. In all cases a nonmesogenic monomer is chosen so as to lower the T_g of a copolymer relative to the T_g of a LC homopolymer. This is achieved due to increase in distance between large mesogenic groups ("mesogenic group dilution") as well as due to the change of the chemical nature of the polymeric backbone. For a majority of copolymers both factors act simultaneously. The shift of T_g is usually accompanied by the lowering of the clearing temperature, i.e. mesophase thermostability is decreased.

On incorporation of nonmesogenic units into LC homopolymers the type of mesophase is not changed as a rule. For instance, the majority of cholesterol-containing copolymers preserve in a broad interval of copolymer compositions a smectic structure characteristic for their homopolymers [36,45]. This was also observed for copolymers containing as side groups phenyl benzoate derivatives of acrylates [139] and polysiloxanes [39,118]. At the same time, among a couple of dozens of copolymers investigated, only three types of copolymers displaying the change of mesophase type on "dilution" of LC homopolymer macromolecules with nonmesogenic units have been described (Table 11). The incorporation of butylacrylate or isoprene units within cholesterol-containing polymers enhances their backbone flexibility and leads to a gradual transformation from a smectic to a nematic polymer. The opposite case, i.e. the formation of a more ordered smectic mesophase for a copolymer compared to a nematic homopolymer, is observed for one of the polysiloxanes (polymer 3.2, compare with polymer 3.5 in Table 11). At the present time unambiguous explanation of the observations is quite difficult. However, the necessity in systematic investigations of copolymers with mesogenic and nonmesogenic components aiming at the establishment of general principles of copolymer mesophase formation is undoubted as it could have led to prediction of the mesophase type of copolymers.

4.3.2.2 Copolymers of Two Mesogenic Monomers

Copolymerization of two mesogenic monomers is, at the present time, the only pathway to obtain polymers with cholesteric mesophase (see Part 4.4). On the other hand, only by copolymerizing smectogenic and nematogenic monomers and investigating the properties of copolymers in a broad interval of compositions, is it possible to establish the principles of formation of each type of mesophase. We demonstrate

Table 11. LC copolymers from mesogenic and nonmesogenic monomers which form mesophases different from the type of homopolymer mesophase [37, 159, 39]

Copolymer No.	Structure of the copolymer unit	Mesogenic monomer, x, mole %	T_g, °C	Mesophase types and transition temperatures, °C
1. Cholesterol containing copolymers				
	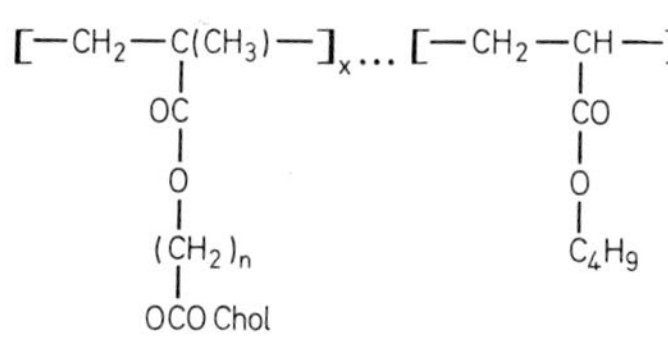			
1.1.	n = 10	100	60	S158I
1.2.	10	58	34	S130I
1.3.	10	20	25	N50I
1.4.	5	38	55	N97I
2.	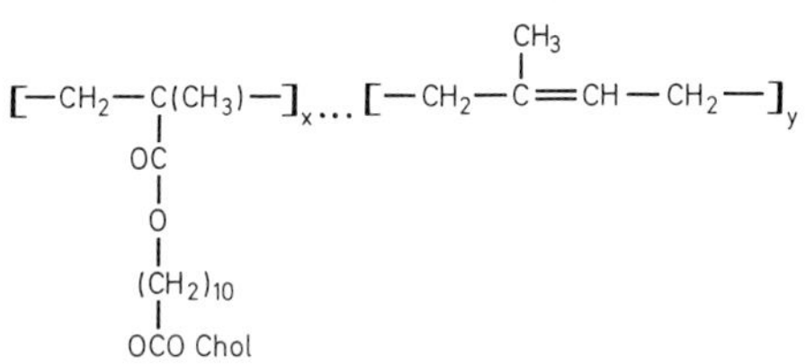			
2.1.		54	33	S83I
2.2.		35	20	N45I
3.	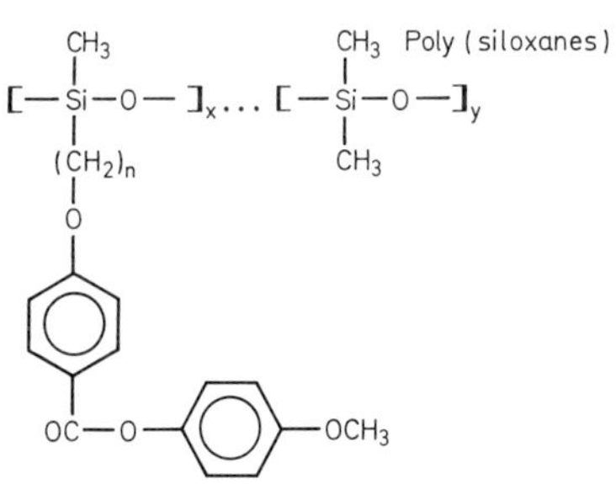			
3.1.	n = 3[a]		15	N61I
3.2.	n = 4[a]		15	N95I
3.3.	n = 5[a]		—	S87I
3.4.	n = 3		3	N21I
3.5.	n = 4	x = y = 60	−6	S50I
3.6.	n = 5	x + y = 120	−10	S73I

[a] Homopolymers (y = 0)

the effectiveness of this approach on copolymers [140] with identical cyanbiphenyl mesogenic groups. The copolymers varied

a) in the length of aliphatic spacer (n) while the polyacrylate chain remained unchanged, (copolymers 5-11):

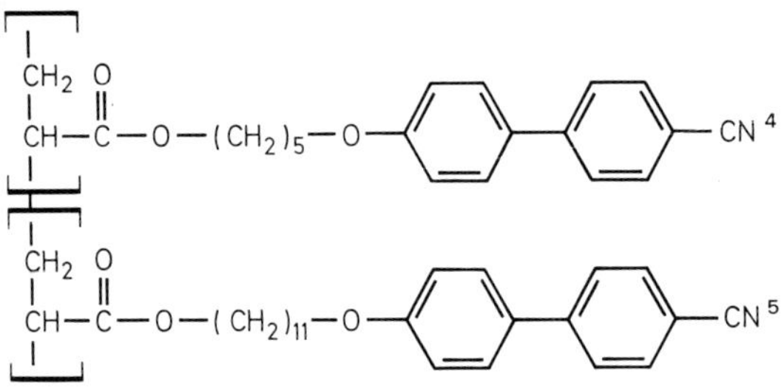

b) in the chemical nature of the main chain, the side groups being the same (copolymers 5-5):

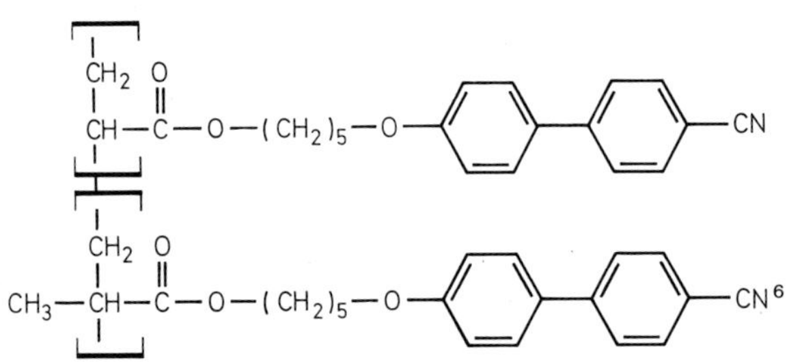

Homopolymers of the comonomers in each case display different structural types– nematic and smectic. Some data on thermal properties of these copolymers are given in Fig. 22. For copolymers 5-11 T_{cl} and ΔH_{cl} gradually increase and reach the values characteristic for smectic homopolymer PM-11, as the content of longer side groups is increased. The transformation of a nematic to a smectic mesophase takes place already at 30% content of the longer spacer groups, as is evident from Fig. 22 and X-ray analysis data. The thickness of smectic layer formed due to overlapping of mesogenic groups increases linearly with the increase of content of smectogenic groups for n = 11. Thus, in the case of relatively flexible polyacrylate backbone the type of mesophase formed is determined, just as for low-molecular liquid crystals, by the length of aliphatic fragments.

The case is quite different for copolymers 5-5. Here, the transformation from a nematic mesophase to a smectic one is a result of gradual decrease in flexibility of the main chain. As seen from Fig. 22a, b, the plots of T_{cl} and ΔH_{cl} on composition have a minimum. This could be the consequence of a more defective packing of side groups due to incorporation within the backbone of units of different chemical nature. A sharp rise in ΔH_{cl}, indicating transformation to a smectic mesophase type, begins only at

4 Homopolymer PA-5
5 Homopolymer PA-11
6 Homopolymer PM-5

methacrylate unit content exceeding 70 mole-%. This implies that in polymers with short side groups (whose low-molecular counterparts form a nematic mesophase) a smectic LC state is feasible only due to a rigid main chain (compare with Chap. 4.1.2). An increase in flexibility on incorporation of relatively small amounts (30 mole-%) of acrylate units leads to the formation by the side groups of a nematic structure.

Glass transition temperatures of these copolymers lie between the T_gs of the respective homopolymers, as is usual for copolymers of mesogenic and nonmesogenic monomers.

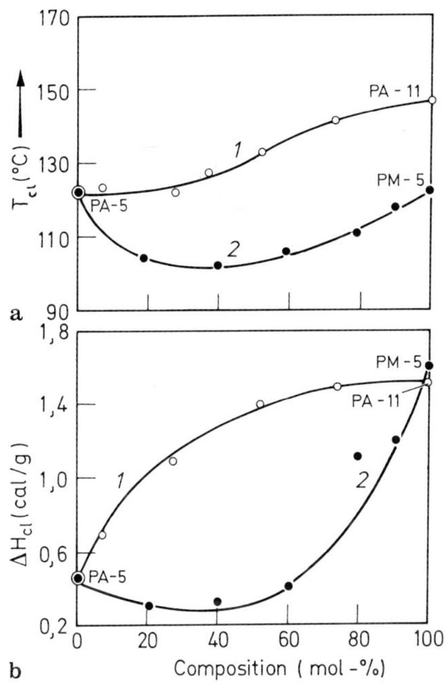

Fig. 22a and b. Dependences of T_{cl} **(a)**, and ΔH_{cl} **(b)** of the copolymers 5–11 (1) and copolymers 5–5 (2) on their composition

The situation is different for copolymers with CN- and alkoxy-substituted mesogenic groups (Table 12) [119]. It is seen that copolymerization leads to a significant increase of smectic mesophase thermostability (from 121 °C for a homopolymer to 171–177 °C for copolymers). An analogous phenomenon is observed for copolymers with Schiff base mesogenic groups as nematogenic comonomer [119]. The increase in stability of a smectic mesophase may be explained by additional donor-acceptor interaction between mesogenic groups with electron-donor (—OCH_3) and electron-acceptor (—CN) substituents as is the case for blends of low-molecular compounds. Copolymerization of monomers containing electron-donor and electron-acceptor substituents reveals new perspectives for increasing polymer smectic mesophase thermostability.

Copolymerization thus presents an effective method for modification of LC polymer properties aimed at diversification of their thermal characteristics as well as at the creation of new types of LC polymers, cholesteric LC polymers in particular.

3.1.	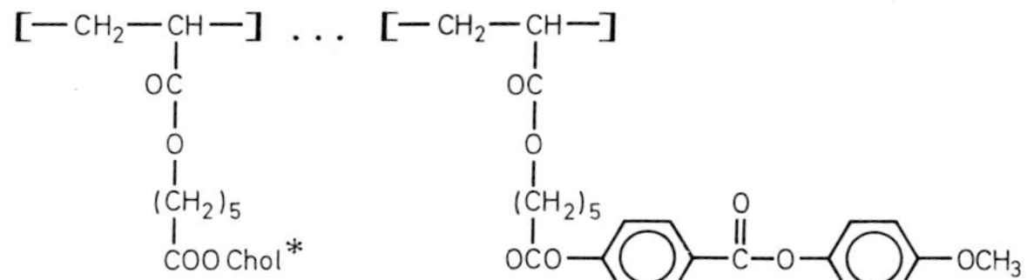		14	25	103	555	
3.2.			35	48	103	495	51)
3.3.			45	40	110	400	
4.1.	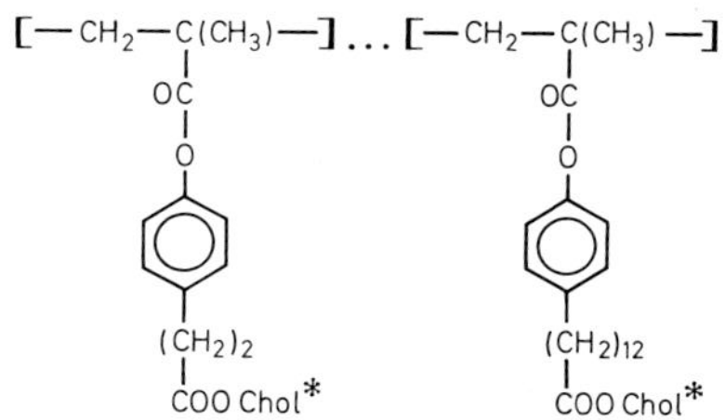		51[b]	—	209	Visible region	145)
5.1.	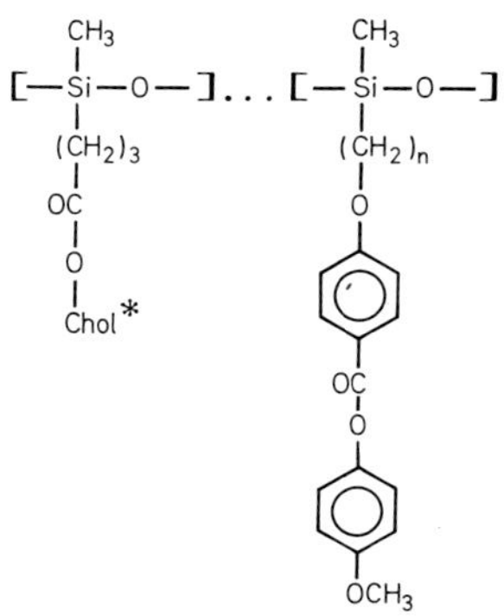	n = 3	3–15[c]	—	—	~400–2000	
5.2.		n = 4	5–15[c]	—	—	~600–2000	105)
5.3.		n = 6	5–15[c]	—	—	~560–2000	

6.1.	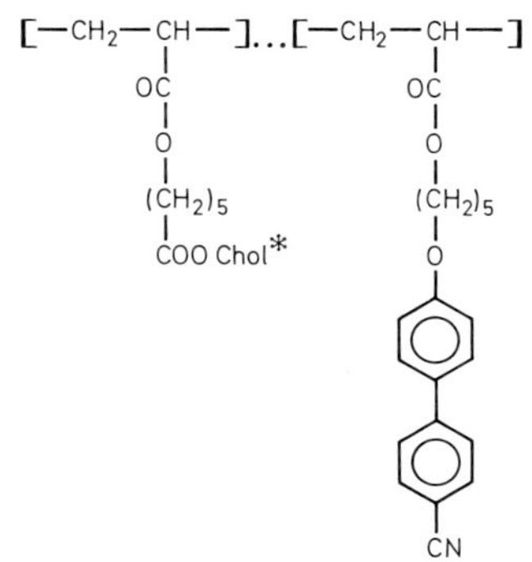	34	50	98	850	146)
6.2.		40	50	102	660	
6.3.		55	55	105	555	
6.4.		65	55	150	500	

[a] Chiral group is marked by a asterisk;
[b] λ_{max} for define temperature region (usually between T_g and T_{cl}) indicated in Ref. is given;
[c] Content of the chiral comonomer in the initial monomer mixture is given

each having a different length of spacer, but forming homopolymers of smectic structure [145] (Polymer 4.1, Table 13).

The main feature identifying a cholesteric mesophase in polymers is the presence of optical texture with selective circularly-polarized light reflection. This indicates the formation of l-helical cholesteric structure in LC copolymers. The X-ray patterns of actually all cholesteric copolymers described (with the exclusion of polymers 3.1 and 4.1, Table 13) correspond to those of nematic and cholesteric low-molecular liquid crystals, which is manifested in a single diffuse reflex at wide scattering angles. At the same time, for copolymers 3.1 and 4.1 (Table 13) small angle reflexes were observed [123], that are usually missing in low-molecular cholesterics.

Consequently, as in the case of nematic LC polymers (see Chap. 4.2), apparently, a greater extent of ordering is typical for cholesteric polymers than for their low-molecular analogs. The enthalpy of melting ΔH_{cl} of polymeric cholesteric is close to that of nematic LC polymers.

As is seen from Table 13, cholesteric copolymers display a maximum of selective light reflection (λ_{max}) in an IR- or a visible part of the spectrum. By varying the composition of a copolymer, it is possible to vary λ_{max}, in accordance with the stipulation $\lambda_{max} = nP$, is proportional to the pitch P of the helical structure of a LC polymer (n — is the refractive index). The pitch of the helix in cholesteric copolymers is usually decreased, when the temperature is raised [105] (at temperatures above T_g), which is equally common for low-molecular cholesterics [142] (Fig. 23a). The observed fact that the helix pitch for LC copolymers 2.1–2.3 (Table 13, Fig. 23b) is increased, is rather unusual but explicable within the theoretical views regarding vibrational movement of macromolecular fragments and their conformational mobility [60].

The macromolecular nature provides an interesting feature of LC polymeric cholesterics, namely the possibility of obtaining monochromic films. Thus for polymeric liquid crystals the helix pitch is practically not altered with temperature below T_g, when a cholesteric phase is frozen in a glassy matrix (Fig. 23a). This implies that fast cooling of polymeric films from a mesomorphic state (shown with arrows) fixes their optical properties, which makes it possible to use them at ordinary temperatures as selective monochromic reflectors. On the other hand, such polymeric films display the extraordinary polarizing properties of cholesterics, i.e. the different absorption

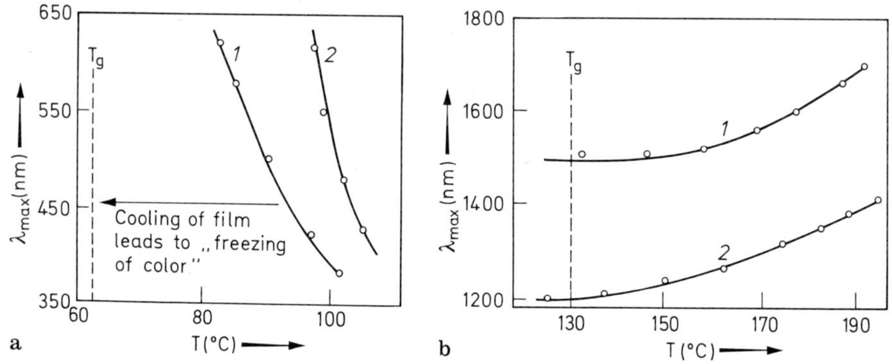

Fig. 23a and b. Relationship between the wavelength of maximum reflection λ_{max} and the temperature for cholesteric copolymers: **(a)** copolymers 6.2 (1) and 6.3 (2) (Table 13); **(b)** copolymers 2.1 (1) and 2.2 (2) (Table 13)

of right — and left — circularly — polarized light. Utilizing such films as circularly — dichroic optical filters possessing all the advantages of polymeric materials is of evident interest[7].

5 Behaviour of Liquid-Crystalline Polymers in Electric Fields

One of the most specific and unique features of low-molecular liquid crystals is their ability for orientation in external fields — mechanical, electric and magnetic. It is this property that establishes wide capabilities for technical application of liquid crystals. Today electric and magnetic optics of liquid crystals are an independent and useful for practics branch of the physics of the condensed state of matter [42,43,147–150].

As regards electrooptical and magnetooptical studies of LC polymers, these works are not more than two–three years old.

First investigations were carried out for LC polymers with mesogenic side groups in the Institute of Organic Chemistry of Mainz [119,151,152] and in Moscow State University [44,89,124,137,153–158]; a little later the works on the effect of electric field on LC polymers began in France and USA with polymers containing mesogenic groups in the backbone [159–162].

It should be noted that in case of low-molecular liquid crystals, electrooptical effects are exhibited only for compounds with large constant dipole moment and high polarizability anisotropy of molecules. A total measure of these two parameters is presented by the anisotropy of dielectric constant $\Delta\varepsilon = \varepsilon_{\|} - \varepsilon_{\perp}$, which is the difference in liquid crystal dielectric constants measured along the external field ($\varepsilon_{\|}$) and perpendicular to it ($\varepsilon_{\perp}$). For molecules of elongated shape and with constant dipole moment directed along the molecular axis, the $\Delta\varepsilon$ value is practically always positive. Among LC polymers synthesized until recently there were none with large $\Delta\varepsilon$. This, apparently, was one of the main reasons retarding the study of electro-optical phenomena in LC polymers.

The example of the above mentioned specific features of low-molecular liquid crystals, capable of orientation in an electric field, has led recently to the synthesis of a series of acrylic, methacrylic and siloxane polymers and copolymers of nematic and smectic types with nitrile end groups in the side branches (see Tables 3, 4, 9, 12):

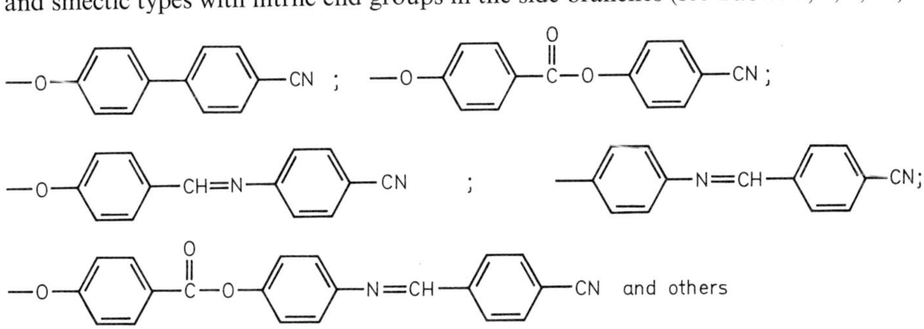

7 A more detailed description of specific properties of polymeric cholesterics is given in a review of Rehage and Finkelmann [27] of this volume

The presence of these groups ensured a large value of the parallel component of the dipole moment in the side chain direction and a positive dielectric anisotropy of these polymers.

All electrooptical effects known to the present time for polymeric liquid crystals may be divided into two groups. First of all there are so called orientational effects, which are due solely to the effect of the electric field (field effect) on LC polymers, but are not a result of a current flowing. The second group of electrooptical effects is attributed to the phenomena ascribed to the anisotropy of electrical conductivity ($\Delta\sigma$) of liquid crystals. These are called electrohydrodynamic effects.

5.1 Orientational Effects

5.1.1 Low-Molecular Liquid Crystals

As a consequence of the anisotropy of dielectric constant ($\Delta\varepsilon$) and diamagnetic susceptibility ($\Delta\chi$), the free energy of the ensemble of nematic liquid crystal molecules reaches a minimum at a strictly defined orientation of the axes of mesogenic molecules in respect to the field. A unit vector n — called a director — characterizes a prevalent orientation of the large axes of mesogenic molecules at each point of a liquid crystal. At positive values of $\Delta\varepsilon$ and $\Delta\chi$ a director tends to align along the field, at negative — perpendicular to it. Elastic forces counteract the orientation of a director as they tend to return the structure of a liquid crystal to an equilibrium state, predetermined by initial conditions (usually, the walls of the electrooptical cell). Competition of the two effects — dielectric and elastic reactions, results in deformation of a layer of liquid crystal. This, in turn, leads to the change of its optical properties, which is easy to monitor through the changes in the degree of light polarization due to the substantial birefringence Δn of a liquid-crystalline compound.

For low-molecular nematic liquid crystals three major kinds of electrooptical effects, shown on Fig. 24 [42], are distinguished.

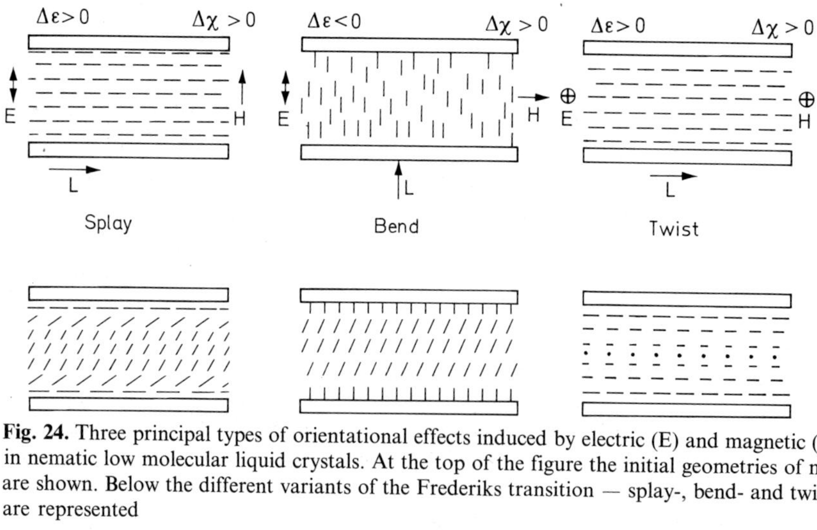

Fig. 24. Three principal types of orientational effects induced by electric (E) and magnetic (H) fields in nematic low molecular liquid crystals. At the top of the figure the initial geometries of molecules are shown. Below the different variants of the Frederiks transition — splay-, bend- and twist-effects are represented

It is important, that for all of the three effects exhibited, there is no need for a current to flow through a layer of a liquid crystal, i.e. the named effects are purely field effects. Besides, as is seen from Fig. 24, the initial molecular orientation is strictly predetermined.

The theory of nematic liquid crystal deformation, forced by an electric field is well developed and permits to establish the relationship between the threshold voltage U_{th}, causing sample orientation, with $\Delta\varepsilon$ and elasticity constants of a liquid crystal (K_{ii}). For the main S and B types of deformation the equation is the following [27]:

$$U_{th} = \pi \sqrt{\frac{4\pi K_{ii}}{\Delta\varepsilon}} \tag{4}$$

where $K_{ii} = K_{11}$ (S-effect) and $K_{ii} = K_{33}$ (B-effect).

5.1.2 Liquid-Crystalline Polymers

5.1.2.1 S-effect (Frederiks Transition)

Preceding the discussion of orientational effects in LC polymers, it is worth mentioning that for a nematic and a smectic phase A of LC polymers only the S-effect was discovered and investigated. This started with works [119,124,137,138], that demonstrated the ability of LC polymers to orient in permanent and alternating electric fields. The structural formulas of some of the polymers and copolymers investigated are given below:

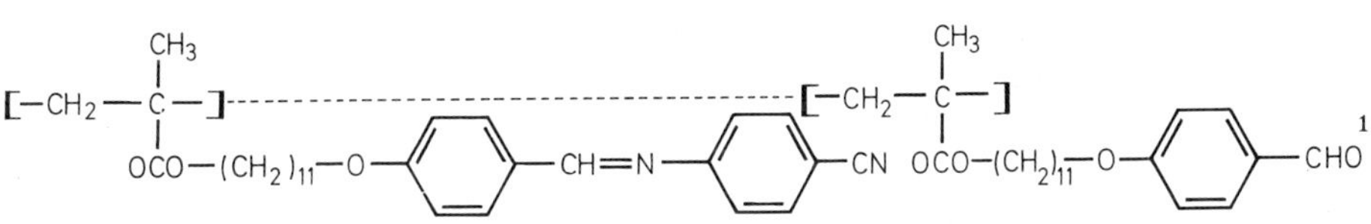
154)

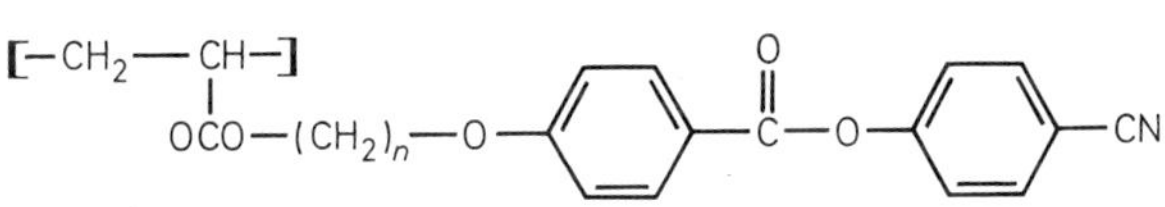
$n = 2, 6$ [119]

Research works carried out with such and other polymers [44,152,158] established that their optical properties are really strongly affected by the electric field.

For instance, for a nematic polymer with positive anisotropy of dielectric constant ($\Delta\varepsilon > 0$) orientation of mesogenic groups along the applied field takes place (homeotropic orientation). The fact of orientation is illustrated in Fig. 25, which shows that under crossed polarizers the optical transmittance I of a film of nematic polymer with optically anisotropic texture (taken for $\sim 100\%$) falls practically to zero when a low-frequency field is switched on.

The rate of light intensity decrease effected by an electric field (f = 50 Hz) is strongly dependent, as is seen from fig. 25a, on the value of the applied effective voltage U_0 (see below), and rises sharply when U is increased. The authors also discovered, that the time value during which the light-transmittance changes (at a fixed value of $I = 50\%$) is inversely proportional to the square of the voltage value (Fig. 25b). The dependence of the threshold voltage U_0 on the frequency of the applied

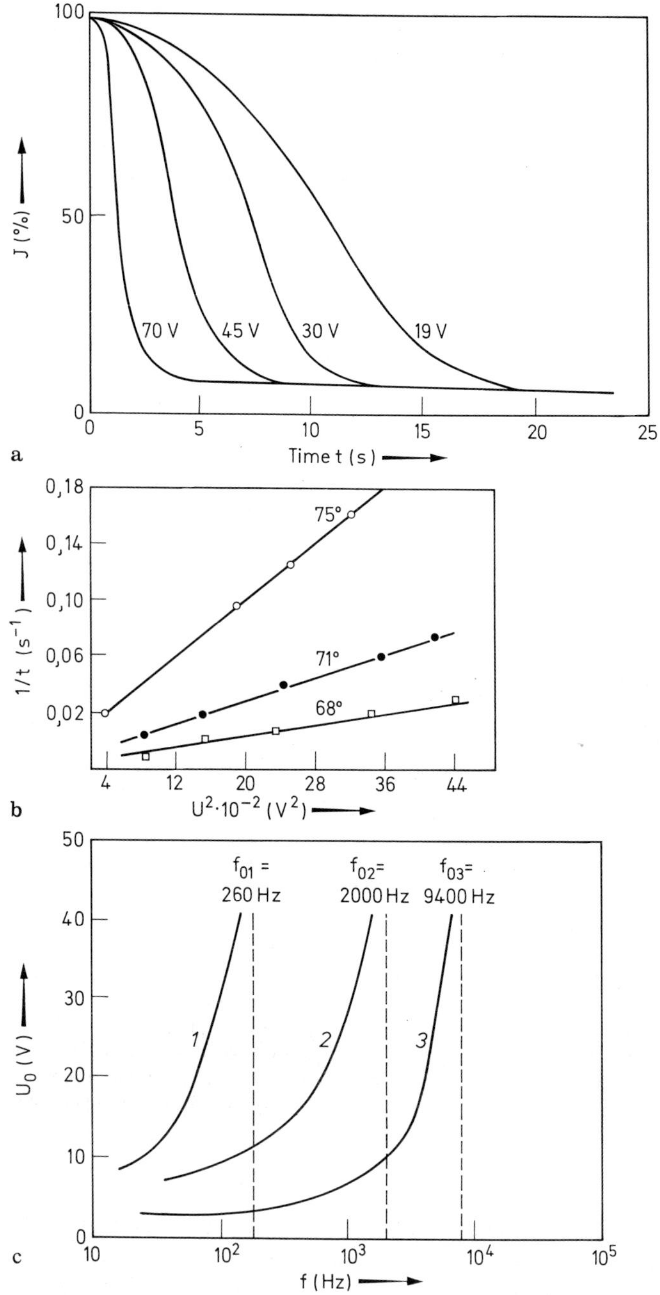

Fig. 25a–c. Electro-optical behaviour of nematic polymer XII ($T_g = 50$ °C; $T_{cl} = 77$ °C; $P_w = 20$): **(a)** Optical transmission I as a function of time t at different voltage (f = 50 Hz; T = 75 °C, crossed polarizers); **(b)** reciprocal rise time as a function of the voltage square at different temperatures; **(c)** threshold voltage U_0 as a function of frequency f at 55 (1); 63 (2) and 68 °C (3)

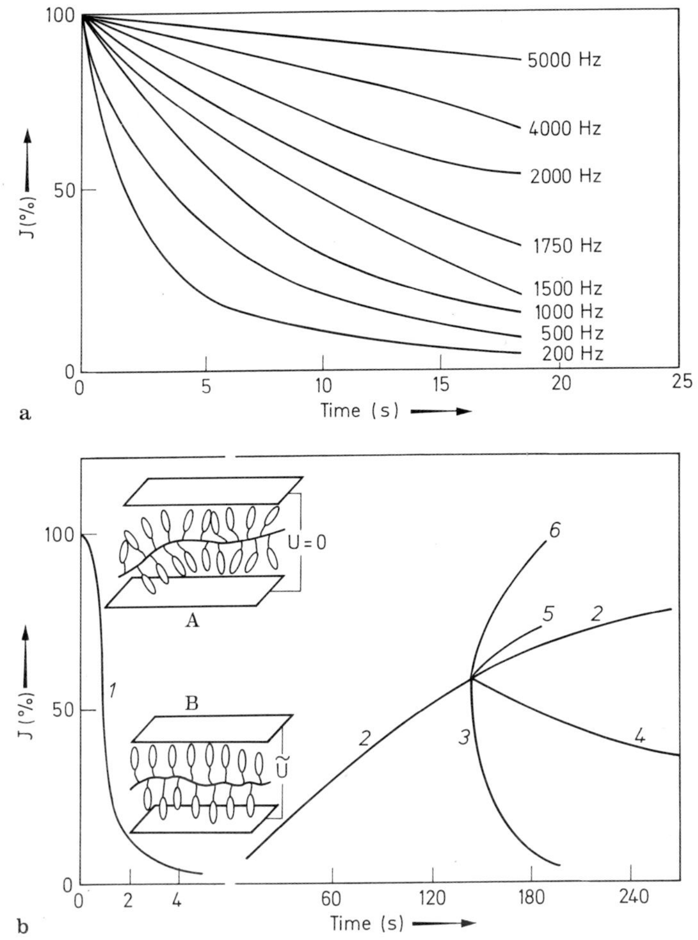

Fig. 26a and b. Influence of the electric field frequency on the electro-optical behaviour of the nematic polymer XII and scheme of mesogenic groups orientation before (A) and after (B) the application of electric field: **(a)** optical transmission as a function of time at different frequencies (U = 30 V; T = 75 °C); **(b)** optical transmission as a function of time upon application of an electric field at U = 85 V (f = 50 Hz) (1); relaxation upon switching the electric field off (2), upon application of an electric field (U = 80 V) of different frequency f = 1 (3); 5 (4); 7 (5) and 20 kHz (6) during the relaxation process

field diverges asymptotically (Fig. 25c). The analysis of these results gave grounds for the analogy of the observed electrooptical effect and the above mentioned S-effect (Frederiks effect), known for low-molecular nematics with $\Delta\varepsilon > 0$ (Fig. 24). In contrast to the latter, however, in a LC polymer the orientation obtained in an electric field may be fixed by cooling the polymer below T_g. The structure of such a film corresponds to the structure of a uniaxial positive monocrystal, the optical axis of the latter coinciding with the direction of mesogenic groups and the field intensity vector (Fig. 26).

The analogy in the behaviour of polymeric and low-molecular liquid crystals is exhibited also in the frequency dependence of the mesogenic group orientation. The appearence of the dependences $U_0 = \varphi(f)$, given in Fig. 25c, indicates the existence of a frequency f_0, at which the sign of $\Delta\varepsilon$ changes, a fact intrinsic to low-molecular liquid crystals with $\Delta\varepsilon > 0$. Figure 26a, b demonstrate the influence of the electric field frequency on the mode of orientation of mesogenic groups. Curve 1 in Fig. 26b corresponds to the process of homeotropic structure formation effected by a low-frequency field ($f = 50$ Hz, $U = 85$ V). Switching off the field leads to demolition of homeotropic orientation — the system is disoriented (Fig. 26b, curve 2). Repeatedly applying the field at various frequencies (at $\tau = 135$ sec) one may observe different orientation of mesogenic groups. For instance, at $f < f_0 = 6$ kHz the side groups are aligned along the field, at $f > f_0$ they tend to be positioned perpendicularly to the field. Such an effect is accounted for by the change of $\Delta\varepsilon$ sign at f_0 and is thus an example of the frequency addressing of the Frederiks effect. Polymeric specificity is exhibited in this case as the decrease of the frequency of the $\Delta\varepsilon$ sign change and is apparently a result of the high viscosity of a LC polymer melt. Consequently, by varying the frequency of the alternating field and its intensity one may vary the mode of side group alignment within the mesophase and then, by cooling the polymeric films below T_g, one can fix their macroscopic structure.

Comparing the effect of an electric field on low-molecular and polymeric liquid crystals it is necessary to stress the following:

a) In a majority of studies on orientational phenomena in LC polymers in electric fields, investigations were carried out on unoriented ("polycrystalline") samples. Until now the attempts to obtain initially homogeneous orientation of mesogenic groups by special treatment of the glass walls of electrooptical cell did not produce reliable positive results. At the same time, a strict quantitative estimation of such significant parameters as the threshold voltage U_{th}, τ_{on} and τ_{off} (disorientation time) times of the Frederiks effect are feasible only for LC polymer samples with homogeneously oriented initial structure [42]. It is thus a necessity to develop experimental techniques allowing to produce homogeneous initial orientation of mesogenic groups. Besides this, it also seems reasonable to introduce some "reduced" parameters, for characterisation of orientational effects in LC polymers with the purpose to standardize the characteristics of the initial structure and to compare the values obtained for U_{th}, τ_{on} and τ_{off} for various LC compounds. For instance, in [155] the effective threshold voltage U_0 was introduced; it was defined as the maximal voltage at which no changes in optical properties of an unoriented LC polymer sample are observed.

b) Another important parameter of orientation processes is the rise-time (τ_{on}) of the orientational effect. For low-molecular liquid crystals the values of τ_{on} are usually of $10^{-3} - 10^{-1}$ sec. For polymeric liquid crystals its value is substantially higher, and, what is most significant, it depends to a large extent on the degree of polymerization (molecular mass) (Table 14). As is seen from the table, a 100 fold increase in the degree of polymerization is accompanied by almost a 200 fold increase of the rise-time. It is worthy of attention that the rise-time continues to increase even when such thermodynamic parameters as T_{cl} and ΔH_{cl} reach their limit.

This indicates that kinetic parameters of orientational process are defined mostly by the macroscopic viscosity of a polymer. The substantial difference in mesophase

Table 14. Influence of the degree of polymerisation P_w on the rise time τ_{on} for polymer XII [44)]

M_w	P_w	T_{cl}, °C	τ_{on} at 100 V and f = 1 kHz (min)
$6.9 \cdot 10^3$	20	77	0.05
$2.4 \cdot 10^4$	70	90	0.5
$7.9 \cdot 10^4$	240	120	1.5
$6.6 \cdot 10^5$	2000	120	10.0

viscosity of LC polymers and low-molecular liquid crystals also apparently determines the significant difference in their rates of orientation in an electric field. To an even greater extent the viscosity of mesophase affects the rate of disordering of the oriented state (orientation relaxation) after the electric field is switched off. At high degrees of polymerization (DP = 2000) the values of orientation relaxation time are so high that even at temperatures around T_{cl} disordering is not observed.

In view of the effect of molecular mass on orientational phenomena the results of [151)] seem to be more explicable. In this work surprisingly low values for threshold voltage (U ≈ 8–10 V) and rise and decay times (τ ≈ 200 msec) were observed for an array of nematic polymers and copolymers. They are close to the corresponding values for low-molecular liquid crystals, which implies presumably that the polymers investigated were of low degrees of polymerization or had a very wide molecular mass distribution.

The discovered dependence of kinetic parameters of orientation processes on the degree of polymerization [44)] is a consequence of the duplex nature of LC polymers — that is the presence of the main chain and of mesogenic side groups. This is why a correct juxtaposition of the kinetic characteristics of orientational processes of low-molecular and polymeric liquid crystals requires an explicit knowledge of the degree of polymerization of a corresponding polymer.

The results published in [120)] should have been analyzed exactly from this viewpoint. The work [120)] presents interesting comparative data on the estimation of threshold voltages U, τ_{on}, τ_{off} for homopolymers XIII–XVI and for a series of copolymers with varying spacer length

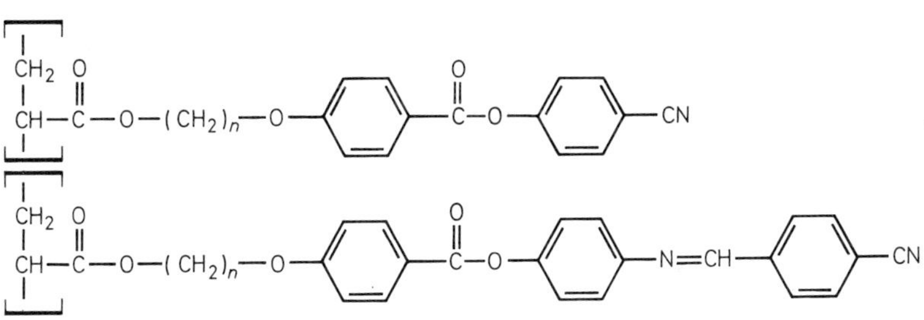

XIII n = 2
XIV n = 6
XV n = 2
XVI n = 6

It was shown that polymers with a short spacer (n = 2) exhibit low threshold voltages of reorientation (4–5 V) which are close to the corresponding values for low-molecular nematics, while for polymers with longer spacer groups (n = 6) the observed

U values are substantially higher (U = 20–60 V). This was interpreted to be a result of significant differences in elastic constants of polymeric nematics with varying spacer lengths.

The ratio of elastic constants K_{11}, calculated for the S-effect according to the equation (4) appeared to be (K_{11} (polymer XIV)/K_{11} (polymer XIII)) $\approx$ 1:100 and (K_{11} (polymer XVI)/K_{11} (polymer XV)) $\approx$ 1:36. Yet, as we have just indicated, taking into account molecular masses of the LC polymers and reducing k_{11} values for various polymers to equal values of DP one may come to substantially different values for ratios of constants presented. It is necessary to note that up to date no quantitative data on the determination of elastic constants of LC polymers has been published (excluding the preliminary results on Leslie viscosity coefficients for LC comb-like polymer [127]). Thus, one of the important tasks today is the investigation of elastic and visco-elastic properties of LC polymers and their quantitative description.

On the other hand, the problem of clarifying the mechanism of orientation processes in LC polymers effected by electric field application is equally important. The complicated structure of polymeric molecules that includes backbone, spacer, and mesogenic group, requires for the movements of all macromolecular fragments in orientation process to be taken into account. Preliminary results obtained by the authors together with R. V. Talrosé and V. V. Sinitsyn reveal that the activation energy of orientation process (E_{or}) is actually independent of the degree of polymerization, but exceeds the published values of E_{or} for S-transition in low-molecular nematics by a factor of 3 to 4. The molecular mechanism of orientation of a polymer in an electric field resembles apparently the mechanism of viscous flow when independent movements of segments lead to an overall displacement of the macromolecule as a whole.

5.1.2.2 *"Guest-Host" Effect and the Order Parameter*

One of the manifestations of orientational effect in LC polymers is presented by a so called "guest-host" effect, which is well-known for low-molecular liquid crystals.

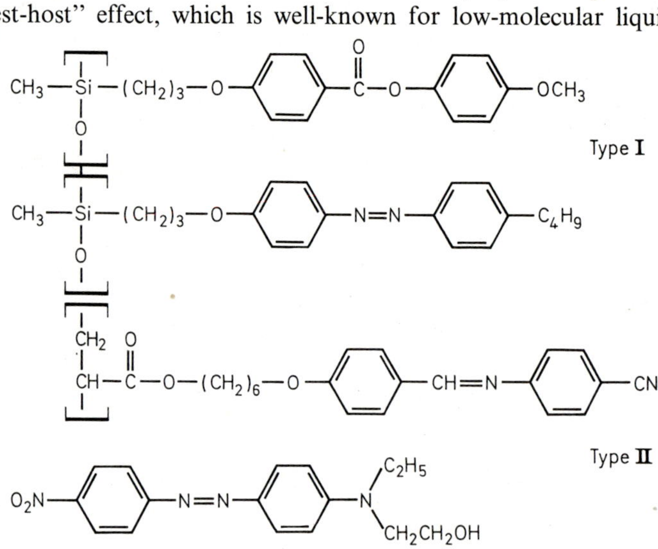

Type I

Type II

In the case of LC polymers, the polymeric matrix performs as a host, while the guest is a dye, whose molecules are elongated in shape, and the absorption oscillator is parallel (or perpendicular) to the big axis of the molecule [65,163–165]. The experiments investigating "guest-host" effect in nematic polymers with dichroic dyes covalently attached to the polymer [163] (type I) and mechanically incorporated [65] (type II) reveal the possibility to obtain regulated color indicators (see page 60).

Mesogenic groups of a polymer, having been oriented in external field (mechanical or electric), enable the dye molecules to orient and thus causing the emergence or the change of color depending on the dye type (sign of $\Delta\varepsilon$) and the parameters of the external field (frequency, intensity). Due to the polymeric character of such liquid crystals, the required structure may be "frozen" within a glassy matrix by cooling the mesophase.

By utilizing a "guest-host" effect, it is also possible to get information on the structural organization of LC polymers. This was done by evaluating the order parameter $S = 1/2(3\cos^2\Theta - 1)$, where Θ is an average angle between the polymer side group direction and the director of a liquid crystalline sample. The order parameter may be stipulated for instance, from the data on IR- and UV-dichroism of added dye, isomorphic to liquid crystal, or from ESR spectra of specially introduced "labels" with paramagnetic spin [166]. The values of order parameter for some LC polymers are given in Table 15. As is seen, the values of S for nematic LC polymers are in the interval 0,45–0,65, and for smectic polymers S is 0,85–0,92. At the same time, a maximum value of S = 0,6–0,8 for low-molecular nematics and it reaches 0,9 for smectics. The comparison of these values shows that the ordering in smectic polymers is very close to the degree of ordering in low-molecular smectics. Nematic polymers are somewhat less ordered than their low-molecular analogues. Taking into account the tendency of comb-like polymers to layer ordering, the first relationship may be regarded as quite evident, while the reasons for the low degree of ordering in nematic polymers are not yet clear. The appearance of the temperature dependence of the order parameter for polymers 1–2 (Table 15) has shown a close analogy to the corresponding dependence for low-molecular liquid crystals. At the same time, preservation prolonged for several years of structural order in a glassy state (below T_g), constitutes an important specific feature of LC polymers.

Thus, the possibility to regulate effectively the orientation of polymer side groups by varying the parameters of the electric field, together with the possibility to fix the oriented structure in a glassy state enables the use of such LC systems for making polymeric materials with required optical properties.

5.1.2.3 *Optical Recording of Information (Thermoaddressing)*

One of the outlets of electrooptical phenomena in LC polymers is the construction of devices for the recording and optical duplication of information. This was first described in the works of Soviet scientists from Moscow State University[8]. Figure 27 illustrates the principle of information recording on oriented layer of a polymeric liquid crystal.

---

8 These are the results of joint work of physicists and chemists under the authors' leadership (see V. P. Shibaev, S. G. Kostromin, N. A. Platé, S. A. Ivanov, V. Ju. Vetrov and I. A. Yakovlev, Polymer Commun., *24*, 364, 1983).

Table 15. Values of order parameter S for some LC polymers determined by different methods

Polymer No.	Structure of monomer unit	Type of mesophase	Order parameter S	Method of determination	Ref.
1.	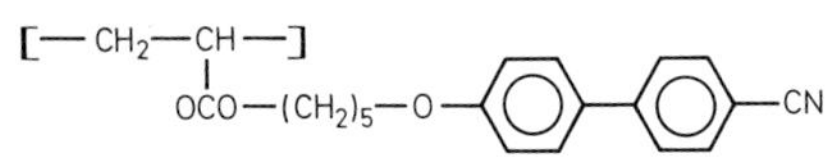	N	0.45	NMR	167)
2.	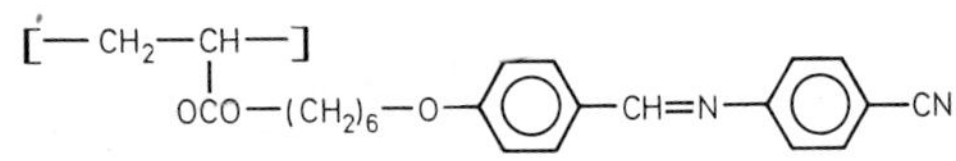	N	0.50	"guest-host" effect	65)
3.	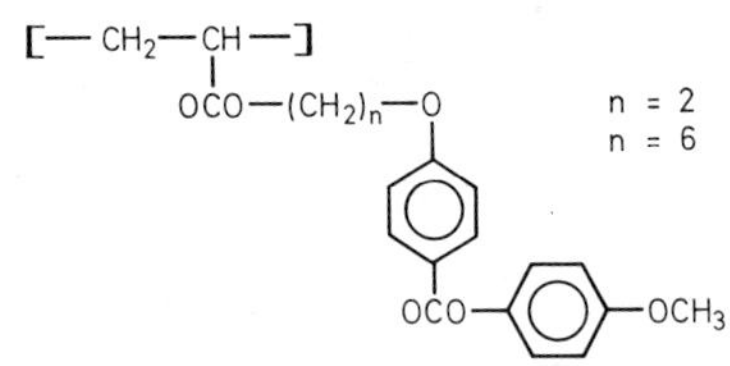	N S	0.65 0.92	ESR with ("marker")	166)
4.	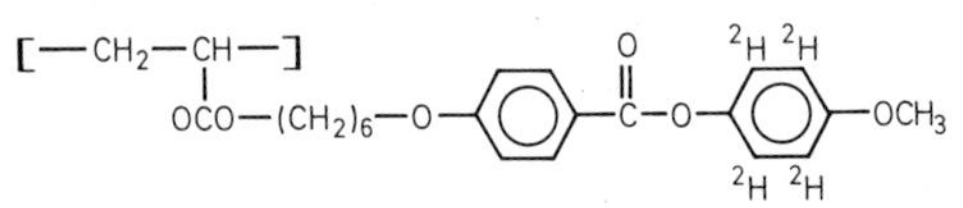	S	0.85 ± 0.05	NMR	168)
5.	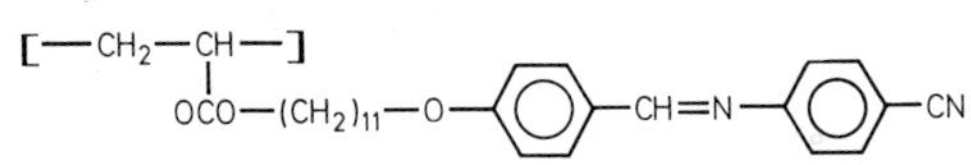	S	0.91	X-ray analysis	65)

A transparent film of homeotropically oriented liquid crystal is subjected to a laser beam which creates regions of overheating. The liquid crystal in these regions passes over to an isotropic melt, its orientation is destroyed (Fig. 27) and instead of a transparent monodomain homeotropic texture a polydomain texture that scatters light is formed on cooling. A certain piece of information being recorded on a transparent film may be "wiped off" by the electric field. Such type of recording procedure is usually called thermorecording or thermoaddressing. For its realization it is necessary for an oriented state to be stable in the absence of an electric field for a rather long time, i.e. it is necessary that the orientation relaxation rate be sufficiently low. Low-molecular nematics do not fulfill this requirement, as they are rapidly disaligned when the electric field is switched off. Accordingly, in thermorecording devices based on low-molecular liquid crystals a "smectic-nematic" transition is most often made use of, as homeotropic texture is sufficiently stable for the smectic LC state.

It has been mentioned above that the rate of disalignment of nematic polymers having sufficiently high degrees of polymerization (DP = 200–2000) is very small even near the clearing temperature, which enables the use of LC polymers as film-matrix for information recording. The recording may be accomplished by subjecting a homeotropically oriented sample at temperatures near T_{cl} to a focused laser beam. On subsequent illumination of a polymeric film by an unfocused laser beam dark

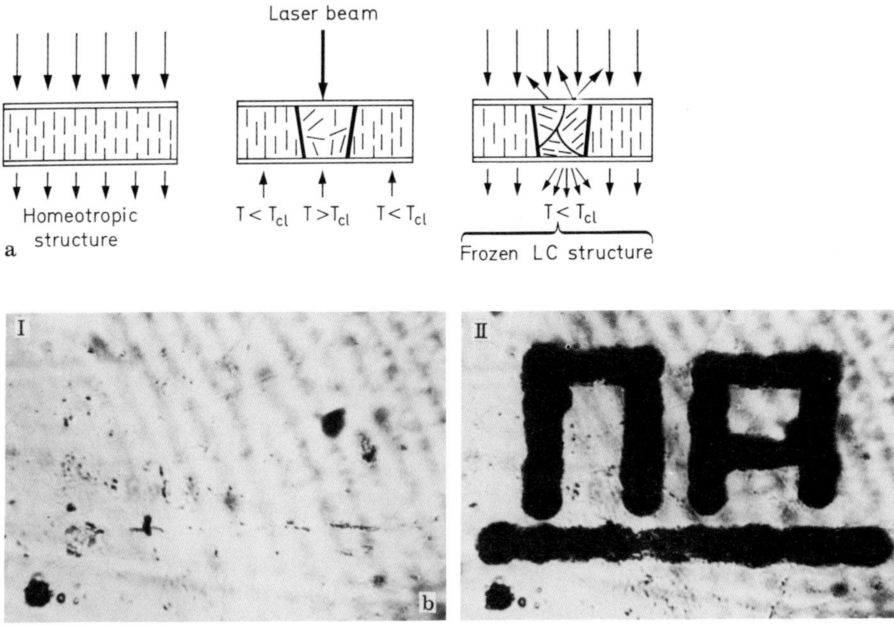

Fig. 27a and b. Scheme of thermorecording using the film of the homeotropic-oriented nematic LC polymer **(a)** and thermorecording of letters on the film of polymer XII at 105,5° (film thickness 40 μm) **(b)**; I — initial homeotropic texture; II — texture with the laser thermo-recorded letters

contrast spots are exhibited on a screen, corresponding to zones of the focused beam action. Figure 27b shows the letters recorded on a film of LC polymer XII. The recorded symbols are completely "wiped off" in 2–3 seconds by an application of alternating electric field.

The laser recorded information is preserved for a long time if the sample is cooled below T_g. From this viewpoint LC polymers differ usefully from low-molecular liquid crystals. The information storage time in the devices based on low-molecular liquid crystals do not usually exceed a few days. The described effects demonstrate the capabilities for the control of structural and optical properties of LC polymeric materials.

5.1.2.4 Electric Field Induced Structural Transition

Besides the electrooptical processes induced by the field (those described above), an entirely new electrooptical effect, lacking analogies among low-molecular liquid crystals, has recently been discovered in LC polymers. For some nematic polymers with $\Delta\varepsilon > 0$ (see Table 9, polymers A.4 and B.3.3.) that were preoriented by an external field, a structural transition was observed on cooling. The same but not preoriented polymers do not exhibit the latter (Fig. 28). In a very narrow temperature interval (at temperature T_1) around the transition region a sharp increase in light transmittance (in crossed polyrazers) is observed, the transparency of the film remaining practically unchanged (Fig. 28). As is seen from Fig. 28, under the applied field the transition is reversible and cooperative. We suppose that the observed transition

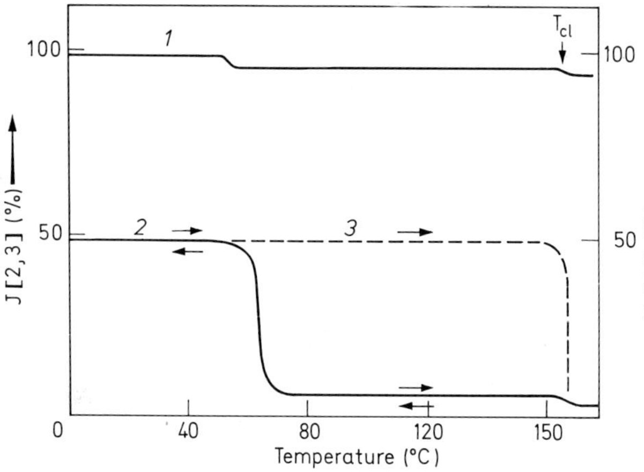

Fig. 28. Temperature dependence of the optical properties of nematic polymer A.4 (Table 9) — transparency Tr (1) and optical transmission in crossed polarizers I (2, 3) upon application of (1, 2), and without (3) an electric field [169]

results from a sharp conformational change in the macromolecules due to a structural rearrangement of mesogenic groups, the whole process being induced by the electric field.

It might be hoped that the subsequent investigation of LC polymer electrooptics will lead to the discovery of a diversity of interesting features in their electrooptical behaviour [170-171].

5.2 Electrohydrodynamic Effects

The described phenomena observed in LC polymers under an applied electric field are not caused by the flow of a current, but are due purely to the effect of the field. At the same time, any ionic current is accompanied by mass transfer, i.e. a hydrodynamic flow, which results in the orientation of a director due to the friction induced moment. As a result of the anisotropy of electric conductivity, complex hydrodynamic flows are developed in a liquid crystal layer, leading to electrohydrodynamic (EHD) instability. EHD instability is characterized by periodic violations of the liquid crystal structure, when the voltage exceeds certain threshold values. On the optical picture this is displayed either as the formation of so called domains (alternating dark and light bands, networks and so on), or as the effect of dynamic light scattering, when the domain picture is followed by the "boiling fluid" picture resulting from turbulent flows.

The investigation of EHD instability in LC polymers has just begun [160-161,170], and the results obtained are of a preliminary character. Below are discussed the specific features of EHD instability, the only investigated cyano-containing polymer of the acrylic series being taken as an example (polymer A.4, Table 9).

When an alternating electric field is applied to a homeotropically oriented layer of LC melt at definite temperatures (close to T_{cl}), a homogeneous orientation is

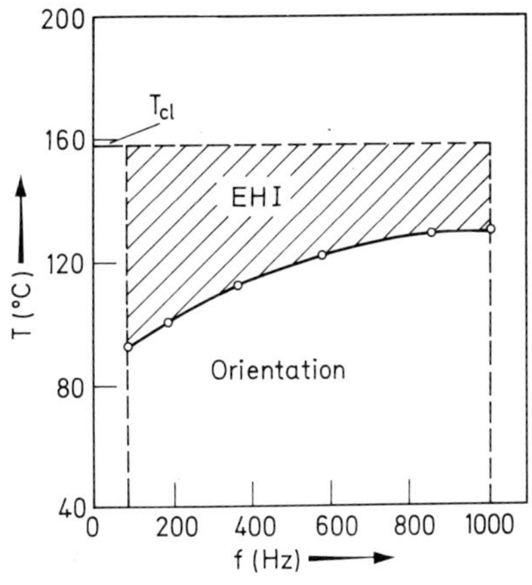

Fig. 29. T_{EHI} of polymer A.4 (Table 9) as a function of electric field frequency (U = 200 V)

disturbed, which is accompanied by a sharp decrease in transparency of a polymeric film. Such violation of orientation is caused by EHD instability resulting from turbulence ("boiling") occuring under these conditions in a LC polymer melt. On cooling a polymer sample the turmoil gradually diminishes and transparency of the sample rises as a homogeneous homeotropic orientation is reformed. The process is reversible, and on heating of a homeotropically oriented film above a definite temperature (T_{EHI}) a reverse transition to EHD instability is observed (Fig. 29). The occurence of EHD instability is characteristic for higher temperatures, at which the mobility of ionic impurities is higher. When the frequency of the electric field is increased, the temperature interval for EHD process becomes narrower. As is seen from Fig. 29, at the increase of frequency T_{EHI} is shifted to higher temperatures. This permits control of optical transparency of polymeric films by stepwise frequency changes (Fig. 30).

Thus, varying the frequency of an applied electric field at a given temperature, one may pass over from homeotropic orientation of the sample (transparent film) to the mode of EHD instability and reverse. This reveals interesting perspectives for controlling the optical properties of polymeric films.

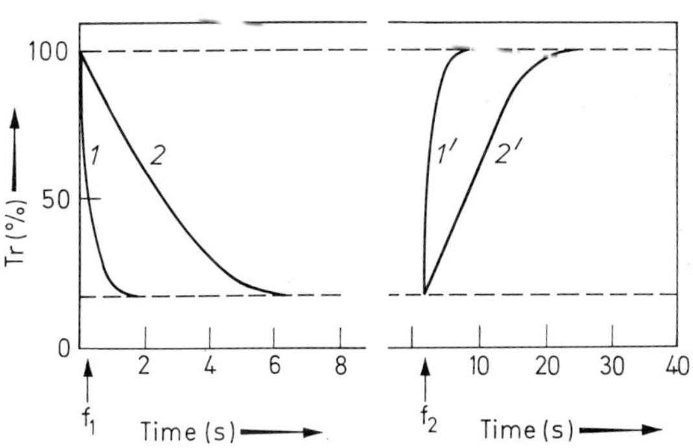

Fig. 30. Regulation of the optical transparency Tr of the films of polymer A.4 (Table 9) via discrete changes of the electric field frequency f_1 = 400 (1), 200 Hz (2) and f_2 = 4 (1′) and 1 kHz (2′). Arrows correspond to switching on the frequency f_1 and f_2

6 The Effect of Magnetic Field on Liquid-Crystalline Polymers

The use of an electric field is not the only effective way to influence the LC polymer structure, magnetic fields displays a closely similar effect [167–168]. It is interesting as a method allowing to orient LC polymers, as well as from the viewpoint of determining some parameters, such as the order parameter, values of magnetic susceptibility, rotational viscosity and others. Some relationships established for LC polymer 1 (Table 15), its blends with low-molecular liquid crystals and partially deuterated polyacrylate (polymer 4, Table 15) specially synthesized for NMR studies can be summarized as follows:

a) Orientation in LC polymers with positive anisotropy of diamagnetic susceptibility $\Delta\chi > 0$ is performed via alignment of mesogenic groups in the direction of the field applied.

b) The limit of orientation is reached substantially slower than on application of the electric field. It takes several dozens of minutes to reach the point.

c) Orientation occurs in a nematic mesophase only (even for normally smectic polymers) and may be freeze-fixed by cooling the polymer below T_g. Smectic mesophases have not yet been oriented.

d) In a frozen LC polymer (below T_g) a restricted motion of mesogenic groups with correlation times $\tau_c = 3 \cdot 10^{-7}$ sec (polymer 4, Table 15) is possible.

e) Order parameters of LC polymers, calculated from NMR measurements indicate a close analogy with those of low-molecular liquid crystals.

Thus the ability of LC polymers to orient in electric and magnetic fields reveals promise for the investigation of the specific features of LC polymer structure, as well as for the study of the mechanism of orientation and structural rearrangement processes in low-molecular liquid crystals, where they are very fast and in some cases are even hard to measure. On the other hand, this provides a method to control the structure of a polymer and thus create new materials with interesting optical properties.

7 Behaviour of Liquid-Crystalline Polymers with Mesogenic Side Groups in Dilute Solutions

It is evident that due to polymeric specificity of LC polymers most of the information on their molecular parameters, i.e. molecular mass, conformational state, polymeric chain flexibility and mobility, optical anisotropy and others, may be obtained from studies of dilute solutions of these compounds. However, taking into account that this branch of polymer science has already been reviewed [134, 172–176] we will here confine our treatment only to the initial steps of LC phase formation in polymer solutions.

A characteristic feature of the solutions of LC polymers with mesogenic side groups is the realization of orientational order of mesomorphic type due to interaction of side chains.

Theoretical and experimental analysis of the dependence of optical anisotropy on the side chain length, performed by Tsvetkov et al. [177] permitted to establish the

Table 16. Segmental anisotropy ($\alpha_1 - \alpha_2$), the Kerr constant K, the relaxation time τ, and the number of monomer units in the Kuhn's segment ν for some comb-like LC polymers in different solvents [172, 177–181]

Polymer No.	Structure of the monomer unit	$(\alpha_1 - \alpha_2) \cdot 10^{25}$ (cm^3) (solvent)	$K \cdot 10^{10}$, ($cm^5 \cdot g^{-1}$) $(V/300)^{-2}$	$\tau \cdot 10^5$ (s)	ν
1.	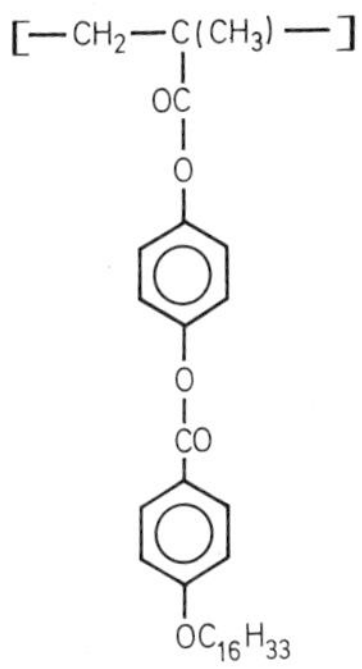	−3100 (tetrachloromethane)	−10	5	24
		−1600 (benzene)	−2,2	—	—
		−4200 (benzene + heptane) (52–48)	−40	—	—
2.	Monomer of Polymer 1	—	+0,21	—	—
3.	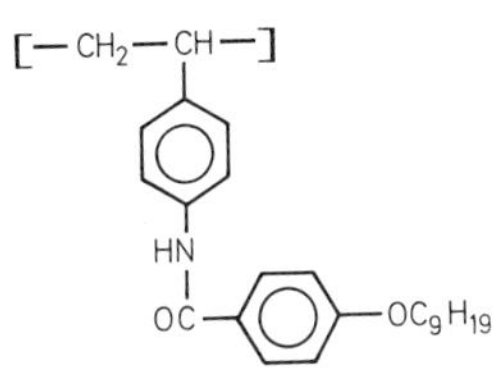	−2500 (benzene)	−140	0,2	20
4.	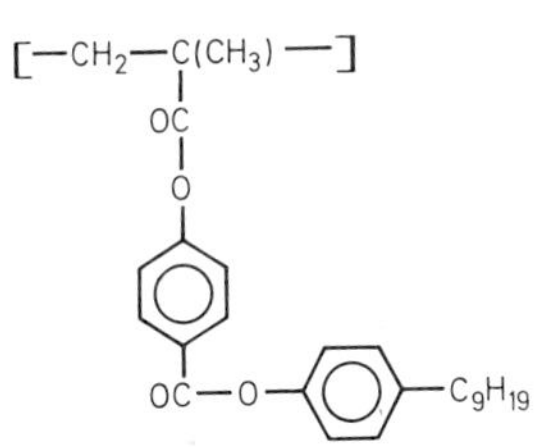	−2700 (tetrachloromethane)	−(8,0 ± 3,0)	—	25
5.	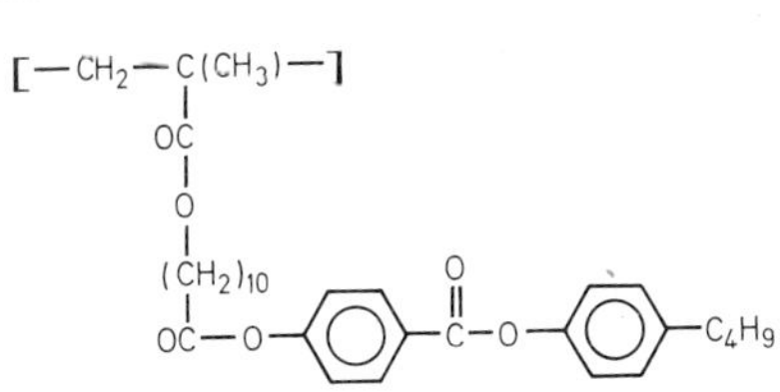	−90 (dioxan)	−(0,8 ± 0,2)	—	26

enhanced equilibrium rigidity of side groups of comb-like polymers (even of those without mesogenic groups — poly-n-alkylacrylates, poly-n-alkylmethacrylates), significantly exceeding the rigidity of the main chains of macromolecules. The increase of side group rigidity and high "crystal-like" order are to a greater extent exhibited by the polymers with mesogenic groups attached directly to the backbone. Investigations of dynamic optical and electrooptical properties of the solutions of such LC polymers (Table 16) indicate extraordinary high segmental anisotropy ($\alpha_1 - \alpha_2$) and dipole moments of macromolecules. Their values usually exceed by more than an order the values for ordinary flexible polymers. A high and negative value of birefringence of the solutions of LC comb-like polymers (Table 16) indicates that not only an axial but also polar order is characteristic for their macromolecules. As is seen from Table 16, the values of Kerr constants for polymers 1, 3, 4 and for monomer 2 differ markedly and even have opposite sign. A high degree of orientationally-polar order at relatively high equilibrium flexibility of the main chain (the Kuhn segment is 20–25 monomeric units) is coexistent with quite peculiar kinetic properties. Relaxation times τ are 10^{-4}–10^{-5} sec, i.e. 5–6 orders higher than the corresponding τ values for flexible polymers. This implies that substantial parts of macromolecules, including hundreds or thousands of monomeric units, are oriented in an electric field as a whole.

Thus, LC polymers of the type considered exhibit intramolecular order of mesomorphic type, mesogenic side groups forming a mobile LC structure.

Electric field induced orientation of polar groups in these polymers occurs via the mechanism of large scale motion, common for rigid polymers. However, the incorporation of a spacer within a macromolecule (polymer 5, Table 16) changes sharply the mode of intramolecular aggregation and the mechanism of orientation in an electric field.

The comparison of optical anisotropy and of Kerr constants for polymers 1, 3, 4, and 5 (Table 16) shows that the spacing out of mesogenic groups from the main chain by "insertion" of a polymethylene spacer leads to a drastic decrease in correlation of side group and backbone orientation, and, consequently, to a worse intramolecular ordering.

A significantly lower Kerr constant value and other factors indicate a small-scale intramolecular mechanism of orientation in an electric field. The size of kinetic units for a polymer with a spacer is significantly lower, and their mobility is significantly higher than for polymers without a spacer group [182]. At the same time, the weakening of correlation between the main chain and mesogenic fragments provides for their autonomy and makes it easier to form a LC order among side groups of distinct macromolecules (see Chap. 4). The data presented in Table 16 for polymer 5 actually give a quantitative confirmation of the independent behaviour of the main chain and side branches that ensure the realization of thermotropic mesomorphism in bulk.

A more serious factor that has direct regard to initial steps of mesophase formation in polymer solutions has to be assessed. That is the interaction between the molecules of LC polymer and of the solvent. The conformation of the macromolecule appears to be sensitive to the thermodynamic quality of the solvent, and this has a very pronounced effect on the mode of intramolecular structure formation. For instance, the folding of the chain in a bad solvent leads to a sharp rise in intramolecular orientational ordering of the side branches. This is manifest as an increase of optical anisotropy and especially of the values of Kerr constants for LC polymer solutions (see ($\alpha_1 - \alpha_2$) and K values for polymer 1 in Table 16).

Thus, varying thermodynamic quality of the solvent (by the use of various solvents and their mixtures or varying the temperature) one may appreciably affect the extent of inter- and/or intramolecular aggregation of the side groups. This, in turn, may definitely influence the formation of supermolecular organization of the mesophase in bulk polymeric samples.

This was qualitatively shown in investigations of conformational behaviour and intramolecular mobility (IMM) of cholesterol-containing polymers in dilute solutions as of a function of solvent quality [134–136, 185–188] and temperature. Polarization luminescence provides one of the most fruitful methods for the evaluation of IMM [175, 176]. The method permits to get direct information about rotational mobility of the macromolecule as a whole, as well as about the mobility of the main chains and side branches. This is achieved via the attachment to macromolecules of so called luminescent markers (LM) — anthracylacyloxymethane groups in the case reported. Below are shown the chain fragments with LM which give information on the mobility of main chains (LM-1) and of side groups (LM-2):

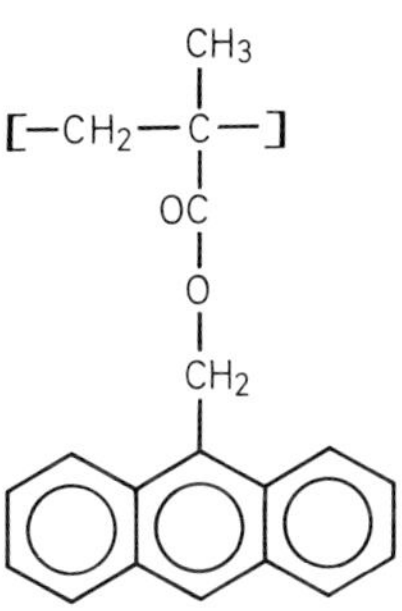

LM-1

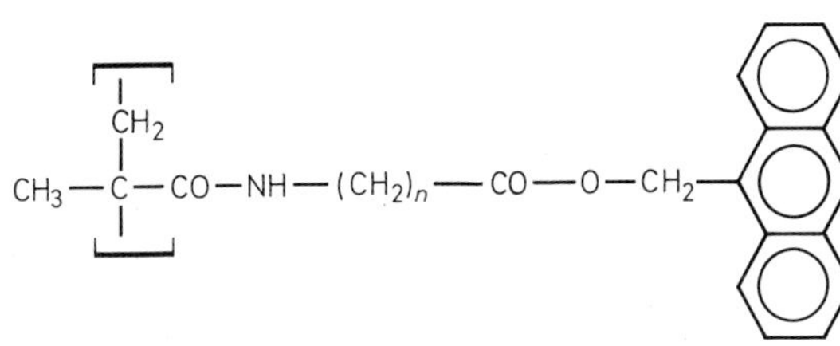

LM-2

The time that characterizes the rotational diffusion of the macromolecule as a whole (τ_m) and the mobility of the main chains (τ_{main}) and of the side branches (τ_{side}) are collected in the following equation

$$\frac{1}{\tau_w} = \frac{1}{\tau_m} + \frac{1}{\tau_{main(side)}}$$

The study of homological series of cholesterol-containing LC polymers with polymethacrylate backbone

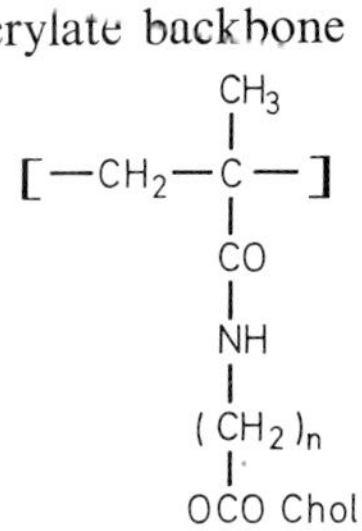

PChMA - *n*

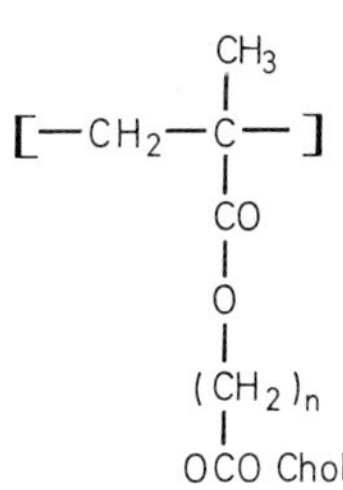

PChMO - *n*

where *n* = 5...14

Table 17. Values of relaxation times τ_w, τ_{main} and τ_{side} for cholesterol containing polymers PChMA-n and PChMO-n in different solvents at 25 °C [135–136, 185–187]

Solvent	PChMA-n			PChMO-n			
	n = 11		n = 6	n = 14	n = 10		n = 5
	τ_{main}, (ns)	τ_{side}, (ns)	τ_w, (ns)	τ_w, (ns)	τ_{main}, (ns)	τ_{side}, (ns)	τ_w, (ns)
Heptane	490	—	490	>400	>600	>200	150
Heptane + 0,2% trifluoroacetic acid	47	—	—	—	—	—	—
Cyclohexane	130		130	50	62	18	200
Toluene	100	3.3	66	28	25	3.2	27
Chloroform	22	1.6	22	16	16	3.0	17

by means of a luminiscent probe method has shown that in dilute solutions intramolecular ordering of cholesterol fragments leads in certain conditions to the formation of mesophase nuclei [135, 136, 175, 176, 185–187].

Table 17 presents relaxation time values characterizing the intramolecular mobility of various fragments of cholesterol-containing polymers with LM in various solvents[9]. As is seen, the values of relaxation times measured for the same polymer in various solvents differ significantly, which reflects the specificity of conformational state and intramolecular organization.

The analysis of these data, together with the results of investigations of optical activity [135, 136], hydrodynamic behaviour and light scattering [186, 188] of the solutions of these polymers, has shown that macromolecules of PChMA-n and PChMO-n dissolved in paraffins (i.e. bad solvents) contain fragments of ordered mesogenic groups (see Table 17, large values for relaxation times in heptane). The formation of even more compact intramolecular structures occurs, as a cooperative coil-globule conformational transition in a sufficiently narrow temperature interval. Figure 31 (curves 1 and 2) and Fig. 32 (curves 1, 2) illustrate how the transition is performed in macromolecules of PChMA-11 and PChMO-10.

The formation of fragments with the ordered alignment of mesogenic (optically active) groups is accompanied by the growth of optical activity of polymer solutions (Fig. 31, curve 1) together with a sharp increase of relaxation times τ_w. The latter reflects the fall of IMM (Fig. 31, curve 1 up to temperature 20 °C). Simultaneously intramolecular retardation of the mobility of the main and side chains increases which is indicated by the increase of τ_{main} and τ_{side} values, as is seen for polymer PChMO-10 for example (Fig. 32, curves 1, 2). At the same time the maximum on the curve of τ_w versus temperature (Fig. 31, curve 2) means that macromolecules tend to accept a compact globular conformation. Subsequent decrease of temperature leads to a decrease of τ_w. This fact cannot be interpreted in terms of IMM only and

9 Here and below relaxation times determined in solvents of different viscosities are reduced to a standard solvent viscosity $\eta_{reduced}$ = 0,38 cP (0,38 × 10^{-3} Ns/m^2).

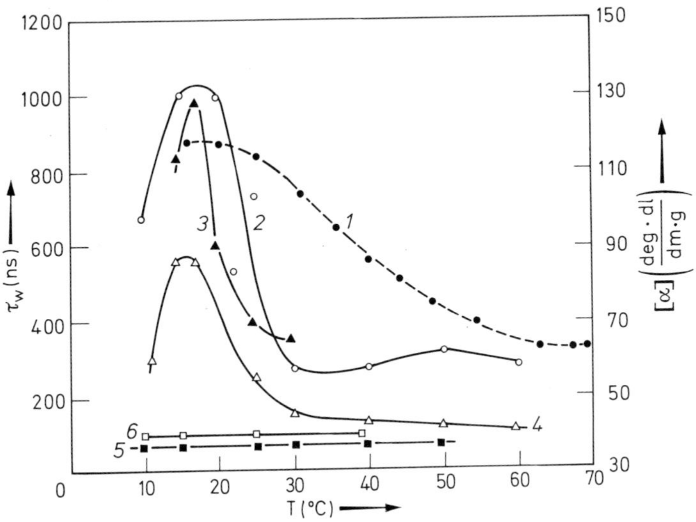

Fig. 31. Temperature dependences of [α] (1) and τ_w (2–6) for solutions of PChMA-11 (1, 2) and copolymers of ChMA-11 with butylmethacrylate (MA-4), containing 10 (3), 25 (4) and 60 mole-% (5) of MA-4 and PCMA-11 (6) in heptane [130]

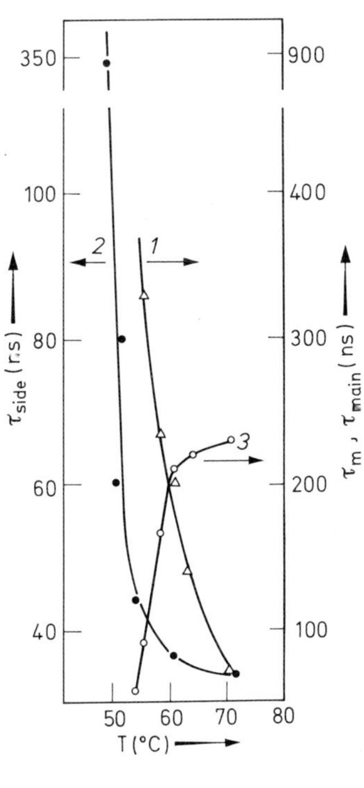

Fig. 32. Effect of temperature on the mobility of the main chain (1), side chain (2) and rotational mobility of the macromolecule as a whole (3) for the polymer PchMO-10 in heptane [187]

needs the mobility of a macromolecule as whole (τ_m) to be taken into account. Compacting of macromolecules of cholesterol-containing polymers is also confirmed by a pronounced decrease of the time characteristic of the rotational mobility of the macromolecule as a whole (Fig. 32, curve 3). This also accounts for a sharp decrease of intrinsic viscosity $[\eta]$ and macromolecule dimensions $(R^{-2})^{1/2}$ (Fig. 33, curves 1, 2) down to the values typical for globular proteins.

The condition necessary for the formation of fragments with ordered mesogenic group alignment, which serve as the mesophase nuclei, is the decrease of side chain mobility (Table 17). If the mobility of side chains is high enough and $\tau_{side} = 2$–3 nsec, structure formation does not occur. If the mobility of side groups is decreased and τ_{side} is one or more orders of magnitude higher, then fragments of ordered mesogenic group sequences and a compact globular structure are formed.

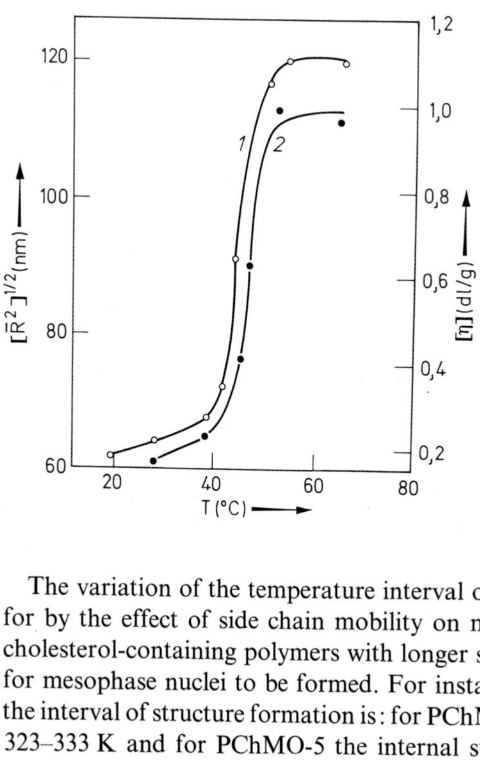

Fig. 33. Temperature dependences of $[\eta]$ (1) and $(R^2)^{1/2}$ (2) of PChMO-10 in heptane [186]

The variation of the temperature interval of structure formation is also accounted for by the effect of side chain mobility on mesophase nucleation. The solutions of cholesterol-containing polymers with longer spacer groups have to be cooled further for mesophase nuclei to be formed. For instance, in a series of PChMO-n polymers the interval of structure formation is: for PChMO-14 — 308–313 K, for PChMO-10 — 323–333 K and for PChMO-5 the internal structure is formed at even higher temperatures [187].

The study of the reasons for the formation of intramolecular structures reveals that the enhanced interaction of cholesterol groups occuring on cooling is very important. The "dilution" of the sequence of cholesterol-containing monomeric units by butylmethacrylate units (copolymers of ChMA-11 with butylmethacrylate) leads gradually to the "degeneration" of the conformational transition (Fig. 31, curves 3–5).

The decisive role of cholesterol groups in the process of compact globular structure formation is confirmed by the temperature dependence of τ_w for solutions of PCMA-11 polymer, which contains paraffinic cetyl groups instead of mesogenic cholesterol groups. As is seen from Fig. 31 (curve 6), a globular structure is not formed in solutions for this polymer.

Intramolecular structures are essentially stabilized by hydrogen bond formation between the amide groups of PChMA-n [188, 189]. For instance, the addition of trifluoroacetic acid (TFAc), which is a strong competitor for hydrogen bond formation, to solutions of PChMA-11 in heptane leads to a sharp increase of IMM (τ_{main} decreases for more than an order at the addition of 0,2% TFAcA) and intramolecular structure is not formed. The data presented manifest the role of kinetic factors in mesophase nucleation in dilute solutions of polymers with mesogenic side groups.

The study of the relaxation of dipole polarization, as well as of the dipole moments of cholesterol-containing polymers and copolymers [128–134, 191–193] presents a sensitive confirmation for the existence of intramolecular structuration of mesogenic groups. This is indicated for instance, by the high values of relaxation times ($\tau^s_{d.p}$) and activation energy ($E^s_{d.p}$) of dipole polarization, as well as by the large values of correlation parameter g, which is a relative measure of the internal rotational retardation in macromolecules (Table 18).

Table 18. Values of correlation parameter g, relaxation time $\tau^s_{d.p}$ and activation energy $E^s_{d.p}$ of a relaxation process of dipole polarization for some comb-like poly(methacrylates) in toluene solutions at 25 °C [190]

$[-CH_2-C(CH_3)-]$ with R attached to C

R	g	$\tau^s_{d.p.}$ (ns)	$E^s_{d.p.}$ (kJ/mole)
$-COO-(CH_2)_{17}-CH_3$	0.6	20	31.0
$-CONH-(CH_2)_{17}-CH_3$	2.0	250	42.0
$-CO-NH(CH_2)_{11}-COO-$Chol	2.6	1090	52.5

The fragments of macromolecules with ordered cholesterol group sequences, that are formed in bad solvents, may serve as nuclei of supermolecular order in films, obtained from these solvents. Structural and optical studies have shown that PChMA-11 films produced by solvent evaporation display different properties: those obtained from chloroform and toluene solutions (small relaxation times, see Table 17) are optically isotropic, and those obtained from heptane solutions (large relaxation times, see Table 17) are optically anisotropic, what reflects the differences in conformational state of polymeric chains in these films. Contrary to the optically isotropic films, a high degree of side branch ordering characterizes optically anisotropic films, which is confirmed by X-ray studies. The observed difference of LC polymer structure in the bulk is thus the consequence of their different conformational state in solution; this reveals some possibilities for the control of LC polymer structure at the initial steps of mesophase nucleation in solutions.

The experimental results regarding the formation of compact globular structures and coil-globule transitions presented above are in good agreement with theoretical works of Grosberg [194)] where a model comprising macromolecules with rod-like mesogenic side groups attached to flexible backbones was chosen for calculations. This work provided theoretical grounds for the formation of intramolecular structure of globular type and coil-globule transition, similar to that investigated in our experimental studies [135, 187)].

The brief data presented in this chapter concerning the initial steps of structure formation in LC polymer solutions, are significant from two viewpoints. On the one hand, the study of these processes provides quantitative information about the molecular parameters and IMM of LC polymers, which is the basis for the understanding and prediction of physico-chemical behaviour of polymeric liquid crystals in bulk. On the other hand, understanding of the features of intramolecular structure formation in dilute solution, reveals broad prospects for the investigation of the formation of lyotropic LC systems of polymers with mesogenic side groups, which is in its infancy [195)].

Summing up, to the present time a substantial amount of data concerning the creation of thermotropic LC polymers has already been accumulated, methods for their synthesis were developed, certain relationships between the structure of the polymer and the mesophase type were established, the structure and some properties of such systems were studied.

However, in a brief survey it is impossible to cover all of the aspects related to LC polymers. This is why such important questions as thermodynamic and dielectric properties, conformational pecularities of LC polymers in solutions and some other subjects were left out.

The contemporary period is characterized by a rapid accumulation of information about thermotropic polymeric liquid crystals. It is yet too diversified and, as a rule, gives only a qualitative description of the observed phenomena. The next step in LC polymer investigations should be aimed at a quantitative description of their behaviour. This approach should necessarily lead to progress in the comprehension of the quite peculiar mesomorphic state of matter as well as in the introduction of LC polymers into technological practice.

Acknowledgement. The authors are greatly indebted to their colleagues and coworkers, especially to Drs. R. V. Talrose, Ya. S. Freidzon, S. G. Kostromin who carried out the synthesis and investigations of LC polymers as well as took part in numerous and very useful discussions.

8 References

1. Uematsu, J., Uematsu, Y.: this issue, p. 37
2. De-Visser, A., De Groot, K. Banties, A.: J. Polymer Sci. *9*, A-1, 1893 (1971)
3. Platé, N. A., Shibaev, V. P., Tal'roze, R. V.: in Uspekhi khimii i fiziki polimerov (Advances in chemistry and physics of polymers), Khimiya, Moscow 1973, p. 127
4. Strzelecki, L., Liebert, L.: Bull. Soc. chim. France, N 2, 605 (1973)
5. Blumstein, A., Blumstein, R., Murphy, G., J. Billard: in Liquid Crystals and Ordered Fluids (Ed.s. Johnson, J. F., Porter, R. S.), Plenum Press, New York 1974, vol. 2, p. 277
6. Shibaev, V. P.: Dissertation for the degree of Doctor of Sciences, Moscow State University, Moscow 1974

7. Perplies, E., Ringsdorf, H., Wendorff, J.: Makromol. Chem. *175*, 553 (1974)
8. Freidzon, Ya. S., Shibaev, V. P., Platé, N. A.: Abstracts of papers at the 3-rd all-Union Conference on Liquid Crystals, p. 214, Ivanovo, 1974
9. Shibaev, V. P., Freidzon, Ya. S., Platé, N. A.: Abstracts of papers at the 11-th Mendeleev Congress on General and Applied Chemistry, vol. 2, p. 164, Nauka, Moscow, 1975
10. Shibaev, V. P., Freidzon, Ya. S., Platé, N. A.: USSR Inventor's Certificate No 525709, Byull. izobreteneiy N 31, 1976
11. Shibaev, V. P., Platé, N. A.: Vysokomolek. Soedin. *A 19*, 923 (1977)
12. Platé, N. A., Shibaev, V. P.: Grebneobraznye polimeri i zhidkie kristally (Comblike polymers and liquid crystals), Khimiya, Moscow, 1980
13. Amerik, Yu. B., Krentzel, B. A.: Khimiya zhidkikh kristallov i mesomorphnyh polimernyh sistem (Chemistry of liquid crystals and mesomophic polymer systems), Nauka, Moscow, 1981
14. Mesomorphic Order in Polymers and Polymerization in Liquid Crystalline Media (Ed. Blumstein, A.). Amer. Chem. Soc. Symposium Series N 74, ACS, Washington, D.C., 1978
15. Liquid Crystalline Order in Polymers (Ed. Blumstein, A.), Academic Press, New York, San Francisco, London, 1978
16. Roviello, A., Sirigu, A.: Polymer Letters *13*, 455 (1975)
17. Tsukruk, V. V., Shilov, V. V., Konstantinov, I. I., Lipatov, Yu. S., Amerik, Yu. B.: Europ. Polymer J. *18*, 1015 (1982)
18. Papkov, S. P., Kulichikhin, V. G.: Zhidko-kristallicheskoe sostoyanie polimerov (Liquid Crystalline state of polymers), Khimiya, Moscow, 1977
19. Mc Intyre, J., Mulburn, A.: British Polymer J. *13*, 5 (1981)
20. Meurisse, P., Noel, C., Monnerie, L., Fayolle, B.: ibid. *13*, 55 (1981)
21. Jackson, W., Kuhfuss, H.: J. Polymer Sci. *14*, 2043 (1976)
22. Polk, M., Bota, K., Akubuiro, E., Phingbodhipakiya, M.: Macromolecules *14*, 1626 (1982)
23. Cox, M., Griffin, B.: British Polymer J. *12*, 147 (1980)
24. Wissbrun, K.: ibid *12*, 163 (1980)
25. Preston, J., in Liquid crystalline order in polymers (Ed. Blumstein, A.): Academic Press, New York, San Francisco, London, 1978, p. 141
26. Mc Intyre, J.: this issue, p. 61
27. Rehage, G., Finkelmann, H.: this issue, p. 99
28. Hardy, Gy, Nyitrai, K., Cser, F.: Abstracts of the fifth International Liquid Crystal Conference of Socialist Countries, Odessa, USSR, 1983, v. 2, part 1, p. 98
29. Kamogawa, H.: J. Polymer Sci. *B 10*, 7 (1972)
30. Paleos, C., Filippakis, S., Margomenou-Leonidopoulou, G.: J. Polym. Sci.-Polym. Chem. Ed. *19*, 1427 (1981)
31. Paleos, G., Margomenou-Leonidopoulou, G., Filippakis, S., Malliaris, A.: J. Polym. Sci. — Polym. Chem. Ed. *20*, 2267 (1982)
32. Finkelmann, H., Rehage, G.: Makromol. Chem.-Rapid Commun. *1*, 31 (1980)
33. Platé, N. A., Shibaev, V. P.: J. Polym. Sci. — Macromolec. Rev. *8*, 117 (1974)
34. Shibaev, V. P., Platé, N. A., Kargin, V. A.: Proceedings of the Third European Regional Conference on Electron Microscopy, Prague, 1964, v. A., p. 415
35. Shibaev, V. P., Petrukhin, B. S., Zubov, Yu. A., Platé, N. A., Kargin, V. A.: Vysokomol. Soedin. *A 10*, 216 (1968)
36. Shibaev, V. P., Freidzon, Ya. S., Platé, N. A.: Vysokomol. Soedin. *A-20*, 82 (1982)
37. Shibaev, V. P., Freidzon, Ya. S., Kharitonov, A. V., Platé, N. A.: Preprints of Short Communications of 26 International Symposium on Macromolecules, Mainz, FRG, 1979, v. III, p. 1571
38. Finkelmann, H., Frenzel, J., Rehage, G.: Abstracts of Commun. 27-th International Symposium on Macromolecules, Strasbourg, v. *2*, 965 (1981)
39. Finkelmann, H., Kokk, H., Rehage, G.: Macromol. Chemie *2*, 317 (1981)
40. Allcock, H., Fuller, T.: Macromolecules *13*, 1338 (1980)
41. Shibaev, V. P., Kulichikhin, V. G., Kostromin, S. G., Vasileva, N. V., Braverman, L. P., Plate, N. A.: Dokl. Akad. Nauk SSSR *263*, 152 (1982)
42. Blinov, L. M.: Elektro- i magnitooptika zhidkih kristallov (Electrooptics and magnitooptics of liquid crystals), Nauka, Moscow, 1978
43. De Gennes, P. G.: The Physics of Liquid Crystals, Clarendon Press, Oxford, 1974

44. Shibaev, V. P., Tal'roze, R. V., Sinitzyn, V. V., Kostromin, S. G., Platé, N. A.: Proceedings of the 28 International Macromolecular Symposium, Amherst, USA, 1982, p. 814
45. Shibaev, V. P.: in Advances in liquid crystal, research and applications (Ed. L. Bata); Oxford: Pergamon Press — Budapest: Akademiai Kiado, v. *2*, 869 (1980)
46. Shibaev, V. P., Freidzon, Ya. S., Platé, N. A.: Dokl. Akad. Nauk SSSR *227*, 1412 (1976)
47. Shibaev, V. P., Kostromin, S. G., Platé, N. A.: Europ. Polym. J. *18*, 651 (1982)
48. Zhidkie kristally (Liquid Crystals) Ed. Zhdanov, S. I., p. 317, Khimiya, Moscow, 1979
49. Kostromin, S. G., Tal'roze, R. V., Shibaev, V. P., Platé, N. A.: Macromol. Chem., Rapid Commun. *3*, 803 (1982)
50. Freidzon, Ya. S., Kharitonov, A. V., Shibaev, V. P., Platé, N. A.: Mol. Cryst., Liq. Cryst. *88*, 87 (1982)
51. Frosini, V., Levita, G., Lupinacci, D., Magagnini, P.: Mol. Cryst., Liq. Cryst. *66*, 341 (1981)
52. Magagnini, P., Marchetti, A., Matera, F., Pizzirani, G., Turchi, G.: Europ. Polym. J. *10*, 585 (1974)
53. Maganini, P.: Makromol. Chemie *4*, 223 (1981)
54. Gray, G. W.: in Liquid Crystals and Plastic Crystals, vol. 1, (Eds. Gray, G. W., Winsor, P. A.), Ellis Horwood, Chichester 1974
55. Matheson, R. R., Flory, P. J.: Macromolecules *14*, 954 (181)
56. Kuznetzov, D. V., Khokhlov, A. R.:Vysokomolek. Soedin. *24*, 418 (1982)
57. Birshtein, T. M., Kolegov, B. I., Goryanov, A. N., Abstracts of the fourth international liquid crystal conference of socialist countries, Tbilisi, USSR, 1981, v. *2*, p. 146
58. Vasilenko, S. V., Khokhlov, A. R., Shibaev, V. P.: Macromol. Chemie Rapid. Commun. *3*, 920 (1982)
59. Vasilenko, S. V., Khokhlov, A. R., Shibaev, V. P.: Abstracts of the First all-Union Symposium on Liquid Crystalline Polymers, Suzdal, USSR, p. 114, 1982
60. Lisezki, L. N.: ibid, p. 26, 1982
61. Flory, P. J.: Proc. Roy. Soc. (London) *234*, 60 (1956)
62. Kostromin, S. G.: Dissertation for the Degree of Candidate of Sciences, Moscow State University, Moscow, 1982
63. Wendorff, I. H., Finkelmann, H., Ringsdorf, H.: J. Polymer Sci-Polymer Symposium *63*, 245 (1978)
64. Shibaev, V. P., Moiseenko, V. M., Freidzon, Ya. S., Platé, N. A.: Europ. Polymer. J. *16*, 277 (1980)
65. Sinitzyn, V. V.: Dissertation for the Degree of Candidate of Sciences, Moscow State University, Moscow, 1982
66. Tal'roze, R. V., Sinitzyn, V. V., Shibaev, V. P., Platé, N. A.: Mol. Cryst., Lig. Cryst. *80*, 211 (1982)
67. Fayolle, B., Noel, C., Billard, J.: J. de Physique *40*, C3-485 (1979)
68. Griffin, A., Havens, S.: Mol. Cryst. Liq. Cryst. *49*, 239 (1979)
69. Griffin, A., Havens, S.: J. Polym. Sci.-Polym. Lett. Ed. *18*, 259 (1980)
70. Cser, F.: J. de Physique Colleque *40*, C3-499 (1979)
71. Cser, F., Hyitrai, K., Hardy, G.: in Advances in liquid crystal, research and applications (Ed. L. Bata), Oxford: Pergamon Press — Budapest: Akademiai Kiado, v. *2*, p. 845, 1980
72. Hardy, G., Cser, F., Hyitrai, K.: Abstracts of the fourth International Liquid Crystal Conference of Socialist Countries, Tbilisi, USSR, 1981, v. 1, p. 371
73. Cser, F., ibid., v. *2*, p. 223, 1981
74. Kozlovski, M. V., Shibaev, V. P.: Abstracts of the First all-Union Symposium on Liquid Crystalline Polymers, Suzdal, USSR, p. 47, 1982
75. Kostromin, S. G., Sinitzyn, V. V., Tal'roze, R. V., Shibaev, V. P., Platé, N. A.: Makromol. Chem., Rapid Commun. *3*, 809 (1982)
76. Lipatov, Yu. S., Tsukruk, V. V., Shilov, V. V., Grebneva, V. S., Konstantinov, I. I., Amerik, Yu. B.: Vysokomol. Soedin. *B23*, 818 (1981)
77. Tsukruk, V. V., Shilov, V. V., Lipatov, Yu. S., Konstantinov, J. J., Amerik, Yu. B.: Acta Polymerica *33*, 63 (1982)
78. Tsukruk, V. V.: Dissertation for the Degree of Candidate of Sciences, Institute of Macromolecular Chemistry, Academy of Sciences of the Ukrainian SSR, Kiev, 1982

79. Wendorff, J. H., Finkelmann, H., Ringsdorf, H.: in Mesomorphic Order in Polymers and Polymerization in Liquid Crystalline Media (Ed. A. Blumstein), ACS Symposium Series, Washington, D. C., American Chemical Society, v. 74, p. 12, 1978
80. Finkelmann, H., Ringsdorf, H., Siol, W., Wendorff, I. H., ibid: v. *74*, 22 (1977)
81. Kelker, H., Wirzing, U.: Mol. Cryst. Liq. Cryst. *49*, 175 (1979)
82. Finkelmann, H., Happ, M., Portugal, M., Ringsdorf, H.: Makromol. Chemie *179*, 2541 (1978)
83. Blumstein, A.: Macromolecules *10*, 872 (1977)
84. Clough, S. B., Blumstein, A., De Vries, A.: Polymer Preprints *18*, 1 (1977)
85. Blumstein, A., Blumstein, R., Clough, S. B., Hsu, E. C.: Macromolecules *8*, 73 (1975)
86. Shibaev, V. P., Kharitonov, A. V., Freidzon, Ya. S., Platé, N. A.: Vysokomol. Soedin. *A 21*, 1849 (1979)
87. Freidzon, Ya. S., Shibaev, V. P., Kharitonov, A. V., Platé, N. A.: in Advances in liquid crystal, research and applications (Ed. L. Bata), Oxford: Pergamon Press — Budapest: Akademiai Kiado, v. 2, p. 899, 1980
88. Perplies, E., Ringsdorf, H., Wendorff, I. H.: Berg. Buns. Phys. Chem. *78*, 921 (1974)
89. Sinitzyn, V. V., Tal'roze, R. V., Shibaev, V. P., Platé, N. A.: Abstracts of the fourth International Liquid Crystal Conference of Socialist Countries, Tbilisi, USSR, 1981, v. 2, 213
90. Konstantinov, I. I., Grebneva, V. S., Amerik, Yu. B., Sitnov, A. A.: ibid, v. 2, p. 183, 1981
91. Shibaev, V. P., Platé, N. A., Freidzon, Ya. S.: J. Polym. Sci., Polym. Chem. Ed. *17*, 1655 (1979)
92. Shibaev, V. P., Platé, N. A., Freidzon, Ya. S.: in Mesomorphic Order in Polymers and Polymerization in Liquid Crystalline Media (Ed. A. Blumstein), ACS Symposium Series, Washington, D.C., American Chemical Society, v. 74, p. 33, 1978
93. Freidzon, Ya. S., Shibaev, V. P., Kustova, N. N., Platé, N. A.: Vysokomol. Soedin. *A 22*, 1083 (1980)
94. Clough, S. B., Blumstein, A., De Vries, A.: Mesomorphic Order in Polymers and Polymerization in Liquid Crystalline Media (Ed. A. Blumstein), ACS Symposium Series, Washington, D.C., American Chem. Soc. v. 74, p. 1, 1978
95. Blumstein, A., Clough, S. B., Patel, L., Blumstein, R. B., Hsu, E. C.: Macromolecules *9*, 243 (1976)
96. Demus, D., Richter, L.: Textures of Liquid Crystals, Leipzig, VEB Deutscher Verlag für Grundstoffindustrie, 1980.
97. Shibaev, V. P., Moiseenko, V. M., Platé, N. A.: Makromol. Chemie *181*, 1381 (1980)
98. Gudkov, V. A., Chistaykov, I. G., Vainshtein, B. K., Shibaev, V. P.: Crystallografia *27*, 537 (1982)
99. Shibaev, V. P., Moiseenko, V. M., Lukin, N. Yu., Platé, N. A.: Dokl. Akad. Nauk SSSR *237*, 401 (1977)
100. Shibaev, V. P., Moiseenko, V. M., Smolyanski, A. L., Platé, N. A.: Vysokomol. Soedin *A 23*, 1969 (1981)
101. Konstantinov, I. I.: J. de Physique *40*, C3-475 (1979)
102. Hahn, B., Wendorff, I. H., Portugall, M., Ringsdorf, H.: Colloid. and Polym. Sci. *259*, 875 (1981)
103. Tal'roze, R. V., Karakhanova, F. I., Borisova, T. I., Burshtein, L. L., Nikanorova, N. I., Shibaev, V. P., Platé, N. A.: Vysokomol. Soedin. *A 20*, 1835 (1978)
104. Shibaev, V. P., Tal'roze, R. V., Karakhanova, F. I., Platé, N. A.: J. Polym. Sci. Polym. Chem. Ed. *A 17*, 1671 (1979)
105. Finkelmann, H., Rehage, G.: Makromol. Chemie, Rapid Commun. *1*, 733 (1980)
106. Lipatov, Yu. S., Tsukruk, V. V., Shilov, V. V., Konstantinov, I. I., Amerik, Yu. B : Vysokomolek. Soedin. *A 23*, 1533 (1981)
107. Lipatov, Yu. S., Tsukruk, V. V., Shilov, V. V., Grebneva, V. S., Konstantinov, I. I., Amerik, Yu. B.: Abstracts of the fourth International Liquid Crystal Conference of Socialist Countries, Tbilisi, USSR, 1981, v. 2, 191
108. Shilov, V. V., Tzukruk, V. V., Lipatov, Yu. S.: Abstracts of the first All-Union Symposium on Liquid Crystalline Polymers, Suzdal, USSR, p. 23, 1982
109. Tsukruk, V. V., Shilov, V. V., Lipatov, Yu. S.: Crystallografia *29*, 520 (1984)
110. Shilov, V. V., Tsukruk, V. V., Bliznyuk, V. N., Lipatov, Yu. S.: Polymer *23*, 484 (1982)
111. Tsukruk, V. V., Shilov, V. V., Lipatov, Yu. S.: Makromol. Chem. *183*, 2009 (1982)

112. Roganova, Z. A., Smoliansky, A. L.: Abstracts of the first All-Union Symposium on Liquid Crystalline Polymers, Suzdal, USSR, p. 28, 1982
113. Clough, S. B., Blumstein, A., Hsu, E. C.: Macromolecules *9*, 123 (1976)
114. Hsu, E., Blumstein, A.: J. Polym. Sci., Letters *15*, 129 (1977)
115. Liebert, L., Strzelecki, L.: Bull. Soc. Chim. France *2*, 603 (1973)
116. Cser, F., Nyitrai, K., Seyfried, E., Hardy, Gy.: European Polym. J. *13*, 678 (1977)
117. Finkelmann, H., Ringsdorf, H., Wendorff, J.: Makromol. Chemie *179*, 273 (1978)
118. Ringsdorf, H., Schneller, A.: British Polym. J. *13*, 43 (1981)
119. Ringsdorf, H., Zentel, R.: Abstracts of Communications of 27-th International Symposium on macromolecules, Strasbourg, France, v. 2, p. 969, 1981
120. Ringsdorf, H., Zentel, R.: Makromol. Chem. *183*, 1245 (1982)
121. Kostromin, S. G., Shibaev, V. P., Platé, N. A.: Abstracts of the fourth International liquid crystal Conference of Socialist countries, Tbilisi, USSR, v. 2, p. 185, 1981
122. Platé, N. A., Shibaev, V. P.: J. Polym. Sci., Polym. Symposium *67*, 1 (1980)
123. Mousa, A. M., Freidzon, Ya. S., Shibaev, V. P., Platé, N. A.: Polymer Bull. *6*, 485 (1982)
124. Shibaev, V. P., Kostromin, S. G., Talroze, R. V., Platé, N. A.: Dokl. Acad. Nauk SSSR *259*, 1147 (1981)
125. Vainshtein, B. K., Gudkov, V. A., Chistyakov, I. G., Shibaev, V. P.: Abstracts of the fourth International liquid crystal Conference of Socialist Countries, Tbilisi, USSR, v. 1, p. 15, 1981
126. Kapustin, A. P.: Eksperimentalnye issledovaniya zhidkih Kristallov (Experimental investigations of liquid crystalls), Moscow, Nauka, 1978, p. 106
127. Blinov, L. M., Yablonski, S. V., Shibaev, V. P., Kostromin, S. G.: Abstracts of Ninth International Conference of Liquid Crystals, Bangalore, India, p. 275, 1982
128. Nikonorova, N. A., Borisova, T. J., Burstein, L. L., Shibaev, V. P.: Abstracts of 12th Europhysics Conference on Macromolecular Physics, Leipzig, GDR, 1981, p. 313
129. Borisova, T. J., Burshtein, L. L., Stepanova, T. P., Freidzon, Ya. S., Shibaev, V. P., Platé, N. A.: Vysokomol. Soedin. *18*, 628 (1976)
130. Nikonorova, N. A., Borisova, T. J.: Abstracts of the first all-Union Symposium on Liquid Crystalline Polymers, Suzdal, USSR, p. 103, 1982
131. Borisova, T. I., Burshtein, L. L., Stepanova, T. P., Shibaev, V. P.: Vysokomol., Soedin. *21*, 829 (1979)
132. Borisova, T. I., Burshtein, L. L., Stepanova, T. P., Freidzon, Ya. S., Shibaev, V. P., Platé, N. A.: Vysokomol. Soedin. *A 24*, 1103 (1982)
133. Borisova, T. I., Burshtein, L. L., Stepanova, T. P., Freidzon, Ya. S., Kharitonov, A. V., Shibaev, V. P.: Vysokomol. Soedin. *24*, 451 (1982)
134. Burshtein, L. L., Shibaev, V. P.: Vysokomol. Soedin. *A 24*, 3 (1982)
135. Shibaev, V. P., Freidzon, Yu. S., Agranovich, I. M., Pautov, V. D., Anufrieva, E. V., Platé, N. A.: Dokl. Akad. Nauk SSSR *232*, 401 (1977)
136. Anufrieva, E. V., Pautov, V. D., Freidzon, Ya. S., Shibaev, V. P.: Vysokomol. Soedin. *A 19*, 755 (1977)
137. Kresse, H., Tal'roze, R. V.: Makromol. Chemie Rapid Commun. *2*, 369 (1981)
138. Kresse, H., Kostromin, S. G., Shibaev, V. P.: Makromol. Chem., Rapid Commun. *3*, 509 (1982)
139. Shibaev, V. P., Moiseenko, V. M., Lukin, N. Yu., Kuznetov, N. A., Roganova, Z. A., Smolyansky, A. L., Platé, N. A.: Vysokomol. Soedin. *A 20*, 2122 (1978)
140. Kostromin, S. G., Kozlova, E. Yu.: Abstracts of the first All-Union Symposium on Liquid Crystalline Polymers, Suzdal, USSR, p. 48, 1982
141. Park, Y., Bac, C., Labes, M.: J. Amer. Chem. Soc. *97*, 4398 (1975)
142. Belaykov, V. A., Sonin, A. S.: Optica kholestericheskikh zidkih kristallov (Optics of cholesteric liquid Crystals), Moscow, Nauka, 1982
143. Finkelmann, H., Koldehoff, Y., Ringsdorf, H.: Angew. Chem. *17*, 935 (1978)
144. Shibaev, V. P., Finkelmann, H., Kharitonov, A. V., Portugall, M., Platé, N. A., Ringsdorf, H.: Vysokomol. Soedin. *A 23*, 919 (1982)
145. Finkelmann, H., Ringsdorf, H., Siol, W., Wendorff, J.: Makromol. Chemie *179*, 829 (1978)
146. Freidzon, Ya. S., Boiko, N. I., Platé, N. A.: Abstracts of the first All-Union Symposium on Liquid Crystalline Polymers, Suzdal, USSR, p. 24, 1982
147. Chistyakov, I. G.: Zhidkie kristally (Liquid Crystals), Moscow, Nauka, 1966

148. Kapustin, A. P.: Electroopticheskie i akysticheskie svoistva zidkikh kristallov (Electrooptical and acoustic properties of Liquid Crystals) Moscow, Nauka, 1973
149. Pikin, S. A.: Stryktyrnye prevrasheniya v zhidkin kristallah (Structural transformations in liquid crystals), Moscow, Nauka, 1981
150. Meier, G., Sackmann, E., Grabmaier, J.: Applications of Liquid Crystals, Springer-Verlag, Berlin, Heidelberg New York, 9, 1975
151. Finkelmann, H., Naegele, D., Ringsdorf, H.: Makromol. Chemie *180*, 803 (1979)
152. Aguilera, G., Ringsdorf, H., Schneller, A., Zehtel, R.: Preprints of International Symposium on Macromolecules, Florence, Italy, Litografia Felici, v. 3, p. 306, 1980
153. Moiseenko, V. M.: Dissertation for the Degree of Candidate of Sciences, Moscow State University, Moscow, 1978
154. Tal'roze, R. V., Sinitzyn, V. V., Shibaev, V. P., Platé, N. A.: in Advances in liquid crystal, research and applications (Ed. L. Bata), Oxford: Pergamon Press — Budapest: Akademiai Kiado, v. 2, p. 915, 1980
155. Tal'roze, R. V., Kostromin, S. G., Shibaev, V. P., Platé, N. A., Kresse, H., Sauer, K., Demus, D.: Makromol. Chem., Rapid Commun. *2*, 305 (1981)
156. Platé, N. A., Tal'roze, R. V., Kostromin, S. G., Shibaev, V. P., Kresse, H.: Abstracts of Commun. 27 International Symposium on macromolecules, Strasbourg, France, 1981, v. 2, p. 978
157. Tal'roze, R. V., Kostromin, S. G., Sinitzyn, V. V., Shibaev, V. P., Platé, N. A.: Abstracts of Commun. 12 Europhysical Conference on Macromolecular Physics, Leipzig, GDR, 1981, p. 313
158. Tal'roze, R. V., Kostromin, S. G., Sinitzyn, V. V., Shibaev, V. P., Plate, N. A.: Abstracts of the fourth International liquid crystal Conference of Socialist countries, Tbilisi, USSR, v. 2, p. 215, 1981
159. Shibaev, V. P., Freidzon, Ya. S., Tal'roze, R. V., Platé, N. A.: Preprints of International Symposium on Macromolecules, Florence, Italy, Litografia Felici., v. 3, p. 310, 1980.
160. Krigbaum, W. R., Lader, H. J., Ciferri, A.: Macromolecules *13*, 554 (1980)
161. Krigbaum, W. R., Lader, H. J.: Mol. Cryst. Liq. Cryst. *2*, 87 (1980)
162. Krigbaum, W. R., Grantham, C. E., Toriumi, H.: Macromolecules *15*, 592 (1982)
163. Benthack, H., Finkelmann, H., Rehage, G.: Abstracts of Communications of 27-th International Symposium on Macromolecules, Strasbourg, France, v. 2, p. 961, 1981
164. Finkelmann, H., Day, D.: Makromol. Chem. *180*, 2269 (1979)
165. Van Dusen, J. G., Williams, D. J., Meredith, G. R.: Proceedings of the 28 International Macromolecular Symposium, Amherst, USA, 1982, p. 814
166. Wassmer, K., Ohmes, E., Kothe, G., Portugall, M., Ringsdorf, H.: Macromol. Chem., Rapid Commun. *3*, 281 (1982)
167. Piskunov, M. V., Kostromin, S. G., Stroganov, L. B., Shibaev, V. P., Platé, N. A.: Makromol. Chem., Rapid Commun. *3*, 443 (1982)
168. Geib, H., Hisgen, B., Pschorn, U., Ringsdorf, H., Spiess, H.: J. Amer. Chem. Soc. *104*, 917 (1982)
169. Tal'roze, R. V., Sinitzyn, V. V., Shibaev, V. P., Platé, N. A.: Polymer Bulletin *6*, 309 (1982)
170. Tal'roze, R. V.: Abstracts of the first All-Union Symposium on Liquid Crystalline Polymers, Suzdal, USSR, 1982, p. 6
171. Shibaev, V. P., Platé, N. A.: Zhurnal Vsesoysnogo Khimicheskogo Obshestva Mendeleeva *28*, 165, 1983
172. Tsvetkov, V. N., Rjumtsev, E. I., Shtennikova, I. N.: in Liquid crystalline order in polymers (Ed. Blumstein, A.), Academic Press, New York, San Francisco, London, 1978, p. 43
173. Tsvetkov, V. N., Shtennikova, I. N.: Abstracts of the first All-Union Symposium on liquid crystalline polymers, Suzdal, USSR, p. 7, 1982
174. Burshtein, L. L.: Dissertation for the Doctor of Sciences, Institute of Macromolecular Compounds, USSR Academy of Sciences, Leningrad, 1982
175. Anufrieva, E. V., Gotlib, Y. Ya.: Advances in Polymer Sciences, Springer-Verlag, v. 40, p. 1, 1981
176. Pautov, V. D.: Dissertation for the Candidate of Sciences, Institute of Macromolecular Compounds, USSR Academy of Sciences, Leningrad, 1978
177. Tsvetkov, V. N.: Vysokomol. Soedin. *4*, 894 (1962); *7*, 1468 (1965); Dokl. Akad. Nauk SSSR *165*, 360 (1965)

178. Tsvetkov, V. N., Andreeva, L. N., Korneeva, E. V., Lavrenko, P. N.: Dokl. Akad. Nauk SSSR *205*, 895 (1972)
179. Andreeva, L. N., Gorbunov, A. A., Didenko, S. A., Korneeva, E. V., Lavrenko, P. N., Platé, N. A., Shibaev, V. P.: Vysokomol. Soedin *15*, 209 (1973)
180. Tsvetkov, V. N., Rjumtsev, E. I., Konstantinov, I. I., Amerik, Yu. B., Krentsel, B. A.: Vysokomol. Soedin. *A 14*, 67 (1972)
181. Tsvetkov, V. N., Rjumtsev, E. I., Shtennikova, I. N., Konstantinov, I. I., Amerik, Yu. B., Krentsel, B. A.: Vysokomol. Soedin. *A 15*, 2270 (1973)
182. Rjumtsev, E. I., Shtennikova, I. N., Pogodina, N. V., Kolbina, G. F., Konstantinov, I. I., Amerik, Yu. B., Vysokomol. Soedin. *A 18*, 439 (1976)
183. Pogodina, N. V., Tsvetkov, V. N., Kolomiez, I. P.: Abstracts of the fourth International liquid crystal Conference of Socialist countries, Tbilisi, USSR, 1981, v. II, p. 203
184. Shtennikova, I. N., Peker, T. V., Kolbina, G. F., Bushin, S. V., Korneeva, E. V.: Abstracts of the fourth International liquid crystal Conference of Socialist countries, Tbilisi, USSR, 1981, v. II, p. 227
185. Freidzon, Ya. S., Shibaev, V. P., Agranovich, I. M., Pautov, V. D., Anufrieva, E. V., Platé, N. A.: Vysokomol. Siedin. *A 20*, 2601 (1978)
186. Freidzon, Ya. S., Shibaev, V. P., Pautov, V. D., Bronich, T. K., Shelukhina, G. D., Kasaikin, V. A., Platé, N. A.: Dokl. Akad. Nauk SSSR *256*, 1435 (1981)
187. Anufrieva, E. V., Pautov, V. D., Freidzon, Ya. S., Shibaev, V. P., Platé, N. A.: Vysokomol. Soedin. *A 24*, 825 (1982)
188. Kasaikin, V. A., Freidzon, Ya. S., Makhaeva, E. E.: Abstracts of the fourth international liquid crystal Conference of Socialist countries, Tbilisi, 1981, v. II, p. 179
189. Smolyansky, A. L., Shibaev, V. P.: Vysokomol. Soedin. *A 21*, 2221 (1979)
190. Shibaev, V. P., Platé, N. A., Smolyansky, A. L., Voloskov, A. Ya.: Makromol. Chem. *181*, 1393 (1980)
191. Borisova, T. I., Burshtein, L. L., Stepanova, T. P., Freidzon, Ya. S., Shibaev, V. P.: Vysokomolek. Soedin. *B 19*, 553 (1977)
192. Borisova, T. I., Burshtein, L. L., Stepanova, T. P., Freidzon, Ya. S., Kharitonov, A. V., Shibaev, V. P.: Vysokomol. Soedin. *B 24*, 451 (1982)
193. Borisova, T. I., Burshtein, L. L., Stepanova, T. P., Freidzon, Ya. S., Shibaev, V. P., Platé, N. A.: Vysokomol. Soedin *A 24*, 1103 (1982)
194. Grosberg, A. Yu.: Vysokomol. Soedin. *A 22*, 96 (1980)
195. Finkelmann, H., Lühmann, B., Rehage, G.: Colloid and Polymer Sci. *260*, 56 (1982)

M. Gordon (Editor)
Received August 19, 1983

Author Index Volumes 1–60/61

Subject Index